BiCMOS/CMOS Systems Design

Other McGraw-Hill Reference Books of Interest

Handbooks

BENSON • *Audio Engineering Handbook*

BENSON • *Television Engineering Handbook*

COOMBS • *Printed Circuits Handbook*

CROFT AND SUMMERS • *American Electricians' Handbook*

DI GIACOMO • *Digital Bus Handbook*

DI GIACOMO • *VLSI Handbook*

FINK AND BEATY • *Standard Handbook for Electrical Engineers*

FINK AND CHRISTIANSEN • *Electronics Engineers' Handbook*

HICKS • *Standard Handbook of Engineering Calculations*

INGLIS • *Electronic Communications Handbook*

JURAN AND GRYNA • *Juran's Quality Control Handbook*

KAUFMAN AND SEIDMAN • *Handbook of Electronics Calculations*

KURTZ • *Handbook of Engineering Economics*

STOUT AND KAUFMAN • *Handbook of Operational Amplifier Circuit Design*

TUMA • *Engineering Mathematics Handbook*

WILLIAMS AND TAYLOR • *Electronic Filter Design Handbook*

Other

ANTOGNETTI AND MASSOBRIO • *Semiconductor Device Modeling with SPICE*

ANTOGNETTI • *Power Integrated Circuits*

BUCHANAN • *CMOS/TTL Digital Systems Design*

CLEMENTS • *68000 Sourcebook*

ELLIOTT • *Integrated Circuits Fabrication Technology*

GEIGER, ALLEN, AND STRADER • *VLSI Design Techniques*

HECHT • *The Laser Guidebook*

INTEL • *i860 Microprocessor Architecture*

INTEL • *386SX Microprocessor Programmer's Reference*

INTEL • *i486 Microprocessor Programmer's Reference*

MUN • *GaAs Integrated Circuits*

SILICONIX • *Designing with Field-Effect Transistors*

SZE • *VLSI Technology*

TSUI • *LSI/VLSI Testability Design*

VAN ZANT • *Microchip Fabrication*

BiCMOS/CMOS Systems Design

James E. Buchanan
Advisory Engineer
Westinghouse Electric Corporation

Drawings by Bert D. Buchanan

McGraw-Hill, Inc.
New York St. Louis San Francisco Auckland Bogotá
Caracas Hamburg Lisbon London Madrid
Mexico Milan Montreal New Delhi Paris
San Juan São Paulo Singapore
Sydney Tokyo Toronto

BiCMOS/CMOS. Systems Design
International Edition 1991

Exclusive rights by McGraw-Hill Book Co-Singapore for manufacture & export. This book cannot be re-exported from the country to which it is consigned by McGraw-Hill.

1 2 3 4 5 6 7 8 9 0 KHL SW 9 7 6 5 4 3 2 1

ISBN 0-07- 008712-1

Not for re-sale in Australia, Canada, Europe, Japan, the United Kingdom and the United States. Export sales may be made only by or with the expressed consent of the publisher.

Library of Congress Cataloging-in-Publication Data

Buchanan, James E. (James Edgar), date.
BiCMOS/CMOS systems design/James E. Buchanan; drawings by Bert D. Buchanan.
p. cm.
Companion vol.: CMOS/TTL digital systems design.
Includes bibliographic references and index.
ISBN 0-07-008712-1
1. Metal oxide semiconductors, Complementary—Design and construction. 2. Transistor-transistor logic circuits—Design and construction. 3. Bipolar integrated circuits—Design and construction. I. Burchanan, Bert D. II. Title.
TK7871.99.M44B83 1991
621.39′732—dc20 90-46798

The sponsoring editor for this book was Daniel A. Gonneau, the editing supervisor was Peggy Lamb, and the production supervisor was Pamela A. Pelton. This book was set in Century Schoolbook. It was composed by McGraw-Hill's Professional Publishing composition unit.

When ordering this title use ISBN 0-07-112541-8

Printed in Singapore.

Contents

Preface

This book is about the system application of state-of-the-art high-speed bipolar combined with complementary metal-oxide semiconductor (BiCMOS) logic devices and advanced CMOS logic devices. It is not about logic design. It is about transistor-transistor logic (TTL)-level compatible BiCMOS and TTL- and CMOS-level advanced CMOS logic devices and the electrical and mechanical environment in which they must function and how to optimize that environment to achieve maximum performance. Their very fast edges, low noise margin, and other nonideal characteristics invite disaster, but with attention to basic electrical and electromagnetic principles, they can be applied successfully.

Unfortunately, an introduction to these basic principles appears to be absent from most digital designers' training. Perhaps the material is considered too simple or obvious—and it is simple and obvious when explained and understood—but without an understanding of the simple electrical issues, there is little likelihood of success when high-speed BiCMOS/CMOS devices are used. Many have learned the hard way in their first encounter with high-speed BiCMOS/CMOS devices that the electrical issues cannot be neglected. The purpose of this book is to help those new to high-speed advanced BiCMOS/CMOS logic design avoid the many pitfalls that await the unsuspecting.

However, a book such as this cannot provide solutions to all BiCMOS/CMOS system design problems. Each design will have its own unique set of circumstances and challenges. The hope is that the book will at least help the reader develop an awareness of the many critical issues that must be addressed in the design of high-speed digital systems. Understanding that an issue is important is the first step to its solution.

This book is intended as a companion to *CMOS/TTL Digital Systems Design* (McGraw-Hill, 1990), which covers advanced Schottky TTL and advanced CMOS devices and their application in systems. This book expands the coverage of advanced CMOS devices to include the application of large-scale integration (LSI) devices, such as programmable logic devices (PLDs), first-in-first (FIFO) devices, random-access memories (RAMs), and application-specific integrated circuits

(ASICs), and covers BiCMOS devices and their application. By necessity, so that each book can stand on its own, each book covers some of the same subjects. Most of the fundamental issues that must be addressed when high-speed logic devices are applied and their solutions are the same regardless of the technology of the devices being applied. Crosstalk, transmission-line effects, and ground and power-supply upsets due to switching currents are universal problems that are controlled by similar techniques.

Advanced CMOS products represent the fastest growing sector of today's semiconductor market, and advanced CMOS process technology is expected to remain the dominant technology in the merchant semiconductor market during the early 1990s. BiCMOS logic devices are relatively new, but BiCMOS is expected to displace CMOS and become the dominant technology of the mid- and late-1990s. Hence, it is crucial that digital designers understand advanced CMOS and BiCMOS logic devices and their application.

Book Arrangement

The book starts with an introduction to advanced BiCMOS/CMOS logic families and their characteristics followed by a brief review of BiCMOS and CMOS logic circuits in Chapter 3. Chapter 4 describes some special problem areas and techniques for overcoming them. Chapters 5 through 12 cover circuit analysis and system implementation techniques that are essential for the successful application of high-speed BiCMOS/CMOS devices. The final three chapters cover some of the special system implementation techniques that must be used when applying advanced BiCMOS/CMOS LSI and VLSI devices, such as memories and ASICs. Each chapter stands on its own so that readers with immediate design tasks can go to specific topics instead of reading the entire book. Chapters 4 through 15 end with a summary of the important points and design techniques discussed.

Summary of Chapters

Chapter 1 stresses that the logic design cannot be separated from the electrical and mechanical design when high-speed BiCMOS/CMOS logic devices are used. The need for power and signal interconnection systems with wide bandwidth is shown.

Chapter 2 presents a brief overview of the relative performance and specifications of the various advanced CMOS and BiCMOS logic families. A number of tables are provided for quick reference and for comparison of characteristics.

Chapter 3 provides a brief review of CMOS and BiCMOS logic circuit operation and introduces circuit modeling techniques that simplify circuit analysis. Several circuit characteristics that designers must understand to successfully apply CMOS and BiCMOS logic circuits are reviewed.

Chapter 4 describes the mechanisms that cause internal device and load dependent transient switching currents and the detrimental effects of inductance in high-speed interconnections. Several examples show the possible magnitude of switching currents and demonstrate the possible adverse effects of excessive inductance. Guidance for minimizing inductance or the effects of inductance in power distribution systems is provided.

Chapter 5 describes the need for low-impedance power and ground distribution networks and describes techniques for achieving them. Techniques for calculating losses in power distribution networks are described. Guidelines for the amount and placement of decoupling capacitors are provided.

Chapter 6 deals with the design of the physical interconnection systems used for high-speed CMOS and BiCMOS devices. Topics include optimum line impedance and how to minimize crosstalk.

Chapter 7 includes a review of transmission-line effects and deals with nonlinear sources and loads as they specifically relate to high-speed advanced CMOS and BiCMOS applications.

Chapter 8 deals with the need for high-quality clock signals in high-speed systems. Techniques for distributing clock signals are described. It is stressed that a sound clock distribution system is one of the keys to a reliable high-speed digital system.

Chapter 9 deals with practical interfacing issues that must be addressed at the device, board, and system levels when high-speed CMOS and BiCMOS devices are used.

Chapter 10 describes the importance of synchronous logic designs and the necessity for using synchronous design practices when high-speed logic devices are used. Critical design issues for synchronous systems are described, such as techniques for synchronizing signals and metastability.

Chapter 11 provides guidance for determining worst-case device and interconnection delays. A typical board-to-board signal path is analyzed for worst-case signal delay.

Chapter 12 describes techniques for generating and distributing system initialization signals and write protection schemes for nonvolatile memory devices, such as EEPROMs, during power switching or transients. Many critical issues that are often overlooked are discussed.

Chapter 13 covers state-of-the-art high-speed memory devices and the design of high-speed memory systems.

Chapter 14 covers some of the special system application concerns associated with programmable array logic (PAL), programmable logic array (PLA), and other LSI advanced CMOS devices.

Chapter 15 covers some of the special system application concerns associated with gate arrays, standard cells, and compiled cells.

Acknowledgments

Many thanks are due my fellow engineers in the Digital Systems Engineering Group at Westinghouse Electric Corporation's Electronic Systems Center, Baltimore, Maryland, who contributed to this book through their efforts as leaders or as participants of the in-house digital design course on which the book is based. Everyone in the Digital Systems Engineering Group contributed in one way or another. Special thanks are due to Michael Juengst and Arden Helland for their contributions and to Jim Hudson and Karl Avellar for their advocacy of the guidelines course. Westinghouse Electric Corporation is thanked for the opportunity and permission to publish this book.

In addition, I want to thank my son Bert for his excellent rendering of the drawings, and Laurie Welms for proofreading the various drafts. My wife Beverly is thanked for her patience and tolerance during the many evenings and weekends that I spent on the book.

James E. Buchanan

Chapter

1

Introduction

Advanced complementary metal-oxide semiconductor (CMOS) and bipolar combined with CMOS (BiCMOS) logic devices represent the state of the art in digital circuits. They are intended for systems that push the state of the art in processing speed and capability, but state-of-the-art system performance does not come without a price. Devices must be pushed to their limits, and the nearer to their limits devices and systems operate, the greater the chance for serious design errors and the less the chance for simple solutions. Advanced BiCMOS/CMOS logic devices are analogues to nitroglycerin: When applied properly and carefully, they offer spectacular performance; when applied carelessly, they invite disaster.

Designers must appreciate that advanced BiCMOS/CMOS devices are not easy to use and that certain fundamental electrical principles cannot be neglected without introducing serious system problems and compromising expected performance. These fundamental electrical principles are not complex, and once explained most seem obvious. However, they are often neglected; perhaps, it is their obvious nature that leads to this neglect.

The two major areas of neglect that lead to most problems encountered in systems built with advanced BiCMOS/CMOS devices are

1. Inadequate timing margin for worst-case device and interconnection delays
2. Neglect of basic electromagnetic principles and Ohm's law

It seems obvious that systems must have adequate timing margin for worst-case component and interconnection delays. Too often, though, systems go into production with the expected operating speed based on naive, optimistic assumptions for device timing parameters and little or no allowance for interconnection delays. Often the fact that a breadboard operates at a given speed is taken as proof of the design, but it is naive to assume that all components in a breadboard have worst-case delays and represent the worst-case combination of conditions that will occur over a production run. Perhaps, in the past when system clock rates were typically very low, it was possible to ignore worst-case timing and interconnection delays without suffering major production disasters. However, when dealing with high-performance systems, in the 20-MHz or greater range, casual approaches to establishing operating speed are sure to lead to systems that are difficult to produce. The nearer a system operates to the limits of present-day device and interconnection technology, the greater is the chance for serious timing errors. Actual in-circuit worst-case device parameters and interconnection delays must be determined and used to establish system operating speeds. Under ideal conditions, individual devices may operate at very high speeds. However, when many high-speed devices must be interconnected, interconnection delays are often responsible for a significant portion of most signal delays. At 20 MHz, interconnection delays are a significant portion of most signal delays; at 40 MHz, interconnection delays are a major portion of most signal delays. In spite of this, interconnection delays are often ignored.

To utilize the potential operating speed of advanced BiCMOS/CMOS devices, the functional design and the interconnection and power distribution system design cannot be separated. Their fast edges cause crosstalk, transmission-line effects, power-supply transients, and ground upset even when the best interconnection design practices are followed. Yet, digital (logic) designers very often fall into the trap of viewing digital circuits as simple functional logic blocks rather than as complex electric circuits. No matter how simple or complex the functional logic being performed is, the actual circuitry consists of an imperfect signal interconnection system, an imperfect power and ground distribution network, and logic devices with nonideal internal circuitry. To ignore these imperfections is to ignore basic electromagnetic principles and Ohm's law.

Nonetheless, most of the effort in most digital designs remains focused on the functional logic design even though logic errors or functional concept errors are generally fixable, but systems with poorly designed or conceived electrical interconnection schemes are often unfixable. The steps required to correct functional logic error may be unpleasant and costly, but they can be accomplished. However, if the electrical system is inadequate, no amount of patching will provide a solid system. One may be able to hand tune each system as it comes off the production line so that it works for the moment, but whether it will work under all the required conditions or for how long will be difficult to establish. Problems with systems with inadequate electrical systems are often never understood. The problems are blamed on various nebulous causes. Very frequently, the problem is blamed on bad or marginal parts rather than on the real cause of the problem: a poor electrical system or system timing that is not compatible with worst-case device parameters. In such systems, very often a given problem will go away with a change to another part or another vendor's part. In the great majority of such cases, however, the part is not the real problem; the problem is that the part is being applied in such a manner that there is very little or no operating margin. Substituting another vendor's part that has a little more margin, or that is a little faster or slower, corrects the immediate, but not the basic, problem.

Logic device and interconnection imperfection are more noticeable when high-speed BiCMOS/CMOS devices are used because signal edge rates, not clock rates, determine the required response of signal and power distribution systems and the onset of transmission-line effects. Today's BiCMOS and CMOS devices typically have edge speeds of 1 to 3 ns. Assuming switching transitions are linear, a switching transition t_r that occurs in 1 ns has a fundamental 3-dB frequency f component of[1,2]

$$f = \frac{0.35}{t_r} \quad \text{in the general case}$$

$$f = \frac{0.35}{1\ \text{ns}} = 350\ \text{MHz} \quad \text{in a given case} \tag{1.1}$$

A linear voltage transition, such as a switching edge, also contains a significant amount of higher-frequency components. The third harmonic composes approximately 10 percent of the amplitude of a linear ramp.

Digital signal interconnections act as low-pass networks (Figure 1.1). Interconnections will attenuate signals to some degree unless they have the bandwidth to pass the third harmonic of the 3-dB frequency content of signal edges. For a signal with a 1-ns edge, the third

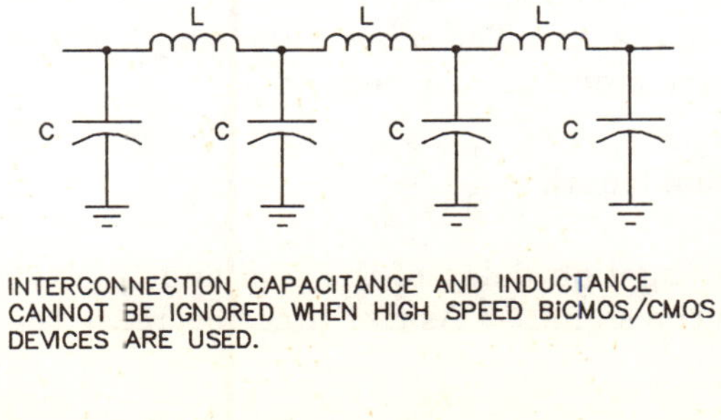

Figure 1.1 All interconnections are low-pass networks when BiCMOS/CMOS devices are used.

harmonic content is approximately 1 GHz (3 × 350 MHz—see given case in Eq. (1.1) above). Not all BiCMOS or advanced CMOS devices have edge rates as fast as 1 ns, but most have edge rates faster than 3.5 ns. A 3.5-ns edge corresponds to 100 MHz (see Table 1.1). Thus, as a minimum, BiCMOS or advanced CMOS systems need interconnection networks with bandwidths in excess of 100 MHz to prevent serious degradation of signals. Digital designers must appreciate that they must deal with frequencies that only a short time ago were considered strictly in the radio frequency (RF) designer's domain.

TABLE 1.1 Rise Time versus Bandwidth

Rise time, t_r, ns	Bandwidth, f^* MHz
1.00	350
1.50	233
2.00	175
2.50	140
3.00	116
3.50	100
4.00	87.5
5.00	70.0
6.00	58.4
7.00	50.0

*f = Upper-bandpass frequency at the 3-dB point for a low-pass filter. The frequency f can also be considered the upper (highest) frequency component (with a significant amount of energy) in the waveform generated when a logic device switches states.

Transmission-line effects, at the circuit board or motherboard level, were of little concern with the older, slower CMOS logic families. Their slow edge rates (5 ns or greater) did not induce transmission-line effects, except on very long lines such as might exist between remote units. However, when advanced BiCMOS/CMOS devices with edge rates that may be as fast as 1 ns are used, transmission-line effects are of concern at the circuit board level. For all practical purposes, all signal lines are transmission lines when advanced BiCMOS/CMOS logic components are used to implement digital systems.

Transmission-line effects begin to appear when the signal rise time is near the propagation delay of the line. The *critical line length* where transmission-line effects become significant is defined as

$$\text{Critical line length} = \frac{1}{2}\left(\frac{t_r}{t'_{\text{pd}}}\right) \tag{1.2}$$

where t'_{pd} is the effective propagation delay of the line and t_r is the rise time (20 to 80 percent) of the signal.[3,4] As rise times decrease relative to line propagation delay, transmission-line effects become more pronounced. Typical propagation delay for track on printed circuit boards is 2 ns/ft. For a signal with a rise time of 1 ns, the critical line length is

$$\text{Critical line length} = \frac{1}{2}\left(\frac{1\text{ ns}}{2\text{ ns/ft}}\right)$$

$$= 0.25\text{ ft or }3\text{ in}$$

which means most lines are transmission lines when BiCMOS/CMOS devices are used. Even on small circuit boards, most lines exceed 3 in, and in many of today's applications boards as large as the one shown in Figure 1.2 are used. Most lines are very long transmission lines on boards as large as the one shown in Figure 1.2; some lines may exceed 2 ft. All the problems associated with transmission lines and transmission-line effects will be present on large boards when BiCMOS/CMOS devices are used. Transmission-line effects that were of no concern in the past at the board level become serious problems when high-speed devices and large boards are combined.

The transient-current generation and noise-tolerance (margin) characteristics of BiCMOS and CMOS devices with either transistor-transistor logic (TTL) or CMOS signal levels are two device properties that must be understood and kept foremost in the system designer's awareness. Neither BiCMOS nor CMOS devices have much noise tolerance, yet they both generate large transient switching currents which cause large amounts of noise. Advanced CMOS devices with "true" CMOS input thresholds have more noise tolerance than BiCMOS or CMOS devices with TTL-level input thresholds, but CMOS rail-to-rail output voltage swings generate more noise. In either case, these two device characteristics tend to combine in a most unfavorable manner. Large transient currents generate large amounts of noise in very close proximity to devices that have very little tolerance for noise. Failure to account for the noise-generating characteristics of high-speed logic devices is certain to lead to the design of unreliable systems.

As clock frequencies move above 20 MHz, BiCMOS and CMOS devices, with either TTL or CMOS signal levels, become increasingly more difficult to use. Not only do interconnection delays become significant, but as frequency goes up, power dissipation goes up, and there is less time for crosstalk and noise to subside. Their large, fast signal swings generate large transient currents. Large transient currents and package pin or interface connector inductance limit the number of signals that can switch simultaneously in a given package or at board or system interfaces without upsetting ground or power levels. Either signal edge speed must be reduced or more package and

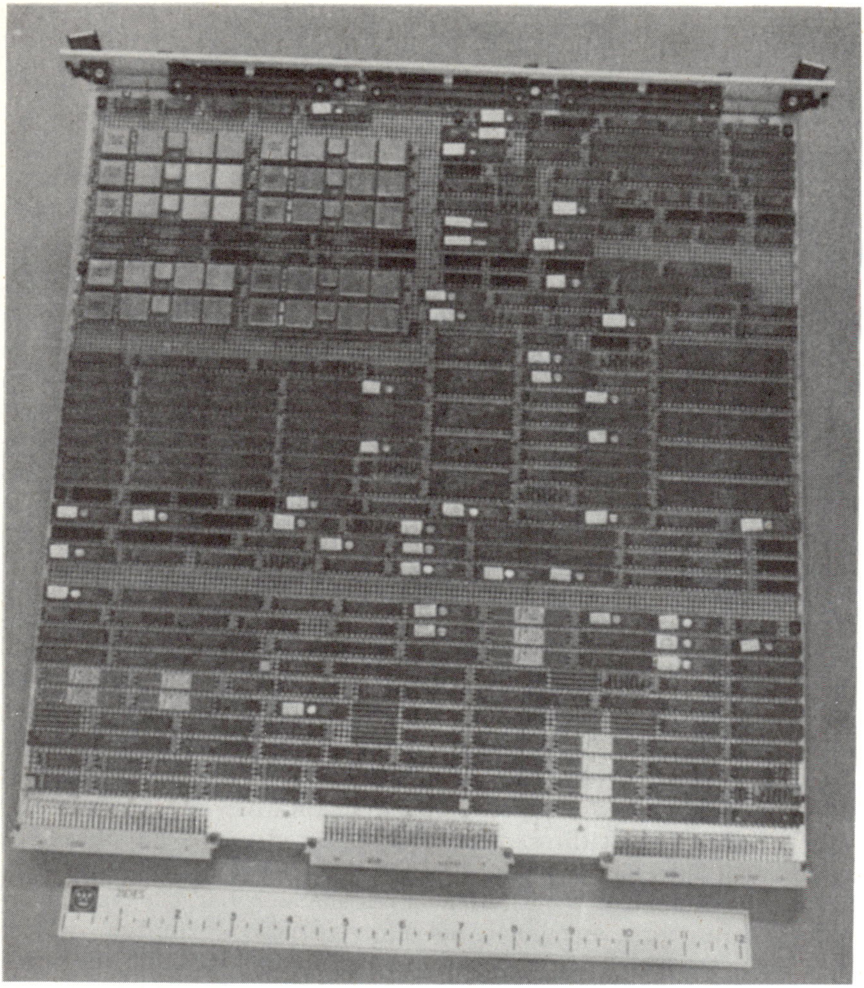

Figure 1.2 Most interconnections must be treated as transmission lines when large circuit boards are used.

interface connector pins must be reserved for power and ground to reduce inductance, but the number of power and ground pins required to make significant speed improvements becomes impractical. No pins are left for signals. Thus, the speed of BiCMOS/CMOS devices and systems with TTL and CMOS signal levels is limited by their interface levels. Devices with lower voltage swings are needed to move to the next level of system performance.[5] High-speed, low-power

BiCMOS devices with emitter-coupled logic (ECL) interfaces offer the promise of a route to the next level of system performance.

In summary, when high-speed BiCMOS and advanced CMOS logic devices are used, the logic design cannot be separated from the electrical and mechanical design. Treating high-speed digital circuitry as functional blocks rather than as a part of a complex electrical network is sure to lead to the design of unreliable systems. Entire areas of concern may be neglected or ignored when devices are applied in block form.

References

1. IEEE Standard 181-1977, "IEEE Standard on Pulse Measurement and Analysis by Objective Techniques," IEEE, 1977.
2. Southard, Robert K.: "High-Speed Signal Pathways from Board to Board," in 1981 WESCON Records, Session 18, September 1981, Paper No. 2.
3. *Applications Handbook,* Cypress Semiconductor Corp., San Jose, Calif., 1989.
4. Blood, W. R.: *MECL System Design Handbook,* 4th ed., Motorola Inc., Phoenix, Ariz., 1988.
5. Lipsett, Ed: "Shift from TTL to ECL," *Electronic Engineering Times,* August 14, 1989, pp. 29–30.

Chapter

2

Advanced CMOS and BiCMOS Logic Families

To make cost- and performance-effective logic device selections for present and future systems, designers need an understanding of the advantages and disadvantages of advanced CMOS and BiCMOS logic devices and logic families. A knowledge of advanced CMOS devices is essential since most of today's state-of-the-art digital products are fabricated using advanced CMOS technology. A knowledge of BiCMOS devices and technology is needed to prepare for the future since BiCMOS technology is expected to displace CMOS and become the dominant digital technology of the mid- and late 1990s.

Each logic device technology has characteristics that must be weighed against system requirements. Tradeoffs must be made based on[1]

Speed requirements

Power consumption

Supply voltage level requirements

Drive capability

Noise margin

Available package styles

Signal level compatibility

Availability

Cost

plus many other considerations unique to each application. Speed is usually a basic selection criterion. If the technology does not have the capability of meeting the speed requirements, there is little point in further consideration of that technology. Most of today's high-performance systems require devices fabricated with advanced Schottky TTL, advanced CMOS, BiCMOS, or ECL-technologies; devices fabricated with the older TTL and CMOS logic technologies are too slow for most present-day applications. In the past, ECL technology was the only choice for high-speed systems. Devices fabricated using ECL technology have outputs that have controlled rise times and that can drive load-terminating resistors. Controlled rise time and control of the match between line impedance and load-terminating resistors provides a means of achieving the signal quality needed for high-performance systems. However, ECL terminating resistors dissipate a great deal of power, increase parts count, and usually require an additional power supply for the termination voltage. Furthermore, ECL circuits inherently dissipate a great deal of power.[2] Advanced CMOS and BiCMOS technologies offer the promise of high-speed performance without the power dissipation of TTL or ECL devices and without the additional components (terminating resistors) required for ECL systems.

When advanced CMOS technology was first introduced, its advertised selling point was high speed with low power consumption. In general, that promised performance has not been attained. The fast rail-to-rail voltage swings of first-generation advanced CMOS devices have compounded most of the problems associated with high-speed logic applications, such as crosstalk, ground bounce, and transmission-line effects. System designers have found advanced CMOS devices very difficult to use (at any speed), and as system operating speeds move into the 30- to 50-MHz range, first-generation advanced CMOS devices with their wide, fast voltage swings may dissipate more power than advanced Schottky TTL devices. Manufacturers of first-generation advanced CMOS devices apparently gave little consideration to how their parts might perform in a system environment. System designers needed fast parts, so design manufacturers built fast (switching) parts, but it was quickly learned that higher device speed did not always translate to higher system speed. In many cases, reducing rise time was found to actually improve system performance when the effects of the interconnection system were considered. With that lesson learned, most second-generation advanced CMOS logic families have output slew rate control, and at least one family has returned to TTL output swings.

Advanced BiCMOS devices are available with both ECL and TTL levels. The discussion in this book is limited to BiCMOS devices and logic families that have TTL input and output levels. BiCMOS devices with traditional TTL logic levels and interface circuitry coupled to a CMOS core offer the best of both CMOS and advanced Schottky bipolar TTL technology.

2.1 Advanced CMOS Logic Families

Advanced CMOS technology is used to build a number of what is customarily called "discrete" logic families as well as several bus logic families. Discrete logic families typically consist of small-scale integration (SSI) devices such as ORs, ANDs, NANDs, and buffers; medium-scale integration (MSI) devices such as decoders, latches, and registers; and perhaps some small pin count large-scale integration (LSI) devices. Bus logic families tend to consist primarily of octal, or wider, buffers, transceivers, latches, and registers. For classification purposes in this book, these logic families are divided into first- and second-generation devices or families of devices. As might be expected by the naming of the two classifications, the grouping is based on the chronological order of development. First-generation devices tend to have rail-to-rail output voltage swings with no control of slew rate which can lead to serious system noise problems. Second-generation devices tend to have output slew rate control or TTL-level output voltages or both that make them easier to use in systems. However, any grouping of first- and second-generation devices is dubious at best. Manufacturers are continually changing and improving the capabilities and characteristics of their devices. Some manufacturers have upgraded devices and families without changing the designation used to identify the family, so it is important to understand that the grouping shown in Table 2.1 is a transitory ranking. Current manufacturers' device specifications should be consulted for the present statuses of devices and logic families.

Discrete advanced CMOS logic devices and logic families are identified by letter designators such as AC, ACT, and FCT (with a few exceptions). Most SSI and MSI advanced CMOS logic devices in the advanced CMOS logic families are pin compatible with equivalent TTL logic functions; most have speed and drive comparable to those of AS or FAST advanced Schottky TTL devices. Some advanced CMOS logic families have both devices with the input switching threshold approximately centered (those with letter designators without a T) and devices with input thresholds shifted so that they are compatible with TTL levels (those with ACT, FCT, etc., designators). However, several advanced CMOS logic series are only offered with TTL input levels. One example is the FCT series; there is no FC series. Most advanced CMOS devices are optimized to operate with TTL supply voltage lev-

TABLE 2.1 Advanced CMOS Logic Families

Logic family	Manufacturers
First Generation	
"Generic" 54/74AC, ACT	Harris, Performance, VTC
FACT 54/74AC, ACT	National, Motorola
54/74AC11, ACT11	Texas Instruments, Signetics
54/74FCT, FCT-A	IDT, VTC, Harris
54/74PCT, PCT-A	Performance
29C800	AMD
Second Generation	
FACT QS 54/74AC, ACT	National
54/74FCT-T, FCT-AT, FCT-CT	IDT

els (5 V ±10 percent), but those with CMOS input levels are specified for operation over a wider supply range, typically between 2 or 3 to 6 V. Those with TTL input levels are specified with normal TTL supply voltage levels of 5 V ±10 percent. Yet, most advanced CMOS devices with TTL input levels will operate over the same range as the CMOS-level devices in the family, but when the supply level is outside the normal TTL range, input and output levels may not meet TTL-level requirements.[3]

Table 2.1 attempts to provide some ordering of the confusing assortment of advanced CMOS logic families. Listed are some of the major subgroups within the two major groupings (at least one manufacturer is shown for each family). Table 2.1 is not an all-inclusive list. It is not practical to provide a comprehensive list of manufacturers and the devices they manufacture in a book of this type. The situation is so dynamic that the list is certain to be obsolete before the book is published. The current *IC MASTER*[4] is recommended as a source for up-to-date information on the availability of devices and manufacturers.

The rationale used for the grouping shown in Table 2.1 is as follows: Generic AC/ACT and FACT are shown as separate logic families since some manufacturers tend to follow that convention. However, to confuse matters, parts for both families are marked the same with AC or ACT letter designators. The AC11/ACT11 logic families with center power and ground pins also use the same basic AC/ACT letter designators with 11 added.[5] The FCT and 29C800 families are not true logic families since they do not have NANDs, NORs, etc. They are often described as "bus interface families." Bus interface families tend to have only octal or wider parts, but the FCT family does have some counters and multiplexers (the 29C800 family does not). There are a number of speed ranges available in the FCT family. They are distinguished by the addition of an A, B, etc. In the second-generation category, FACT QS and the improved FCT-T families are separated because of slight differences in speed. Even though the FACT QS family is advertised as having a 15 percent speed improvement over the original FACT family, it is still slower than the improved FCT-T families. One feature that distinguishes the FCT-T families from other advanced CMOS families is the return to true TTL output levels.

Acceptance of advanced CMOS parts has been slow because of the many system level complications caused by the fast edges and wide voltage swings of first-generation devices. Acceptance has also been hindered because several of the major semiconductor manufacturers introduced AC parts that could not be mixed with TTL devices first introduced and were slow in expanding the number of ACT devices offered. Not only did that preclude a gradual introduction of advanced CMOS parts into systems, it also limited the complexity of systems that could be built with advanced CMOS parts since most random-access memories (RAMs), read-only memories (ROMs), programmable array logics (PALs), etc., are only available with TTL input and output levels. Most system designers found the risks associated with advanced CMOS devices which include no alternate parts to fall back on (in case of problems with "new" AC parts) unacceptable. They could not simply substitute an advanced Schottky TTL part because of input-output level incompatibility. To further complicate matters, some manufacturers changed to packages with power and ground pins located in the center of the package (to solve some of the problems caused by fast edges). The new package again created a situation where system designers had no parts to fall back on. Thus, unless faced with some very compelling reason to do otherwise, in the great majority of cases system designers have found it more prudent to continue to use advanced Schottky TTL devices.[6] Second-generation advanced CMOS devices with controlled slew rates and in some cases lower voltage swings promise to change that attitude.

2.2 BiCMOS Logic Families

BiCMOS is expected to be "the key technology of the nineties." It offers the best features of bipolar and CMOS technologies by combining advanced Schottky bipolar technology and advanced CMOS technology. Typically, BiCMOS devices have CMOS input structures and CMOS internal logic combined with bipolar output structures. However, some BiCMOS devices also have bipolar input structures as well as outputs, and some BiCMOS ASICs and LSI devices use bipolar

TABLE 2.2 BiCMOS Logic Families

Logic family	Manufacturers
74BC	Motorola, Toshiba
54/74BCT	Texas Instruments
54/74FBT	IDT

drive stages throughout the internal logic. The combination of technologies provides high speed, high drive, and low power. Speeds and drive capabilities of BiCMOS devices are comparable to those of advanced Schottky devices.

A wide variety of BiCMOS logic devices are available. They include application-specific integrated circuits (ASICs), RAMs, programmable logic devices (PLDs), and discrete logic devices. A BiCMOS logic family that includes SSI devices such as NANDs, ANDs, and ORs is available from Motorola and Toshiba, and BiCMOS bus interface families are available from Texas Instruments and IDT (Table 2.2). All the BiCMOS logic families listed in Table 2.2 have TTL-level output swings and TTL-level input thresholds. Traditional TTL logic levels and interface circuitry improves overall system signal quality. TTL levels cause less crosstalk, less dynamic power dissipation, less transient-current generation, and less ground bounce than CMOS levels. TTL levels also provide compatibility with LSI devices, such as RAMs, and with test equipment. The advantage of BiCMOS devices over bipolar TTL devices is less static power dissipation.

2.3 Advanced BiCMOS/CMOS Timing Specifications

Most SSI and MSI advanced BiCMOS/CMOS device data sheets list typical, maximum, and minimum +25°C timing for devices intended for both the commercial and the military markets, maximum and minimum timing for commercially rated devices, and maximum and minimum timing for military-rated devices as shown in Figure 2.1. Traditionally, commercially rated devices are specified for operating conditions of 0 to +70°C and ±5 percent from nominal power-supply limits. However, there is little consistency in the temperature range or power-supply range limits for commercially rated advanced BiCMOS/CMOS parts. Some are specified for the −40 to +85°C range which is closer to what has traditionally been called the industrial range. Some commercially rated advanced CMOS parts are specified with ±10 percent power-supply limits instead of ±5 percent. Specification limits are much more consistent for military-rated devices.

AC Electrical Characteristics

Symbol	Parameter	V_{cc}* (V)	74AC T_A – +25°C C_L = 50 pF			54AC T_A – −55°C to +125°C C_L = 50 pF		74AC T_A – −40°C to +85°C C_L = 50 pF		Units
			Min	Typ	Max	Min	Max	Min	Max	
t_{PLH}	Propagation Delay	3.3	2.0	7.0	9.5	1.0	11.0	2.0	10.0	ns
		5.0	1.5	6.0	8.0	1.0	8.5	1.5	8.5	
t_{PHL}	Propagation Delay	3.3	1.5	5.5	8.0	1.0	9.0	1.0	8.5	ns
		5.0	1.5	4.5	6.5	1.0	7.0	1.0	7.0	

*Voltage Range 3.3 is 3.3V ±0.3V.
Voltage Range 5.0 is 5.0V ±0.5V.

(a)

AC Electrical Characteristics

Symbol	Parameter	V_{cc}* (V)	74ACT T_A = +25°C C_L = 50 pF			54ACT T_A = −55°C to +125°C C_L = 50 pF		74ACT T_A = −40°C to +85°C C_L = 50 pF		Units
			Min	Typ	Max	Min	Max	Min	Max	
t_{PLH}	Propagation Delay	5.0	1.5	5.5	9.0	1.0	9.5	1.0	9.5	ns
t_{PHL}	Propagation Delay	5.0	1.5	4.0	7.0	1.0	8.0	1.0	8.0	ns

*Voltage Range 5.0 is 5.0V ±0.5V.

(b)

Figure 2.1 (*a*) FACT 54AC/74AC00; (*b*) 54ACT/74ACT00 dynamic (ac) characteristics. (*Reprinted by permission of National Semiconductor Corp.*)

Most military-rated devices are specified for operating conditions of −55 to +125°C and ±10 percent power-supply variations from nominal. The timing information supplied for LSI and VLSI devices is usually less complete than that for SSI and MSI devices; in many cases no typical or minimum data are supplied.

Device timing specifications are tied to loading. Unfortunately, not all manufacturers use the same load conditions when specifying timing. Load resistance does not have a large impact on device timing. However, variations in load resistance from manufacturer to manufacturer make interpreting and comparing timing specifications difficult when differences are small, e.g., when a logic family is claimed to be 1 or 2 ns faster than another family. When devices are tested to TTL thresholds (1.5 V), loading networks that limit *high* levels tend to improve *high*-to-*low* propagation times. It takes less time to make a transition over a shorter distance. Load capacitance has a major effect on output timing, but fortunately there is more standardization of load capacitance than there is of load resistance. Most SSI and MSI advanced BiCMOS/CMOS logic devices are specified with 50-pF loads. Most LSI and VLSI devices, such as memories, are specified with 30- or 35-pF loads. If the actual load varies significantly from the load used to specify the device timing, the timing must be adjusted to reflect the actual load conditions. That is not necessary in most cases. Signals used for control functions or local point-to-point data paths seldom exceed 50 pF. Thus, in most cases, manufacturers' maximum and minimum specified timing can be used without adjustments. Buses are exceptions. Backpanel bus capacitance will usually exceed 50 pF. However, most buses are long line, and thus are not lumped loads and should not be treated as such (they should be treated as transmission lines—see Chapter 7).

2.3.1 Comparison of dynamic operational characteristics

The advanced CMOS logic families have never achieved the standardization that exists for TTL logic families. Advanced CMOS devices from different manufacturers tend to have different alternating current (ac) characteristics even though they have the same letter designation. For example, Table 2.3 shows the variation in specified worst-case operating speed at +125°C with a 5-V power supply for AC 240 buffers and 374 registers from several different manufacturers. Table 2.4 does the same for ACT 240 buffers and 374 registers (in both tables, FACT, AC11, and generic AC devices are included). The variation in specified speeds of AC/ACT devices complicates the process of classifying AC/ACT parts on a generic

TABLE 2.3 54AC240 Buffer and 54AC374 Register Worst-Case Speed at +125°C and 5 V ±10 Percent Supply Levels

		Buffer speed,* ns	Register speed,* ns	
Family	Manufacturer	t_{prop}	t_{prop}	t_{setup}
AC11	Texas Instruments	9.0	10.9	2.5
FACT AC	National	8.5	12.0	5.0
AC	VTC	9.0	11.0	1.0
AC	Harris	7.6	10.8	2.0

*The slower of t_{PHL} or t_{PLH} is shown.

TABLE 2.4 54ACT240 Buffer and 54ACT374 Register Worst-Case Speed at +125°C and 5 V ±10 Percent Supply Levels

		Buffer speed,* ns	Register speed,* ns	
Family	Manufacturer	t_{prop}	t_{prop}	t_{setup}
ACT11	Texas Instruments	11.1	10.9	2.5
FACT ACT	National	9.5	12.0	8.5
ACT	VTC	10.5	13.0	3.0
ACT	Harris	8.6	11.2	2.0

*The slower of t_{PHL} or t_{PLH} is shown.

basis to other advanced CMOS and BiCMOS families. Table 2.5 compares the relative dynamic performance of the major advanced CMOS and BiCMOS logic families by using the best-case and worst-case AC and ACT speeds from Tables 2.3 and 2.4. Table 2.5 is intended as a general guide to the relative performance of the different families. The times listed must not be interpreted as potential system operating speeds. Actual in-circuit operating speed is a complicated function of device input-output characteristics, the arrangement in which devices are interconnected, and the characteristics of the interconnections (see Chapter 11).

Tables 2.3 to 2.5 show that there is not a great deal of difference in the speeds of the various advanced CMOS and BiCMOS logic families. The most notable differences are those for the FCT-A and PCT-A families which are typically advertised as being up to 50 percent faster than standard FCT or PCT families. Devices in the generic AC and ACT and FACT AC and ACT first-generation advanced CMOS logic families tend to have propagation delays

TABLE 2.5 54XX240 Buffer and 54XX374 Register Worst-Case Speeds at +125°C and 5 V ±10 Percent Supply Levels

Family	Buffer speed,* ns	Register speed,* ns	
	t_{prop}	t_{prop}	t_{setup}
AC (best case)	7.6	10.8	2.0
AC (worst case)	9.0	12.0	5.0
ACT (best case)	8.6	10.9	2.5
ACT (worst case)	11.1	13.0	3.0
FCT	9.0	11.0	2.5
FCT-A	5.1	7.2	2.0
FCT-T	NA†	NA†	NA†
PCT	7.5	9.0	2.5
PCT-A	5.1	7.2	2.0
FBT	NA†	NA†	NA†
BC‡	7.5	10.0	2.0
BCT	6.4	11.6	6.5

*The slower of t_{PHL} or t_{PLH} is shown.
†NA: not available.
‡Worst-case speeds at +85°C.

slightly slower than devices built using advanced Schottky technologies. One manufacturer equates ACT speeds to those of ALS.[7] AC11/ACT11 speeds are comparable to those of generic AC/ACT devices. FCT and PCT speeds are typically equated to those of FAST parts.[7,8] FCT-A and PCT-A speeds are slightly faster than those of FAST parts.

2.3.2 Timing parameter adjustments for worst-case conditions

If worst-case device timing is not specified, the system designer must adjust the timing information supplied to reflect actual worst-case operating conditions and process variations. The manufacturer is the best source of information on parameter limits when they are missing and should be contacted for the missing information. However, for those cases where the manufacturer does not have the needed specific parameter limits or is unwilling to supply them, the following rules of thumb can be used to estimate timing parameter limits (for more exact means of estimating timing parameter limits see Chapter 11).

Rules of Thumb for Timing Parameter Adjustments for Worst-Case Conditions
To convert typical +25°C timing parameters to worst case over the commercial temperature range, multiply typical +25°C timing parameters by a factor of 1.5.
To convert typical +25°C timing parameters to worst case over the military temperature range, multiply typical +25°C timing parameters by a factor of 2.
To convert maximum (or minimum) +25°C timing parameters to worst case over the commercial temperature range, multiply maximum (or minimum) +25°C timing parameters by a factor of 1.25.
To convert maximum (or minimum) +25°C timing parameters to worst case over the military temperature range, multiply maximum (or minimum) +25°C timing parameters by a factor of 1.5.

2.3.3 Caution—beware of F_{MAX}

The parameter F_{MAX}, which is listed as the maximum toggle rate or maximum clock rate on data sheets for clocked devices such as counters, flip-flops, and shift registers, should never be used as an indication of the useful speed of a device.[9] It is a measure of what might be achieved with an individual part under ideal conditions with no restrictions on input pulse widths or load conditions.[9] Since most digital systems must operate under conditions that vary greatly from ideal, and since devices must communicate with other devices to serve a useful purpose, F_{MAX} is of little use for actual system timing. Actual signal path propagation times must be used to determine the maximum operating speed of systems, not F_{MAX}.

2.4 Static Input-Output Characteristics

The static [direct current (dc)] parametric specifications provided in data books for advanced BiCMOS/CMOS logic families tend to be more complete than those of many of the older logic families. Most device data sheets have three columns of dc characteristics as shown in Figure 2.2. In most cases, input-output current and voltage limits are specified for typical operating conditions as well as for worst-case conditions over the extremes of both the commercial and military operating conditions (see Section 2.3 for temperature and power-supply voltage limits). In addition, output-voltage limits are typically specified for several load currents.

DC Characteristics for 'ACT Family Devices

Symbol	Parameter	V_{cc} (V)	74ACT		54ACT	74ACT	Units	Conditions
			$T_A = +25°C$		$T_A = -55°C$ to $+125°C$	$T_A = -40°C$ to $+85°C$		
			Typ	Guaranteed Limits				
V_{IH}	Mininum High Level Input Voltage	4.5 5.5	1.5 1.5	2.0 2.0	2.0 2.0	2.0 2.0	V	$V_{OUT} = 0.1V$ or $V_{CC} - 0.1V$
V_{IL}	Maximum Low Level Input Voltage	4.5 5.5	1.5 1.5	0.8 0.8	0.8 0.8	0.8 0.8	V	$V_{OUT} = 0.1V$ or $V_{CC} - 0.1V$
V_{OH}	Mininum High Level Output Voltage	4.5 5.5	4.49 5.49	4.4 5.4	4.4 5.4	4.4 5.4	V	$I_{OUT} = -50\mu A$ $I_{OUT} = -50\mu A$
		4.5 5.5		3.86 4.86	3.70 4.70	3.76 4.76	V	$*V_{IN} = V_{IL}$ or V_{IH} I_{OH} - 24 mA I_{OH} - 24 mA
V_{OL}	Maximum Low Level Output Voltage	4.5 5.5	0.001 0.001	0.1 0.1	0.1 0.1	0.1 0.1	V	$I_{OUT} = 50\ \mu A$ $I_{OUT} = 50\ \mu A$
		4.5 5.5		0.36 0.36	0.50 0.50	0.44 0.44	V	$*V_{IN} = V_{IL}$ or V_{IH} I_{OL} 24 mA I_{OL} 24 mA
I_{IN}	Maximum Input Leakage Current	5.5		± 0.1	± 1.0	± 1.0	µA	$V_I = V_{CC}$, GND
I_{CCT}	Maximum I_{CC}/Input	5.5	0.6		1.6	1.5	mA	$V_I = V_{CC} - 2.1V$
I_{OLD}	†Minimum Dynamic Output Current	5.5			50	75	mA	$V_{OLD} = 1.65V$ Max
I_{OHD}		5.5			- 50	- 75	mA	$V_{OHD} = 3.85V$ Min
I_{CC}	Maximum Quiescent Supply Current	5.5		4.0	80.0	40.0	µA	$V_{IN} = V_{CC}$ or GND

*All outputs loaded; thresholds on input associated with output under test.
†Maximum test duration 2.0 ms, one output loaded at a time.
NOTE: I_{CC} for 54ACT @ 25°C is identical to 74ACT@ 25°C.

Figure 2.2 54ACT/74ACT00 static (dc) characteristics. *(Reprinted by permission of National Semiconductor Corp.)*

2.4.1 BiCMOS and CMOS input-voltage levels

Input-voltage levels (i.e., switching thresholds) of advanced CMOS logic devices fall into two distinct categories, those with true CMOS input levels (AC) and those with TTL input levels (ACT, FCT). Advanced CMOS devices with true CMOS input levels (e.g., AC devices) have input thresholds at approximately 30 and 70 percent of actual device V_{cc}, as shown in Figure 2.3. Thus, CMOS devices with CMOS input levels recognize input levels less than 30 percent of V_{cc} as logic *low* inputs, and those greater than 70 percent of V_{cc} as logic *high* inputs. Levels in the 30 to 70 percent range produce undefined responses.

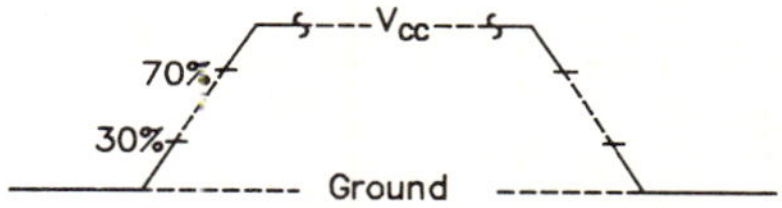

Figure 2.3 Advanced CMOS devices with CMOS input thresholds recognize input levels less than 30 percent of V_{cc} as logic *lows* and input levels greater than 70 percent of V_{cc} as logic *highs*.

In contrast to TTL thresholds, CMOS device threshold levels are a function of V_{cc} levels. Lowering or raising V_{cc} lowers or raises both *high* and *low* input thresholds. For standardization, most CMOS device input levels are specified with V_{cc} at 4.5 V. For example, AC parts are specified to have a minimum *high* input $V_{IH\ MIN}$ of 3.15 V which is 70 percent of 4.5 V, and a maximum *low* input $V_{IL\ MAX}$ of 1.35 V which is 30 percent of 4.5 V.

Worst-case input level limits for CMOS and BiCMOS devices with TTL input levels (e.g., ACT, BCT, FCT) are specified as 0.8 V maximum for *low* inputs $V_{IL\ MAX}$ and 2.0 V minimum for *high* inputs $V_{IH\ MIN}$ (i.e., the same as those of Schottky TTL devices[10]). However, a subtle point often overlooked is that most CMOS and BiCMOS devices with CMOS input structures only meet normal TTL input limits when V_{cc} is equal to 4.5 V. At higher V_{cc} levels, $V_{IH\ MIN}$ may not meet 2.0 V. Both $V_{IL\ MAX}$ and $V_{IH\ MIN}$ tend to increase slightly with increased V_{cc}.

2.4.2 BiCMOS and CMOS input currents

Table 2.6 lists input currents for NANDs and most standard parts for several of the more common advanced CMOS and BiCMOS logic families.

Table 2.7 lists input currents for 54XX240 buffers for several of the more common advanced CMOS and BiCMOS bus interface logic fam-

TABLE 2.6 54/74XX00 NAND Gate Input Current* at the *Low* and *High* Input-Voltage Levels Listed

Family	I_{IL}, μA	@	V_{IL}, V	I_{IH}, μA	@	V_{IH}, V
AC	±1	@	GND†	±1	@	V_{cc}
ACT	±1	@	GND	±1	@	V_{cc}
AC11	±1	@	GND	±1	@	V_{cc}
ACT11	±1	@	GND	±1	@	V_{cc}
BC‡	±1	@	GND	±1	@	V_{cc}

*Current out of a terminal is given a negative sign.
†GND: ground.
‡Input current for 74BC00 is listed; 54BC00 data is not available at this time.

TABLE 2.7 54/74XX240 Buffer Input Current* at the *Low* and *High* Input-Voltage Levels Listed

Family	I_{IL}, μA	@	V_{IL}, V	I_{IH}, μA	@	V_{IH}, V
AC	±1	@	GND†	±1	@	V_{cc}
ACT	±1	@	GND	±1	@	V_{cc}
AC11	±1	@	GND	±1	@	V_{cc}
ACT11	±1	@	GND	±1	@	V_{cc}
FCT	−5	@	GND	+5	@	V_{cc}
FBT	−5	@	GND	+5	@	V_{cc}
BCT	−1000	@	0.5	+20	@	2.7

*Current out of a terminal is given a negative sign.
†GND: ground.

ilies. Most bus interface devices in a given family will have similar input currents. Note that AC/ACT and AC11/ACT11 buffers have the same input specifications as NAND gates (in the respective family).

2.4.3 BiCMOS and CMOS output-voltage levels

Output levels of advanced CMOS devices vary greatly with load conditions and temperature. Under static conditions, AC devices with complementary CMOS output structures go to V_{cc} or ground when driving high-impedance loads such as CMOS inputs. However, AC CMOS output levels are generally not specified with ground or V_{cc} output levels but instead are typically specified with very light 50- or 100-μA loads. ACT and FCT device output levels are specified with currents that range from 1 to 24 mA for standard devices and up to 64 mA for bus interface devices. Thus, it is difficult to pick a standard set of conditions for defining advanced CMOS output levels. However, for the purposes of this book, and in most CMOS application information, CMOS output levels are typically specified as follows:

1. A worst-case true CMOS *high* output is defined as being between V_{cc} and 0.1 V less than V_{cc} with V_{cc} at 4.5 V, and a worst-case CMOS *low* output is between ground and 0.1 V above ground.[3,5,7] Output levels within 0.1 V of power or ground are only possible when devices are lightly loaded (100 μA or less), i.e., when driving high-impedance CMOS inputs.
2. A worst-case TTL-compatible ACT CMOS minimum *high* output is defined as 3.7 V, and a worst-case maximum *low* output is 0.5 V (in some cases 0.55 V) for 54ACT devices over the full range of military operating conditions.[3,5,7] In most cases, ACT devices, with TTL-compatible inputs, have their output levels specified under load current conditions similar to those used to specify AS output levels (see Tables 2.8 and 2.9).
3. A worst-case TTL-compatible FCT (including FCT-T) CMOS minimum *high* output is defined as 2.4 V, and a worst-case maximum *low* output is 0.55 V for 54FCT devices over the full range of military operating conditions.[8] In most cases, FCT devices, with TTL-compatible inputs, have their output levels specified under load current conditions similar to those used to specify advanced Schottky output levels (see Tables 2.8 and 2.9).

Output levels of BiCMOS devices tend to be more stable with changing load conditions and temperature than CMOS outputs. BiCMOS outputs behave much the same as advanced Schottky TTL output. A worst-case BiCMOS minimum *high* output is typically 2.4 V, and a worst-case maximum *low* output is 0.5 V or 0.55 V.[11,12] In most cases, BiCMOS devices have their output levels specified under load current conditions similar to those used to specify advanced Schottky output levels (see Tables 2.8 and 2.9).

TABLE 2.8 54XX00 NAND Gate Output Voltage at the Load Currents* Shown and with V_{cc} Equal to 4.5 V

Family	V_{OL}, V	@	I_{OL}, mA	V_{OH}, V	@	I_{OH}, mA
54AC00	0.1	@	0.05	4.4	@	− 0.05
	0.5	@	24	3.7	@	− 24
54ACT00	0.1	@	0.05	4.4	@	− 0.05
	0.5	@	24	3.7	@	− 24
54AC11000	0.1	@	0.05	4.4	@	− 0.05
	0.5	@	24	3.7	@	− 24
54ACT11000	0.1	@	0.05	4.4	@	− 0.05
	0.5	@	24	3.7	@	− 24
74BC00†	0.5	@	20	2.5	@	− 1

*Current out of a terminal is given a negative sign.
†Input current for 74BC00 is listed; 54BC00 data is not available at this time.

TABLE 2.9 54XX240 Buffer Output Voltage at the Load Currents* Shown and with V_{cc} Equal to 4.5 V

Family	V_{OL}, V	@	I_{OL}, mA	V_{OH}, V	@	I_{OH}, mA
54AC240	0.1	@	0.05	4.4	@	−0.05
	0.5	@	24	3.7	@	−24
54ACT240	0.1	@	0.05	4.4	@	−0.05
	0.5	@	24	3.7	@	−24
54AC11240	0.1	@	0.05	4.4	@	−0.05
	0.5	@	24	3.7	@	−24
54ACT11240	0.1	@	0.05	4.4	@	−0.05
	0.5	@	24	3.7	@	−24
54FCT240	0.55	@	48	2.4	@	−12
54FBT240	0.55	@	48	2.4	@	−12
54BCT240	0.55	@	48	2.4	@	−3

*Current out of a terminal is given a negative sign.

Tables 2.8 and 2.9 list dc output voltage versus current for several of the more common advanced CMOS and BiCMOS logic families. Table 2.8 is for 54XX00 NAND gates, and Table 2.9 for 54XX240 buffers. In general, the drive and output levels shown for each family's 54XX00 NAND gate are representative of all standard devices in that family. However, caution must be exercised because some devices within a family may have different characteristics, and *drivers,* by definition, are expected to have more drive than NAND gates, but that is not always the case for the advanced CMOS logic families—they may have the same ratings. The drive and load levels shown for each family's 54XX240 buffers are representative of all buffers and bus interface devices in the family, but again caution must be exercised because some bus interface devices within a family may have different characteristics. The device specifications for a given device should always be consulted and closely studied before the device is applied. Failure to read and understand device specifications remains one of the major sources of design error. Data sheet specifications should be read over and over. A great deal of attention should be given to the notes and fine print. For complex devices it is best to find and consult someone that has some experience with that part of interest, but having talked to someone about a part is not a very good excuse for having not studied the data sheet when the part does not work as expected.

2.5 Compatibility of Logic Families

The various advanced TTL, CMOS, and BiCMOS logic families *with TTL input levels* are voltage level compatible and can be mixed as long as current drive specifications are not exceeded.[3,13] Advanced CMOS logic devices that have inputs compatible with TTL levels have

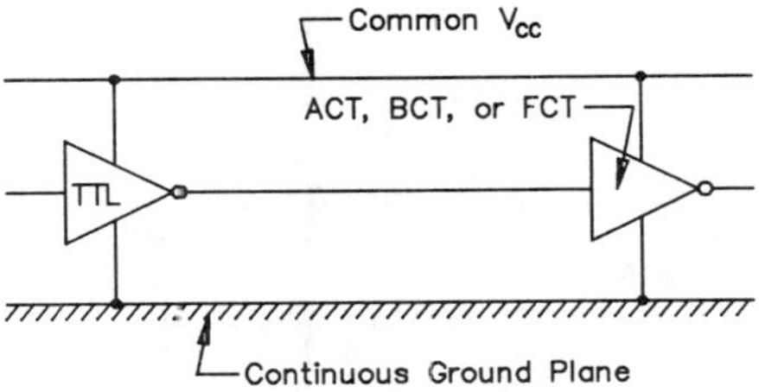

Figure 2.4 No additional circuitry is required to transfer signal from devices with TTL output levels to ACT, BCT, or FCT devices.

logic family letter designators that end with T. For example, the ACT and FCT families have input thresholds that are designed to be compatible with TTL levels. No additional circuitry is required to transfer signals from TTL or BiCMOS sources to ACT, BCT, FCT, or HCT devices (see Figure 2.4). Current drive capability of TTL-input-compatible CMOS families are similar to those of advanced Schottky TTL and BiCMOS families, and CMOS input currents are of such a low magnitude that all TTL and BiCMOS families can source or sink several orders of magnitude more dc current than is needed to drive a very large number of CMOS inputs. In general, dc drive limits should never be exceeded in mixed advanced BiCMOS/CMOS systems since input currents are typically only a few microamperes and output drive is typically 20 mA or more. *Caution:* Texas Instruments BCT bus interface family has 1-mA input current sink requirements. Extra care must be taken to ensure that signals are not overloaded in systems with mixed logic families. Particular care must be taken in those systems where devices with high input current requirements are mixed with low-drive LSI devices, ASICs, or some of the older logic families. In all applications, an audit of signal loading should be performed as one of the final design tasks.

The various CMOS and BiCMOS families with TTL output levels and CMOS families with CMOS input levels cannot be mixed without the use of additional interface circuitry for TTL- to CMOS-level signal transfers. The output voltage levels of all advanced CMOS devices are compatible with TTL input levels, but CMOS input levels are not compatible with TTL output levels. Thus, additional circuitry is required to transfer signals from TTL to CMOS voltage levels, but not in the other direction. Devices with CMOS input levels are TTL *low*-level compatible, but are not TTL *high*-level compatible. The minimum *high*-output specification of TTL, CMOS, and BiCMOS devices with TTL-level outputs is 2.4 or 2.5 V, and the minimum *high*-input requirement for CMOS devices is 3.15 V (see Tables 2.8 and 2.9). Thus, when signals must be transferred from devices with TTL output levels to CMOS devices with CMOS thresholds, level shifters or pull-up resistors are required (see Figure 2.5).[3,13] A number of problems are as-

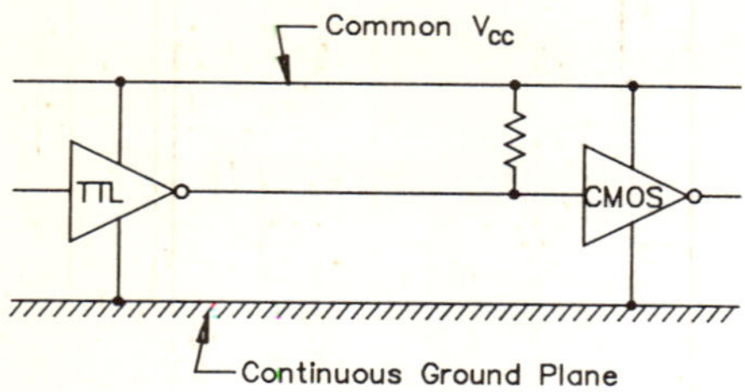

Figure 2.5 When signals must be transferred from devices with TTL output levels to CMOS devices with CMOS input levels, pull-up resistors or level-shifting circuits are required.

sociated with pull-ups such as degraded rise times, extra power dissipation, and additional current-sinking requirements for drivers. The best means of transferring from TTL levels to CMOS levels is to use ACT CMOS devices with TTL input levels and CMOS output levels.

Advanced CMOS logic devices have output levels that are compatible with TTL input levels when powered from a common source. *Low* CMOS output levels are approximately the same as TTL output *low* levels. *High* CMOS output levels are greater than normal TTL *high* levels, but do not exceed maximum TTL input limits. In many applications, higher *high* levels are of benefit. In noisy applications, they provide more noise margin. Higher *high* levels do have some disadvantages. They generate more system noise due to larger signal transitions and increase *high*-to-*low* signal response since they require more time to reach *low* switching thresholds. However, in most applications, transferring signals from CMOS devices to TTL devices requires no special circuitry provided that all devices are powered with the same power supply (see Figure 2.6).[3,13]

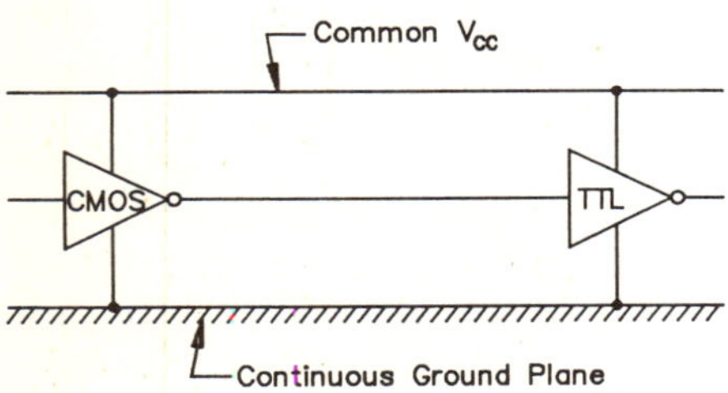

Figure 2.6 Transferring signals from CMOS devices with CMOS output levels to devices with TTL input levels requires no special circuitry provided that all devices are powered with the same power supply.

Dynamic as well as static interface compatibility must be considered when logic families are mixed. Mixing slow and fast devices can cause problems. For example, in synchronous systems, signals originating from fast devices may go away before they can be captured by slow devices (see hold-time requirements discussion in Chapter 10).

2.6 CMOS or TTL Levels?

The system designer has the choice of using devices with either TTL or CMOS levels. In general, devices with TTL input and output levels are the best choice; TTL output swings create less noise than CMOS rail-to-rail output swings and dissipate less dynamic power. As operating speed increases there is less time for noise to die out, and dynamic power becomes more and more of a concern since it is a direct function of frequency (see Chapter 3). Other advantages of TTL levels include: Many LSI and VLSI parts, such as memories and PLDs, are not available with CMOS inputs or output levels. Most standard test equipment is designed for TTL levels. The use of TTL levels allows a fallback to bipolar TTL parts if it is suddenly discovered that the preferred parts are not available. Only where maximum noise margin is needed, such as for clock signals, should CMOS levels be considered.

Once either TTL or CMOS interface levels are selected, the system designer's task is to select devices or drivers with adequate drive. In general, the drive characteristic needed depends upon whether lightly loaded control signals or heavily loaded buses are being driven. Standard advanced BiCMOS/CMOS logic functions typically have drive levels of 20 to 24 mA, which are adequate for most lightly loaded applications. However, drive requirements in high-performance systems are a complicated function of dynamic loading and transmission-line driving requirements and whether first incident wave switching is required (see Chapters 6 and 7). Static (dc) drive is usually not an issue in advanced BiCMOS/CMOS systems since the dc input current for devices with CMOS input stage is typically less than $\pm 10\ \mu A$, but caution must always be exercised—some BiCMOS bus interface parts have 1-mA input currents.

2.7 Noise Margins of BiCMOS/CMOS Logic Families

Noise margin is the difference between the worst-case output level of a driving device and the worst-case input level at which the receiving device no longer recognizes the input as the intended *high* or *low* logic level. Both static and dynamic noise margins are of concern in high-speed systems. Static noise margins define the low-frequency safety margins built into device input-output levels. Dynamic noise margin relates to the sensitivity of logic devices to noise spikes. For example, Figure 2.7 illustrates static input-output voltage level limits and static noise margins for most BiCMOS and advanced CMOS devices with TTL levels. Notice that *high* and *low* noise margins are different. Most BiCMOS devices and most CMOS devices with TTL input levels have more *high* than *low* noise margin. Most CMOS devices with true CMOS inputs have equal *high* and *low* noise margins. Advanced CMOS devices with true CMOS levels also have more noise margin than CMOS or BiCMOS devices with TTL levels, but they tend to need

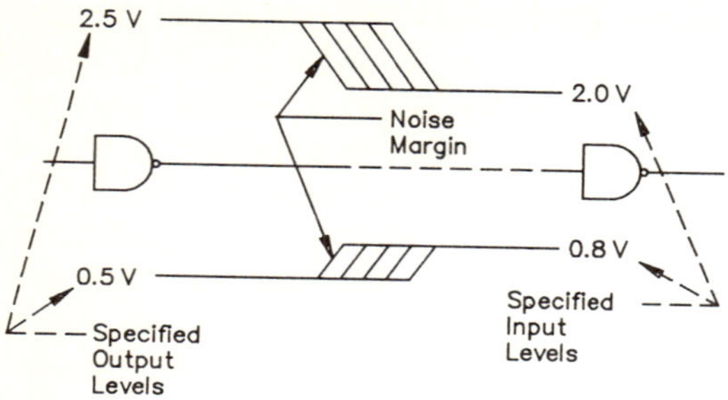

Figure 2.7 Static TTL input-output voltage-level limits used to define static noise margins for most BiCMOS and advanced CMOS devices with TTL input and output levels.

more noise margin since their wider, and in many cases faster, voltage swings generate more noise.

Device noise margin and system noise immunity are of great concern when advanced BiCMOS/CMOS devices are used; their fast edges generate much more noise than digital system designers are accustomed to dealing with. Yet, their static noise margins are approximately the same as those of older, slower bipolar and CMOS devices. Their dynamic noise margin tends to be less than that of the older logic families. Because they are designed for high-speed operation, they are more apt to react to narrow noise spikes.

An understanding of the static and dynamic noise margin of each device and each logic family used in a system is needed so that maximum advantage is taken of the existing noise margin. For example, most devices with TTL input and output levels have more *high*-level noise margin than *low*-level noise margin. That additional *high*-level noise margin should be taken advantage of by arranging the polarity of critical signals that have low duty cycles so that they are in the *high* state most of the time. Thus, critical control signals, such as resets, should be *high* when inactive. Only devices that clock on *low*-to-*high* transitions should be used so that clock signals are *high* during the noisy time immediately following clock switching edges.

In general, systems using advanced CMOS devices with true CMOS input-output levels have greater noise margins than systems using TTL, CMOS, or BiCMOS devices with TTL input-output levels. However, it is not clear whether there is any system advantage since advanced CMOS devices tend to generate more noise. Thus, CMOS devices need more noise margin than TTL devices. Where advanced CMOS devices with TTL input thresholds are connected to CMOS devices with CMOS output levels, *high*-level noise margin is greater, but

TABLE 2.10 Specified Worst-Case Input-Output Voltage Levels

Family	Input levels		Output levels	
	$V_{IL\ MAX}$, V	$V_{IH\ MIN}$, V	$V_{OL\ MAX}$, V	$V_{OH\ MIN}$, V
AC	1.35	3.15	0.1	4.4
ACT	0.8	2.0	0.5	3.7
AC11	1.35	3.15	0.1	4.4
ACT11	0.8	2.0	0.5	3.7
FCT	0.8	2.0	0.55	2.4
BC*	0.8	2.0	0.5	2.5
FBT	0.8	2.0	0.55	2.4
BCT	0.8	2.0	0.55	2.4

*Worst-case levels at +85°C.

low-level noise margin is approximately the same as in systems where advanced Schottky devices are interconnected. CMOS device input-output levels vary greatly with load conditions, and thus it is difficult to define generic noise margins (as can be done for TTL circuits). Actual loaded input-output levels must be determined to establish the actual noise margin for a given application.

Table 2.10 shows the worst-case specified input and output voltage levels for the input and output currents listed in Tables 2.6 through 2.9. The difference between the worst-case input and output levels is the tolerance that a given set of devices, or a given system built with a particular logic family, has for noise and other imperfection in the reference and interconnection systems.

2.7.1 Static noise margins of advanced CMOS logic families

Table 2.11 lists static *high* and *low* noise margins for the AC, ACT, and FCT logic families based on the above defined input-output limits.

TABLE 2.11 Static Noise Margins of Advanced CMOS Logic Families

Family	*Low*-level noise margin, V	*High*-level noise margin, V
AC	1.25	1.25
ACT	0.3	1.7
AC11	1.25	1.25
ACT11	0.3	1.7
FCT	0.25	0.4

2.7.2 Static noise margins of BiCMOS logic families

The BiCMOS logic families have TTL input-output levels and thus have static noise margins equivalent to TTL devices. Table 2.12 lists worst-case static noise margins for the various BiCMOS families over the military temperature range, with V_{cc} equal to 4.5 V, and static output current conditions (typically 3 mA or less). The values listed in Table 2.12 can be derived from the specified worst-case input and output voltage levels listed in Table 2.10 (also see References 8, 11, and 12). Note, that worst-case *high* and worst-case *low* noise margins are different for all the families. For all the BiCMOS logic families, *low* noise margin is either 250 or 300 mV and *high* noise margin is either 400 or 500 mV. A 150- or 200-mV difference between *high* and *low* noise margin may not seem significant, but advantage should be taken of higher noise margin where it exists. In many applications, 150 or 200 mV of additional noise margin may significantly enhance system reliability.

TABLE 2.12 Noise Margins of BiCMOS Logic Families

Family	*Low*-level noise margin, mV	*High*-level noise margin, mV
BC	300	500
BCT	250	400
FBT	250	400

2.7.3 Dynamic noise immunity

Dynamic noise immunity refers to the sensitivity of logic devices to narrow pulses. Most of the older, slower logic families were insensitive to narrow pulses of only a few nanoseconds. That is not the case with advanced BiCMOS/CMOS devices. They tend to react to pulses as narrow as 2 or 3 ns. Thus, narrow ground bounce and cross-coupling spikes caused by the fast edges of advanced BiCMOS/CMOS devices are of great concern (see Chapter 4). Some manufacturers' data books have information on pulse-rejection characteristics, but the information tends to be on a generic logic family basis rather than for specific devices. Most advanced BiCMOS/CMOS devices with TTL-level inputs have *low*-level pulse-rejection characteristics similar to that shown in Figure 2.8. Most advanced CMOS devices with true CMOS input levels have pulse-rejection characteristics similar to those shown in Figures 2.9 and 2.10. As Figures 2.8 through 2.10 show, most devices have some additional immunity to upset or output changes from very narrow input spikes. Most pulse-rejection curves are "typical" data. Thus, pulse-rejection curves should be viewed as general guidelines as

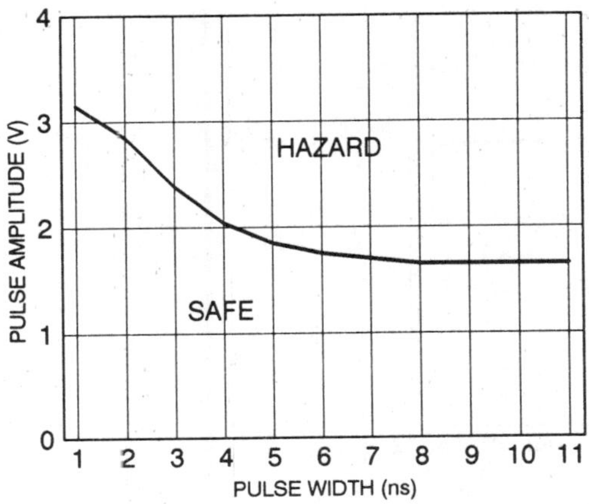

Figure 2.8 *Low*-state pulse-rejection characteristics of IDT FCT244 when operating with V_{cc} equal to 5 V. Pulse width measured at 50 percent amplitude. Most advanced BiCMOS/CMOS devices with TTL-input thresholds have similar characteristics. (*Reprinted by permission of Integrated Device Technology, Inc.*)

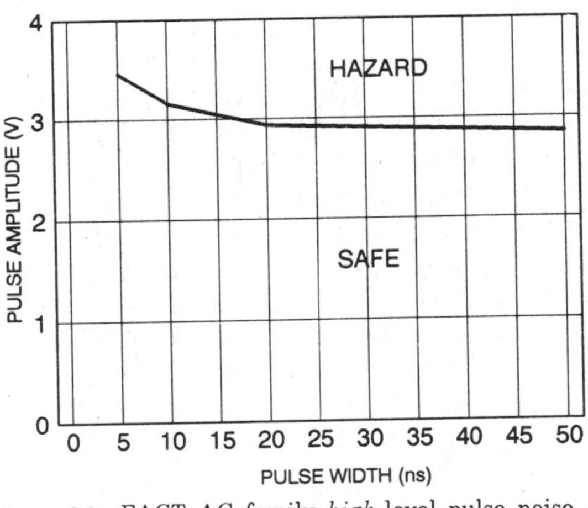

Figure 2.9 FACT AC family *high*-level pulse noise margin with V_{cc} equal to 5 V. Pulse amplitude is measured from V_{cc}. (*Reprinted by permission of National Semiconductor Corp.*)

to expected performance—not as an absolute definition. Device pulse-rejection characteristics are difficult to precisely define for all conditions. Clock and asynchronous circuits should not rely on pulse-rejection characteristics of devices to prevent malfunctions. Clock circuits and control signals to asynchronous inputs must be designed to be free of stray pulses.

2.7.4 Noise margin summary

In general, advanced CMOS devices with CMOS input thresholds have more noise margin than advanced CMOS and BiCMOS devices

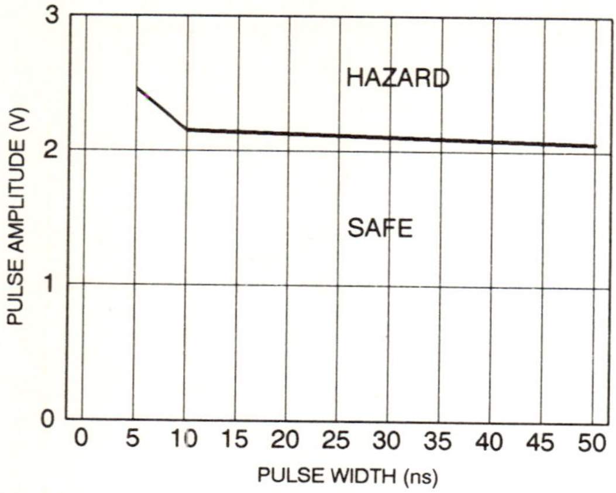

Figure 2.10 FACT AC family *low*-level pulse noise margin with V_{cc} equal to 5 V. Pulse amplitude is measured from ground. (*Reprinted by permission of National Semiconductor Corp.*)

with TTL input and output levels. However, more noise margin is needed since the large voltage swings and fast edges of first-generation advanced CMOS devices with full rail-to-rail output swings generate more noise than second-generation advanced CMOS and BiCMOS devices with TTL-level output swings. There are subtle system implications with respect to CMOS input thresholds that are often overlooked. Both CMOS and BiCMOS (with CMOS input structures) input threshold levels are a function of V_{cc} where TTL device thresholds are a function of voltage drops across diodes or base-emitter junctions that are referenced to ground. Since CMOS thresholds fluctuate with V_{cc} changes, where CMOS devices must interface with other devices (i.e., not other CMOS devices), extra care must be taken to ensure compatibility over worst-case V_{cc} levels.

Because of the noise-generating characteristics of advanced BiCMOS/CMOS devices, system noise immunity must be optimized by selecting signal polarity with regard to optimum noise rejection. Most TTL, CMOS, and BiCMOS devices with TTL input-output levels have more *high*-level noise margin than *low*-level noise margin (the FCT family is an exception). Thus, in those systems that use CMOS and BiCMOS (or TTL) devices with TTL input levels, critical signals that have low-duty cycles should be arranged so that they are in the *high* state most of the time. Only clocked devices that clock on *low*-to-*high* transitions should be used so that clock signals are *high* when system noise is at a maximum (noise is a maximum immediately following clock switching edges). The same is true when CMOS devices with rail-to-rail outputs are used to drive CMOS or BiCMOS (or TTL) devices with TTL inputs; *high*-level noise margin is greater than *low*-level noise margin by even a greater ratio (than when all devices have TTL levels). When CMOS or BiCMOS devices with TTL input levels are driven with TTL devices, *high* signals have maximum noise margin, and the same considerations apply with regard to optimum signal polarity for maximum noise rejection.

Where CMOS devices with true CMOS input levels are used, *high* and *low* noise margin specifications are the same. Thus, either signal polarity is acceptable for critical low-duty cycle signals (i.e., reset signals). However, for commonality with parts with TTL levels, it is best to arrange critical low-duty cycle signals so that they are in the *high* state most of the time since most systems will have a mix of devices with both TTL and CMOS levels.

2.8 References

1. Tharalson, Gary: "Which Logic Family is Best for You?" *Electronic Products,* May 1989, pp. 53–57.
2. Matthews, Paul L.: *Choosing and Using ECL,* McGraw-Hill, New York, 1984.
3. *FACT—Advanced CMOS Logic Databook,* National Semiconductor Corp., Santa Clara, Calif., 1988.
4. *IC MASTER,* Hearst Business Communications Inc., Garden City, N.Y., Current year.
5. *Advanced CMOS Logic Data Book,* Texas Instruments Inc., Dallas, Tex., 1988.
6. Baker, Stan: "Extending TTL," *Electronic Engineering Times,* October 16, 1989, pp. 41, 52.
7. *Advanced CMOS Logic Data Book,* VTC Inc., Bloomington, Minn., 1988.
8. *High Performance CMOS Data Book Supplement,* Integrated Device Technology Inc., Santa Clara, Calif., 1989.
9. *Bipolar Microprocessor Logic and Interface Data Book,* Advanced Micro Devices Inc., Sunnyvale, Calif., 1981.
10. *FAST—Advanced Schottky TTL Databook,* National Semiconductor Corp., Santa Clara, Calif., 1988.
11. *Bi-CMOS Logic Data,* Motorola Inc., Phoenix, Ariz., 1989.
12. *BiCMOS Bus Interface Logic Data Book,* Texas Instruments Inc., Dallas, Tex., 1989.
13. *RCA High-Speed CMOS Logic ICs Databook,* RCA Corp., Sommerville, N.J., 1986.

2.9 Bibliography

Advanced CMOS Logic Designer's Handbook, Texas Instruments Inc., Dallas, Tex., 1987.

Alvarez, Antonio R.: *BiCMOS Technology and Applications,* Kluwer Academic Publishers, Norwell, Mass., 1989.

Bus Interface Products 1988 Data Book, Advanced Micro Devices Inc., Sunnyvale, Calif., 1987.

FCT—Fast, CMOS, TTL-Compatible Logic TECH NOTE, Integrated Device Technology Inc., Santa Clara, Calif., 1986.

Frederiksen, Thomas M.: *Intuitive CMOS Electronics,* McGraw-Hill, New York, 1989.

Funk, Dick: "Design Guidelines for CMOS Logic Systems," *Electronic Products,* March 28, 1984, pp. 75–79.

Tuck, Barbara: "TI's BiCMOS Bus Interface ICs Slash Standby Current," *Electronic Products,* June 15, 1987, pp. 17–19.

Wong, Thomas: "Not All BiCMOS is Created Equal," *Electronic Engineering Times,* December 25, 1989, pp. 19, 22, 24.

Chapter

3

Advanced BiCMOS and CMOS Logic Circuits

High-speed digital devices cannot be applied successfully when the design effort is limited to functional logic considerations only. The electrical characteristics and the limitations of the devices being applied must be understood. A knowledge of the minute details of the internal structure of logic devices is not necessary, but an understanding of device interface circuitry is needed to understand device and interconnection system interaction. However, schematics and circuit details of today's logic devices, which are needed to understand interface operations, are not easy to come by. That has not always been the case; most early SSI circuit logic device data sheets included schematics that showed transistor-level device implementations, but current data sheets seldom include schematics. Sometimes schematics of basic SSI devices, such as gates and drivers, can be found in the introductory section of data books. However, for MSI circuit devices, such as 4-bit counters, and for LSI circuit devices, such as ALUs, the most that is provided is a functional-block-level representation. It would be impractical to show detailed internal schematics for MSI and LSI devices in data books and detailed information on internal circuitry would be of little use to system designers, but interface circuitry or limits which are needed are seldom provided either. This chapter's purpose is to provide some insight into BiCMOS and CMOS logic device circuit operation.

3.1 Advanced CMOS Logic Circuits

Advanced CMOS logic devices are built with *n*-channel and *p*-channel enhancement-mode metal-oxide-semiconductor field-effect transistors

(MOSFETs). To understand CMOS logic circuit operation, designers must have a basic understanding of how MOSFETs function. The symbols and terminal nomenclature for enhancement-mode FETs are shown in Figure 3.1.[1] Enhancement-mode n-channel MOSFETs oper-

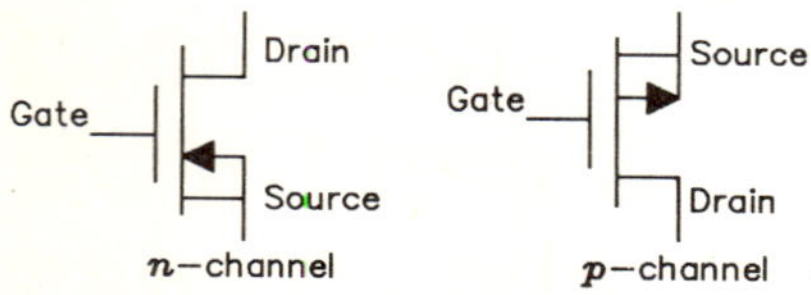

Figure 3.1 MOSFET schematic symbols.

ate somewhat analogous to *npn* bipolar transistors, and enhancement-mode p-channel MOSFETs operate somewhat analogous to *pnp* bipolar transistors (see Figures 3.2 and 3.3).

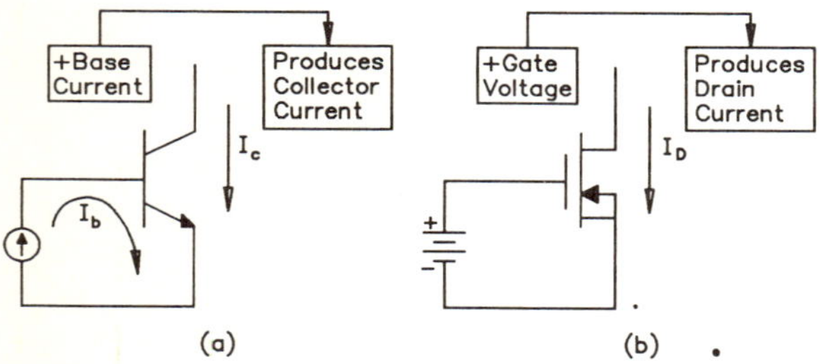

Figure 3.2 Bipolar *npn*- and n-channel MOSFET equivalency.

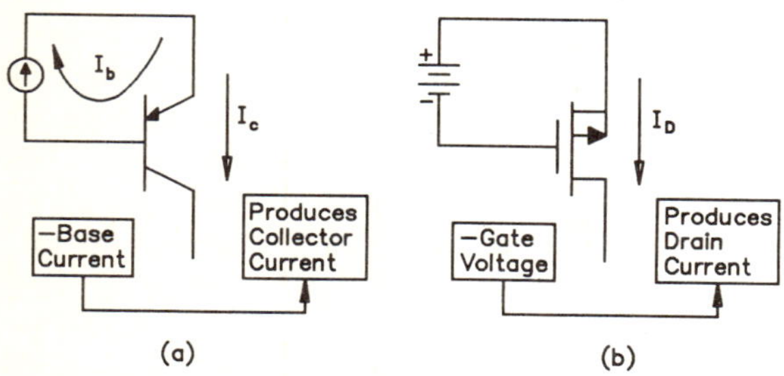

Figure 3.3 Bipolar *pnp*- and p-channel MOSFET equivalency.

Analyzing and understanding CMOS circuit operation is greatly simplified by substituting simple static equivalent circuits or models for the internal MOSFETs. The equivalent circuits or models shown and described in Figure 3.4 are not exact, but they are adequate for most static analysis. The equivalent circuit for an ON MOSFET is a resistor between the drain and source and an open circuit for the gate connection (see Figure 3.4). The equivalent circuit for an OFF MOSFET is an open circuit between all terminals. Typical ON resistance of the MOSFETs used in advanced CMOS devices is near 10 Ω. The OFF impedance of most MOSFETs is extremely high; OFF MOSFETs have some leakage current, but in most digital applications OFF leakage currents are of such a low magnitude that OFF MOSFETs can be considered open circuits. The same is true of the gate input impedance; in most cases it is extremely high.

Most advanced CMOS logic devices are made with MOSFET circuits arranged in complementary n-channel and p-channel pairs.[1] The complementary pairs use p-channel MOSFETs for pull-ups and n-channel MOSFETs for pull-downs. Logic devices implemented with complementary pairs have no dc paths between V_{cc} and ground; one MOSFET in each complementary pair is ON and the other OFF at all times. Hence, CMOS devices dissipate little dc power. A typical complementary inverter stage and its equivalent circuit[2] are shown in Figure 3.5. The

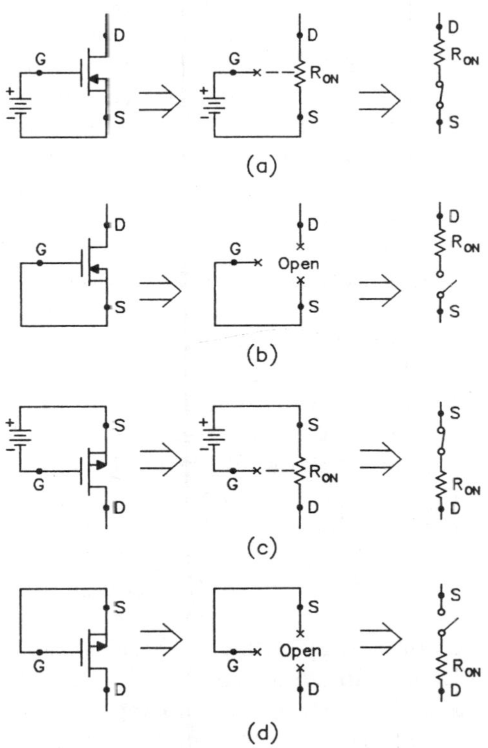

Figure 3.4 Equivalent circuits for ON and OFF MOSFETs.

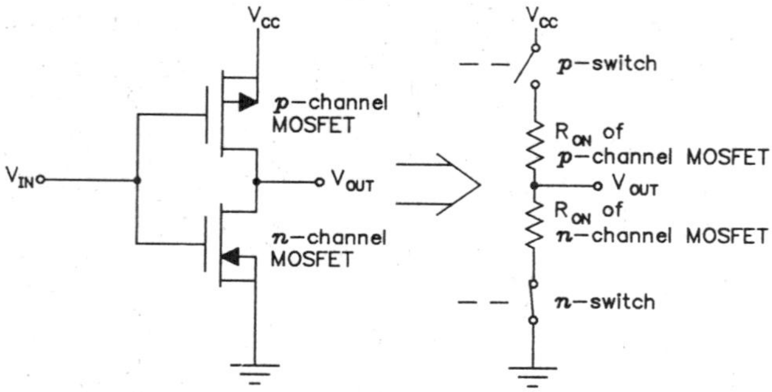

Figure 3.5 CMOS inverter and equivalent circuits.

equivalent circuit for the inverter consists of two resistors and two ideal switches in series between V_{cc} and ground. In the equivalent circuit, the switching action of the MOSFETs is represented by the ideal *p* switch and *n* switch. The two resistors represent the ON resistance of the two MOSFETs. The inverter functions as follows: When the input level is near V_{cc}, the *p*-channel pull-up MOSFET is OFF and the *n*-channel pull-down MOSFET is ON. When the input level is near ground, the *n*-channel pull-down MOSFET is OFF and the *p*-channel pull-up MOSFET is ON. When the *n*-channel pull-down MOSFET is ON, the output level is near ground. When the *p*-channel pull-up MOSFET is ON, the output level is near V_{cc}. Most first-generation advanced CMOS logic devices (see Chapter 2 for the definition of first-generation devices) have complementary outputs that operate in a similar fashion and have signal levels that switch between ground and V_{cc}. Signals that switch between ground and V_{cc} are sometimes described as rail-to-rail levels.

Advanced CMOS technology's chief selling point has been low power consumption and cool system operation, but as system operating speeds move into the 30- to 50-MHz range, first-generation advanced CMOS with its wide, fast, rail-to-rail voltage swings may dissipate more power than advanced Schottky TTL devices. In addition, rail-to-rail voltage swings intensify most of the problems associated with high-speed logic applications, such as crosstalk, ground bounce, and transmission-line effects. Thus, some second-generation advanced CMOS logic families, notably the FCT-T family, have returned to TTL-level output voltages. One means of limiting output voltage swings is to use source follower pull-up circuits in output stages (see Figure 3.6). Source follower pull-up circuits cut off when the voltage across the pull-up MOSFET is approximately equal to the threshold voltage of the MOSFET. MOSFETs used in advanced CMOS logic circuits tend to have threshold voltages in the 1 V or less range. Thus, source follower outputs tend to cut off and cease to supply current at

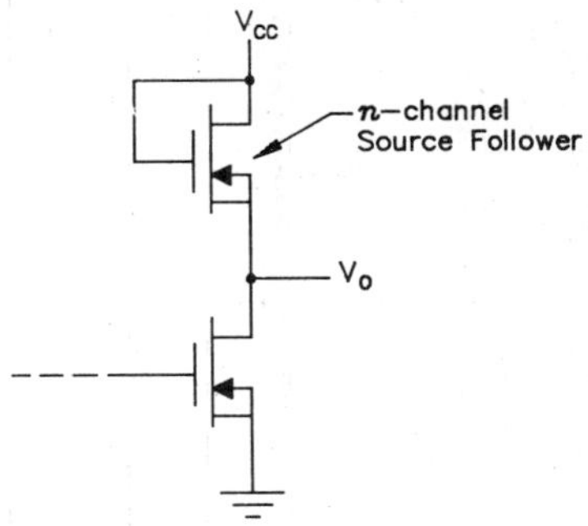

Figure 3.6 CMOS output stage with a source follower pull-up.

about 4 V (under typical conditions) which means devices with source follower outputs typically have *high* outputs that are near 4 V. Thus, advanced CMOS devices with source follower pull-up output stages have output levels approximately the same as advanced Schottky TTL devices. Limiting signal swings to TTL levels reduces dynamic power dissipation, crosstalk, and ground bounce and greatly simplifies the system designer's overall task. Logic families with source follower outputs by necessity have TTL-level input thresholds since under worst-case conditions, *high*-output levels may not meet true CMOS *high*-input level requirements.

The input impedance of advanced CMOS logic devices is very high when signals are between V_{cc} and ground (i.e., within the rails). MOSFETs have very high gate input impedance which is ideal from a signal loading standpoint, but MOSFET gate impedance is so high that all inputs must have some form of electrostatic damage (ESD) protection circuitry. The typical ESD network consists of diodes connected between signal lines and power and ground (see Section 3.4). Some inputs may also have high-speed clamp diodes to limit signal overshoots to improve signal quality in transmission-line environments (see Chapter 7). Under normal signal conditions, these ESD and signal clamp diodes are reverse-biased; they come into play only when signals exceed the rails.

Input threshold levels are typically set by controlling the relative size of the complementary input MOSFETs. If they are equal, the threshold is centered between power and ground. If one device is larger than the other, the threshold shifts in that direction. Thus, no special level-shifting input circuitry is required for advanced CMOS devices with TTL input levels; they have the same high-input impedance of advanced CMOS devices with true CMOS input levels.

CMOS logic structures. Logic functions, such as NORs and NANDs, are implemented using multiple complementary structures.[1,2] For ex-

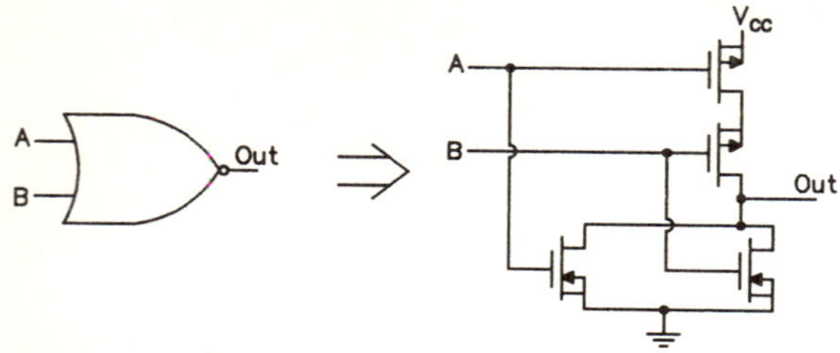

Figure 3.7 Two-input CMOS NOR gate.

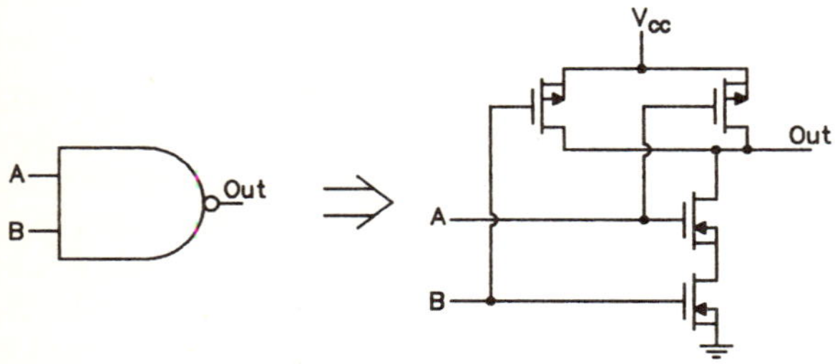

Figure 3.8 Two-input CMOS NAND gate.

ample, two-input CMOS NOR gates are implemented with two series pull-up and two parallel pull-down MOSFETs as shown in Figure 3.7. The opposite arrangement, parallel pull-ups and series pull-downs, is used to implement CMOS NAND gates. For example, two-input NAND gates are implemented as shown in Figure 3.8.

To implement logic functions with multiple inputs, the number of complementary pairs of series and parallel MOSFETs can be increased as needed (up to some practical limit). However, as the number of series and parallel devices increases, output *high* and *low* impedances become increasingly more unbalanced. For example, the variable output impedance of an unbuffered two-input NOR gate (Figure 3.7) is shown in Figure 3.9. Advanced CMOS devices use output buffers to isolate internal logic function structures from line and load capacitances (which tend to be much larger than internal device node capacitances) to balance output drive and response.

3.2 BiCMOS Logic Circuits

BiCMOS logic circuits are built with a semiconductor technology that combines CMOS and bipolar transistors on the same chip. Most discrete BiCMOS logic devices are built with CMOS cores and bipolar output circuits although some discrete BiCMOS devices also have bipolar input structures. Some BiCMOS LSI and ASIC devices use bipo-

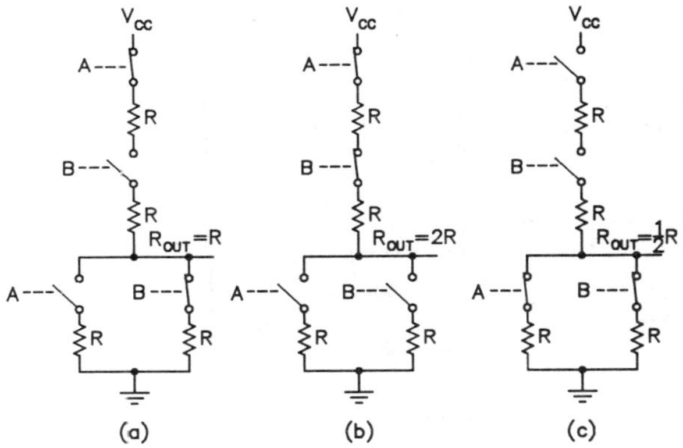

Figure 3.9 Variable output impedance of a CMOS two-input NOR gate.

lar transistors in the core as well as on external interfaces.[3] Bipolar outputs and internal drive stages provide higher drive and more stable drive characteristics than can be achieved with CMOS technology alone. BiCMOS technology can provide either ECL or TTL interfaces. However, the discussion of BiCMOS circuits in this chapter is limited to those with TTL-level interfaces. BiCMOS logic devices, with TTL logic levels and interface circuitry coupled to a CMOS core, offer the best of both CMOS and bipolar advanced Schottky TTL technology. BiCMOS has the following advantages over CMOS: improved signal quality, less crosstalk, less dynamic power dissipation, and less transient-current generation which means less ground bounce. The advantage of BiCMOS over bipolar TTL is less static power dissipation.

Figure 3.10 shows the internal circuitry of a BiCMOS logic device.[4] Typically, the input and core logic of BiCMOS logic devices (those that operate with TTL levels) are implemented using advanced CMOS technology and conventional CMOS circuit techniques. Output stages are implemented using bipolar technology. Most outputs are implemented with conventional totem-pole output stages[5] (as shown in Figure 3.11) that are similar in structure and operation to those used in advanced Schottky TTL devices.[6]

Bipolar totem-pole output stages have several system advantages:

1. Totem-pole stage output-voltage (signal) swing is typically about 3.5 V instead of near 5 V as is the case for normal CMOS outputs. Less output-voltage swing means less dynamic power (see Section 3.6), less crosstalk, and less transient switching currents which means fewer of the problems, such as ground bounce and power and

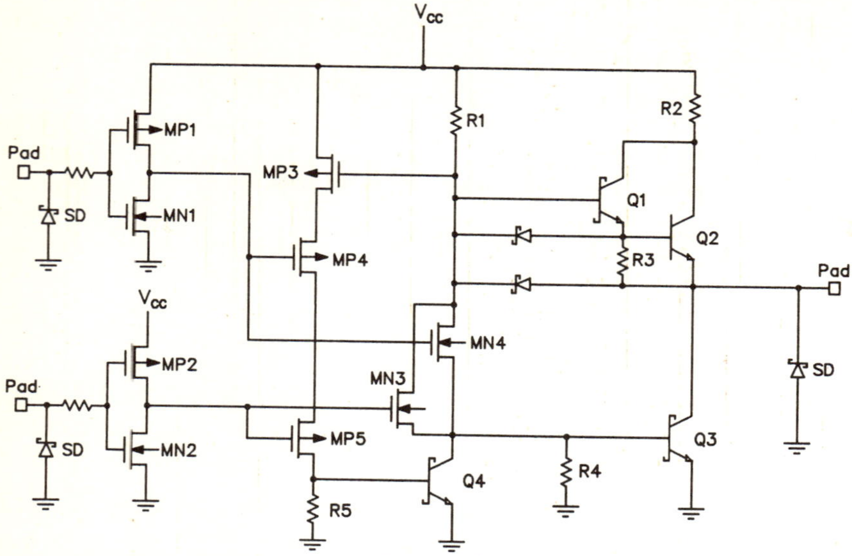

Figure 3.10 Internal circuitry of a BiCMOS 74BC08 AND gate—MP1 and MP2 are high-threshold *p*-channel MOSFETs; MP3, MP4, and MP5 are *p*-channel MOSFETs; MN1, MN2, MN3, and MN4 are *n*-channel MOSFETs. (*Reprinted by permission of Toshiba Inc.*)

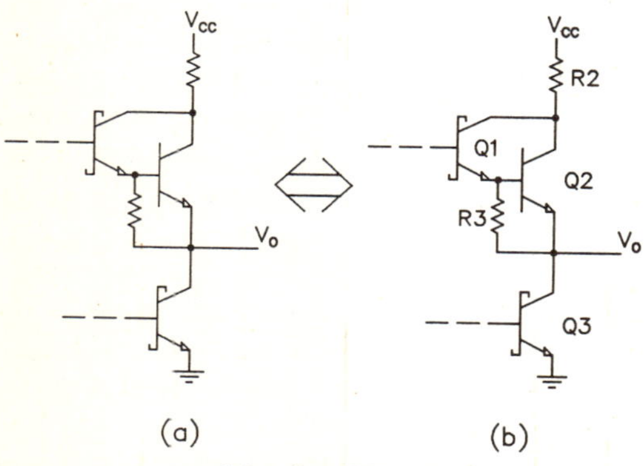

Figure 3.11 (*a*) Schottky TTL totem-pole output stage; (*b*) BiCMOS totem-pole output stage.

ground noise, that come with large transient currents (see Chapter 4).

2. The resistor between the pull-up transistor and V_{cc} in a totem-pole output stage limits the output current drive during *low*- to *high*-transitions and thereby reduces *low*-to-*high* transient switching currents. The resistor also tends to match the *low*-to-*high* output

impedance to that of the typical printed circuit (pc) board characteristic impedance and thereby controls ringing and overshoot of *low*-to-*high* transitions.

3. Bipolar transistor drive characteristics tend to be more stable with temperature than MOSFETs. Thus, bipolar output-stage drive characteristics can be optimized for a given application and will remain optimized over the entire operating temperature range. In contrast, MOSFETs ON impedance changes by a factor of approximately 0.6 percent per degree Celsius which translates to an approximate 0.3 percent per degree Celsius change in speed. Thus, CMOS devices slow down significantly at high temperature and speed up significantly at low temperature. This large variability in output impedance with temperature, approximately four to one when process and power-supply effects are added, makes it impossible to optimize CMOS output impedance for wide-temperature-range operation. Devices that barely have adequate drive at 125°C have excessive drive at −55°C and cause severe ringing unless special precautions are taken. BiCMOS devices do not have such a large variability in drive and speed with temperature. Thus, BiCMOS parts can be made that maintain high drive at high temperature without having excessive drive at low temperature.

Totem-pole output-stage voltage swing is limited to approximately 3.5 V because the Darlington pull-up output transistor no longer has drive above approximately 3.5 V. Darlington transistors require about 1.5 V of base-emitter voltage V_{BE} for base current. Without base current, the output (emitter) can no longer supply load current which causes the output *high* level to stabilize at approximately 3.5 V with V_{cc} equal to 5 V. The actual magnitude of the output voltage V_{out} *high* depends on the value of the load and the output current. There are two sources of output *high* current, I_c through the collector-emitter of $Q2$ and I_b through the base-emitter of $Q2$ as shown in Figure 3.12. Base current I_b must exist for $Q2$ to be ON. Thus, the voltage drop across $R1$ due to I_b establishes a limiting value for V_{out} *high* which is (assuming I_c is very small)

$$V_{out} = V_{cc} - V_{BE}(Q2) - (I_b)(R1) \tag{3.1}$$

As I_b goes to zero (infinite load impedance), V_{out} approaches 3.5 V

$$V_{out} = 5\text{ V} - 1.5\text{ V} - (0)(R1) = 3.5\text{ V}$$

assuming V_{cc} is equal to 5 V and V_{BE} is equal to 1.5 V.

In high load current situations, such as when an output is switching from *low*-to-*high*, the collector-emitter current path I_c through $Q2$

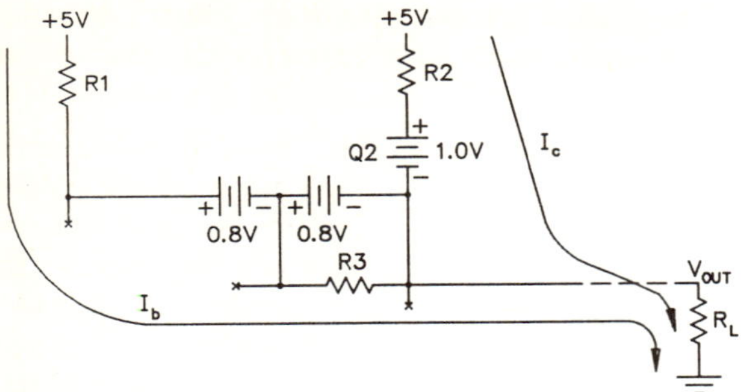

Figure 3.12 BiCMOS totem-pole output stage equivalent circuit for a *high* output.

may establish the output *high* level rather than the current path through the base I_b. The output voltage V_{out} in the collector path limited case can be determined using a simple voltage divider calculation as shown in Equation (3.2).

$$V_{\text{out}} = \left(\frac{R_L}{R_L + R2}\right)(5\text{ V} - V_{\text{CE}} \text{ of } Q2) \tag{3.2}$$

In most static situations, BiCMOS and CMOS inputs have such high impedance that little output current flows which means little base or collector current flows. Thus, in most static situations the presumed conditions of Equation (3.1) are met, which means *high* outputs are limited to about 3.5 V when V_{cc} is equal to 5 V. Yet, most BiCMOS signals observed in the lab have *high* levels that are above 3.5 V and are often in the range of 4.0 to 4.7 V. Why the discrepancy? Signals tend to be higher than 3.5 V because in many cases when signals switch, they overshoot and stabilize above 3.5 V and in most cases leakage currents tend to pull signals up rather than down.

3.3 BiCMOS/CMOS Input Circuit Operation

BiCMOS/CMOS inputs cannot tolerate indeterminate logic levels. Most BiCMOS/CMOS devices have complementary MOSFET input structures [see Figure 3.13(*a*)]. Both MOSFETs in complementary input structures turn ON when input levels are in the intermediate or indeterminate logic level region. When both input transistors are ON, I_{cc} may go up to a dangerous level. The typical profile for supply current versus input voltage for a BiCMOS/CMOS input stage is shown in Figure 3.14.[7–9] The equivalent circuit for a complementary stage with an intermediate-level input is two resistors between V_{cc} and ground [see Figure 3.13(*b*)]. The two resistors represent the ON im-

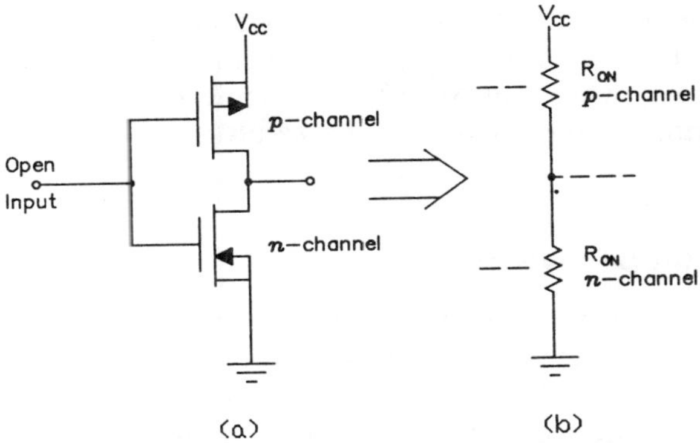

Figure 3.13 Complementary CMOS input stage and equivalent circuit when the input voltage is in the intermediate range between a valid logic *high* and valid logic *low*.

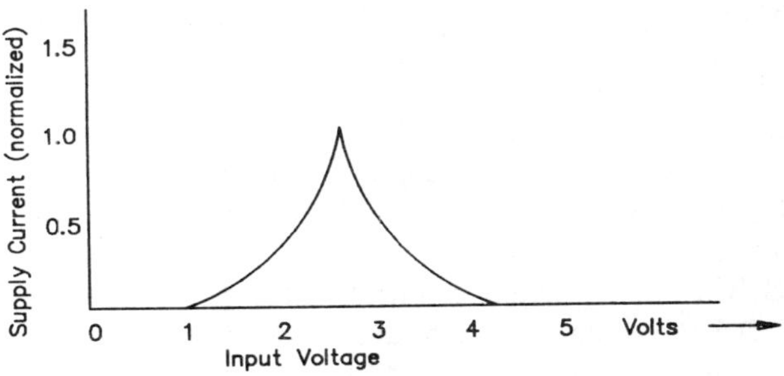

Figure 3.14 Typical profile of supply current versus input voltage for CMOS complementary input stages.

pedance of the two MOSFETs. Since the ON impedance of the MOSFETs used in advanced BiCMOS/CMOS devices is relatively low,[10] internal paths initiated by open or intermediate-level inputs can cause excessive internal device currents. Excessive internal current as a result of intermediate-level inputs can cause overheating and destruction of parts. Even if the intermediate-level input condition only lasts for a very short time, device temperature is increased because of extra current flow. High device temperature increases the susceptibility of BiCMOS/CMOS parts to latch-up which leads to additional high current flow and a greater possibility of device destruction. The time a part can endure with an intermediate-level input depends on the internal impedance of the device and other considerations such as how deep into the internal circuitry the intermediate logic level propagates. In general, intermediate-level input conditions should not be allowed to last longer than a few microseconds. However, some BiCMOS/CMOS parts have specified minimum rise time requirements that translate into more stringent requirements. In all applications, when BiCMOS/CMOS logic devices are used, minimum rise time requirements must be established and adhered to in the design.[11]

3.4 Latch-Up and Latch-Up Prevention

All BiCMOS and CMOS devices are intrinsically susceptible to latch-up.[12] Latch-up occurs when internal parasitic silicon controlled rectifiers (SCRs), structures that are inherent in CMOS (and the CMOS portion of BiCMOS) integrated circuits (ICs), are triggered ON.[13] When triggered ON, parasitic SCRs cause low-impedance paths between V_{cc} and ground that remain ON until V_{cc} is removed or the part is destroyed. Latch-up usually disrupts the functional operation of a part and in many cases will cause permanent damage even though some parts may return to normal operation after power is cycled OFF and then back ON. Latch-up may be initiated by voltage overshoots or undershoots that cause substrate currents that exceed device ratings at one or a combination of device inputs, outputs, or supply terminals. Thus, latch-up can be prevented by limiting both static and transient input and output currents.

All BiCMOS/CMOS ICs have parasitic four-layer *pnpn* structures, as shown in Figure 3.15, that can be triggered into a regenerative switching mode if sufficient current is injected into the appropriate points. Four-layer structures associated with input-output circuits are most exposed to transient currents and are most likely to latch-up. Manufacturers use a number of techniques to minimize the chance for latch-up: increased spacing between parasitic devices to reduce gain, guard rings around diffusion areas,[14] low-impedance substrates,[15] and special doping[16] to prevent injected currents from developing sufficient potential to trigger parasitic structures, to name a few. However, all CMOS ICs have parasitic SCR structures that can be triggered ON if sufficient current is injected. Most advanced BiCMOS/CMOS logic

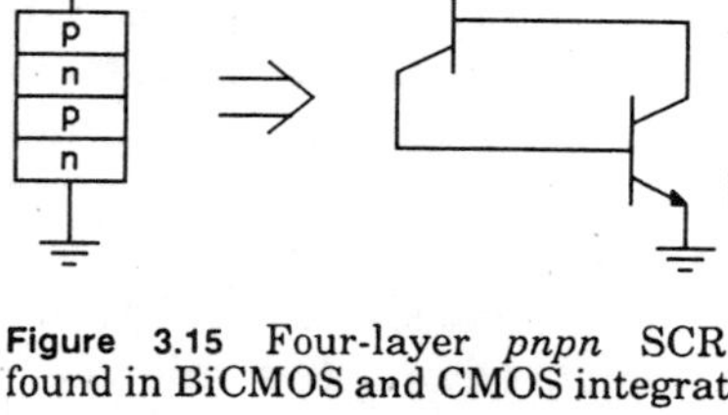

Figure 3.15 Four-layer *pnpn* SCR structure found in BiCMOS and CMOS integrated circuits.

devices have latch-up immunity for injected current levels of 100 to 200 mA depending upon device and device manufacturer. Some of the older devices may latch-up with injected current levels as low as 10 mA.

Military specification MIL-M-38510/606A[17] attempts to bring some standardization to BiCMOS/CMOS device latch-up immunity, but the test required in MIL-M-38510/606A is very limited. It only requires one of each input and output circuit type on a given chip to be subjected to a current pulse of ±150 mA for a duration of 500 ms with the part at V_{cc} equal to 5.5 V and at an ambient temperature of +125°C. Device I_{cc} is monitored to ensure latch-up does not occur. The test implies that parts that meet the specification (i.e., 38510) should not latch-up under worst-case environmental conditions with ±150 mA injected into inputs or outputs. However, there is no assurance that all inputs and outputs on a chip will have the same latch-up immunity or that there is not some accumulative effects if more than one input or output are subjected to injected currents at the same time. Injecting current into more than one input or output at one time is a very real possibility in bus driver-receiver applications. Bus signals may ring and overshoot and inject current into multiple inputs or outputs on byte-wide drivers and receivers. Since MIL-M-38510 does not require a test for that condition and, in general, manufacturers do not test more than one input at a time, there is no assurance that a part will not latch-up at a lower current when current is injected into multiple inputs. Thus, when using BiCMOS/CMOS parts, it is best to limit injected current where possible and practical. For example, unused CMOS inputs should not be connected directly to V_{cc} or ground; some form of current-limiting should be used to prevent noise or transient spikes on power or ground or fast power-supply turn ON from injecting sufficient current to trigger on parasitic SCRs associated with input circuits.

The fact that a system is free from latch-up at room temperature does not guarantee that it will be free from latch-up at high temperature. High temperature increases the susceptibility of BiCMOS/CMOS devices to latch-up. Thus, extra precautions must be taken to limit injected substrate currents at high temperature, and conditions that unnecessarily increase chip temperature must be avoided. Inputs must not be allowed to float or have slow transitions since floating inputs or inputs with slow transitions increase device operating temperature and increase the chance for latch-up.

Where signals cross board or system unit boundaries, which is where they are most exposed to transient conditions, current-limiting should be provided to ensure that device current and voltage ratings are not exceeded. Where BiCMOS/CMOS sources and loads are not powered by the same source, interconnecting signal lines must be current-limited or clamped to prevent injected substrate currents in unpowered parts. Unpowered parts with substrate currents caused by powered input signals may latch-up when local V_{cc} is applied.

Series resistors provide the simplest means of current-limiting BiCMOS/CMOS interfaces. Current-limiting resistors should be as large as possible, but resistor size must be balanced against excessive speed degradation. Under most board-to-board and other internal system conditions, special current-limiting is not required when advanced BiCMOS/CMOS devices are used. However, if long external lines with a high probability of being exposed to abnormal conditions must interface with BiCMOS/CMOS devices, or if some of the older CMOS logic families or special custom devices are being used, great care must be taken to limit interface currents.

3.5 ESD Protection

Oxide isolated gate FETs, such as those used in BiCMOS and CMOS IC logic devices, have extremely large gate-to-drain and gate-to-source impedances of 100 MΩ or more.[18] To prevent damage from electrostatic buildup, some form of low-impedance bypass circuitry must be added to exposed external device terminals[19] to limit gate-to-drain and gate-to-source voltages (due to electrostatic charge) to less than the dielectric breakdown voltage of the gate oxide.

Various combinations of diodes, transistors, and zener diodes are used to implement low-impedance clamp or discharge paths for advanced BiCMOS/CMOS devices. Most input-output protection networks implemented with diodes are similar to the network shown in Figure 3.16. The diodes are arranged to clamp either positive or negative electrostatic voltage excursions to levels below MOSFET gate oxide breakdown limits. In addition to diodes or transistors, most input-protection networks include some resistance[20] to limit current to protect the diodes, but most output clamps do not have current-limiting resistors (see Figure 3.16). Series output resistance would im-

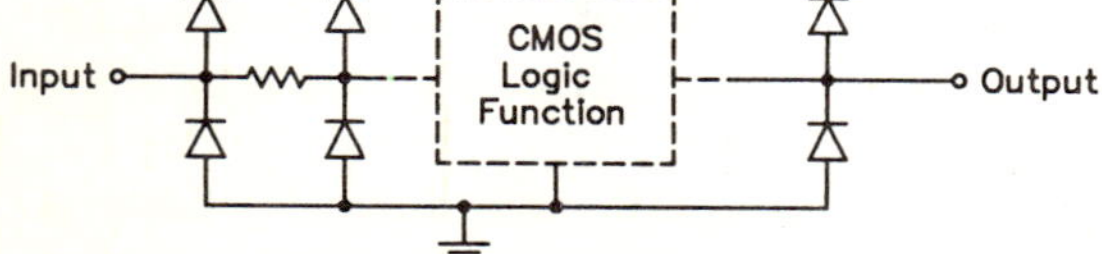

Figure 3.16 Typical electrostatic protection network used on BiCMOS and CMOS logic device inputs and outputs.

pact normal operation, and ideally protection networks protect against electrostatic damage without interfering with normal operation when signal levels are between V_{cc} and ground. However, when signal levels exceed V_{cc} or ground due to overshoots caused by transmission-line effects, some clamping action may occur depending on the speed of response of the diodes or transistors used in the protection networks. In some cases, input-protection networks provide effective dynamic clamping, but input- or output-protection diodes should not be relied upon to provide effective clamping of high-frequency signals. In most instances, input-protection network impedance is too high and diode response too slow for effective dynamic signal clamping. Different manufacturers use different protection networks for the same generic part, so care must be exercised if dynamic response is important. Each manufacturer's parts must be evaluated for the particular application at hand (see Chapter 7). Protection networks reduce but do not eliminate the susceptibility of parts to ESD. Also, the effectiveness of protection circuitry varies from manufacturer to manufacturer, so the ESD sensitivity of the same generic part can be different.

In all applications, large static and dynamic currents must be prevented from flowing in input and output ESD protection networks to guard against possible latch-up of parasitic SCRs that are inherent in CMOS ICs (see Section 3.4). Even normal signals that overshoot or undershoot V_{cc} or ground can induce latch-up if they have excessive energy. Series current-limiting resistors offer the best means of controlling static and transient currents injected into protection networks. Series resistors should be used at system interfaces to limit current in protection networks in applications where signal sources may be powered when receiving devices are not. In all BiCMOS/CMOS interface applications, input currents under abnormal or worst-case transient conditions must be kept below actual specified device limits. Absolute maximum dc input-current limits for most BiCMOS/CMOS logic devices are in the 20- to 30-mA range. Transient limits for most advanced BiCMOS/CMOS devices are typically above 100 mA, but caution must be exercised since some CMOS devices may latch-up with injected currents as low as 10 mA and output circuits are often more susceptible to latch-up than inputs. Thus, all inputs and outputs should have series current-limiting resistors to prevent excessive input or output currents.

Isolated system interface signals that connect directly to CMOS inputs or outputs are an invitation to ESD problems.[21] All external system signals that connect to CMOS inputs should have shunt resistors to V_{cc} or ground and series current-limiting resistors (see Figure 3.17).[22] Shunt resistors help prevent electrostatic buildup when input

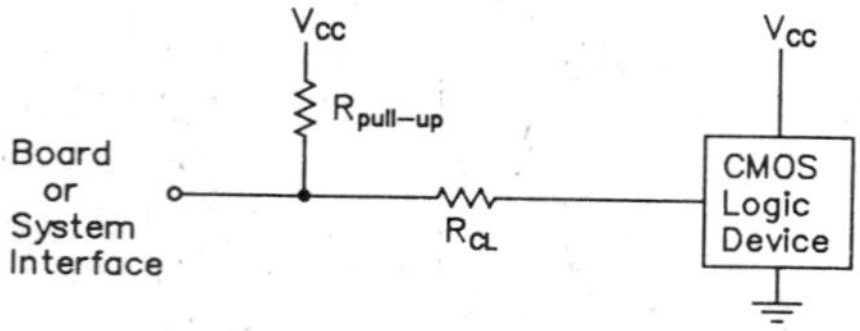

Figure 3.17 ESD and latch-up protection network for exposed BiCMOS and CMOS inputs.

lines are disconnected and prevent open inputs from floating (and perhaps overheating—see Section 3.3). In applications where high-speed signals must cross external unit boundaries, series current-limiting resistors may cause excessive *RC* delays. However, if it is not possible to use series current-limiting resistors, and the possibility of excessive static or transient input currents exists, other steps must be taken to control the situation.

Good low-impedance chassis grounds are essential for ESD control. Chassis of electronic equipment must be solidly connected to ground with a rugged low-impedance ground line.[21] Green lines in power cords should not be relied on to provide chassis ground. Likewise, green lines in power cords should not be relied on to ensure that units under test and test equipment are at the same potential. Units under test and test equipment must be grounded together and to earth ground with visible ground cables before test probes or other test connections are made to units under test. Chassis not only must be grounded, they must also provide good shielding, and internal circuitry must be solidly referenced to chassis ground (generally at one point to prevent ground loops) to prevent electrostatic differences between internal electronic components and chassis. Ideally all pc boards and motherboards are built with multiple power and ground planes so that all signal lines are enclosed by power and ground planes to provide ESD shielding of signal interconnections. Ground planes must extend to all areas of pc boards to maximize the ESD shielding.

Electrostatic stress damage to ICs is responsible for a good share of system failures.[23] A U.S. Department of Defense analysis of field failures of ICs indicates that over 50 percent are due to ESD.[24] It seems safe to assume that ESD is responsible for even a larger percentage of IC failures in commercial equipment since perhaps more care is taken to prevent ESD damage in military systems than in commercial equipment. Because of mounting evidence of widespread failures of ICs occurring as a result of ESD damage, the U.S. Department of Defense requires semiconductor manufacturers to classify military ICs according to their electrostatic discharge tolerance. The three classifications are

Class I: Sensitivity range 0 to 1999 V

Class II: Sensitivity range 2000 to 3999 V

Class III: Sensitivity range 4000 to 15,999 V

Semiconductor manufacturers must indicate the ESD rating of military ICs on the packages. The ESD rating is indicated by

Class I: A single triangle

Class II: Two triangles

Class III: No triangle marking

When parts are available from multiple sources, builders of military equipment are required to buy the parts with the highest ESD classification. Most advanced BiCMOS/CMOS parts are rated as class II devices. Table 3.1 lists ESD ratings for several advanced CMOS logic families.

TABLE 3.1 ESD Ratings of Some Representative Logic Families

Manufacturer	Logic family	ESD rating,* V
National	FACT	> 2000
Motorola	FACT	> 2000
Texas Instruments	BCT	> 2000
VTC	AC and ACT	> 2000

*See respective data books for current ESD specifications.

ESD is a serious concern when advanced BiCMOS/CMOS devices are used. Smaller-geometry devices are inherently more susceptible to ESD, and smaller geometries make it more difficult to build in robust input and output ESD protection networks. In most cases, ESD ratings of ASICs and LSI devices is less than that of SSI devices. Many ASICs fall into the class I category.

For those readers needing more information on ESD and ESD control, Reference 25, *Electrostatic Discharge Control* by Owen McAteer, is recommended.

3.6 Power Dissipation Calculations

Calculating the power dissipation of BiCMOS/CMOS systems is an extremely difficult task. Quiescent or static power dissipation of most BiCMOS/CMOS devices is insignificant. The majority of the power dissipated in high-speed BiCMOS/CMOS systems is a function of signal switching rates which are difficult to determine. In contrast, in TTL systems most of the system power dissipation is due to intrinsic device dissipation. System dc power dissipation is easy to calculate from device dc current specifications which are typically provided on data sheets. System dynamic power dissipation, on the other hand, is difficult to calculate since dynamic power is dependent on system application rather than on inherent device characteristics. Dynamic power dissipation is a function of node frequency, capacitance, and signal voltage swing. Thus, the operating conditions of each system signal, or node, is needed to calculate system dynamic power. Even though individual device dc dissipation tends to be insignificant, in a large system, overall dc dissipation "adds up," so to calculate BiCMOS/CMOS device and system power both the quiescent power and dynamic power of all devices must be determined.

$$P_{\text{total}} = P_{\text{quiescent}} + P_{\text{dynamic}} \tag{3.3}$$

3.6.1 Quiescent power dissipation calculations

Most advanced CMOS SSI and MSI logic devices in the AC and ACT logic families are built with complementary circuits (see Figures 3.5, 3.7, and 3.8) that have no dc current path between V_{cc} and ground, and as a result most AC and ACT advanced CMOS devices dissipate very little dc power. The same is true of BiCMOS SSI and MSI devices in the BC logic family. In this family, the core and input stages are built with complementary CMOS circuits and the output stages are built with bipolar totem-pole structures (see Figure 3.11) that likewise have no dc current path between V_{cc} and ground. When devices are built exclusively with complementary and totem-pole structures, device dc or quiescent power is a function of internal-device leakage current and any dc load current that may exist.

Quiescent power P_q is the product of the supply voltage V_{cc} and the quiescent supply current I_{cc}.

$$P_q = V_{cc} \times I_{cc} \tag{3.4}$$

Quiescent supply current I_{cc} is listed on most data sheets and is typically listed as a maximum value. It is usually specified at +25°C and at the worst-case high operating temperature (typically +125°C), with V_{cc} at 5.5 V in both cases. Worst-case quiescent I_{cc} current for AC and AC11 parts is shown as 4 μA at +25°C and 40 μA at +125°C with all inputs at either V_{cc} or ground.[7,8] (*Caution:* I_{cc} is higher when inputs are at TTL levels—see below.) When I_{cc} is strictly due to leakage current, its value at other temperatures can be calculated when needed since it is known that leakage currents in semiconductors approximately double for each 10°C increase in temperature.[26] That is

$$I_{cc} \text{ at } T_F = (I_{cc} \text{ at } T_I)^{2 \times N} \tag{3.5}$$

for positive changes (i.e., increases) in temperature and by

$$I_{cc} \text{ at } T_F = (I_{cc} \text{ at } T_I)^{1/(2 \times N)} \tag{3.6}$$

for negative changes in temperature where

$$N = \left| \frac{T_F - T_I}{10} \right| \tag{3.7}$$

in Equations (3.5) and (3.6).

Quiescent current is not low in all BiCMOS/CMOS devices. For example, most BiCMOS/CMOS memories and PALs dissipate significant dc power. Texas Instruments' BCT bus interface family data sheets show quiescent I_{cc} currents as high as 80 mA for some BCT parts depending upon input conditions. Most BiCMOS and CMOS devices with TTL-compatible input structures, i.e., ACT and FCT devices, dissipate significantly more power when driven with TTL-level signals than when driven with rail-to-rail CMOS-level signals. Complementary input structures even when optimized for TTL levels may not cut completely OFF when driven with TTL-level signals; V_{IH} may not be sufficiently high. No operational problem is caused, but I_{cc} may be higher than expected unless the system designer carefully reads all the fine print on data sheets. Typically, maximum I_{cc} current is on the order of 1.5 mA for ACT parts when driven with TTL *high*-level signals. Most devices in the FCT logic families have quiescent I_{cc} currents in the 2.0-mA region when driven with nominal TTL *high*-level inputs (i.e., 3.4 V).

3.6.2 Dynamic power dissipation calculations

The equation for calculating dynamic power P_d is[7]

$$P_d = (C_L + C_{pd})(\Delta V_s)^2 F \tag{3.8}$$

where C_L = line and device load (both input and output) capacitance
C_{pd} = internal device capacitance
ΔV_s = signal swing ($\approx V_{cc}$ for CMOS devices with CMOS output levels)
F = node toggle frequency

Signal voltage swing is easily established and C_{pd}, which represents internal device capacitance, is provided on most CMOS and BiCMOS data sheets, but line and load capacitance and the frequency of each node are usually difficult to establish. Yet, if a reasonably accurate estimate of system power is needed, a reasonably accurate estimate of the toggle rate and capacitance of each node is needed to calculate system power dissipation. In most situations, the best that can be done is to neglect all low-frequency control signals and concentrate on data paths where some assumptions as to average toggle rate and load capacitance can be made.

It is generally assumed that CMOS systems dissipate less power than equivalent TTL systems. Perhaps that is true at low frequencies since certain CMOS devices dissipate little dc power, but as system operating speeds increase, the power advantage of first-generation advanced CMOS systems decreases. First-generation CMOS devices have greater output-voltage swings than TTL, BiCMOS, and second-generation advanced CMOS devices with TTL output levels. Thus, dynamic power dissipation is greater in CMOS systems with true CMOS levels than in TTL-level systems since dynamic power is a function of the signal voltage swing squared [see Equation (3.8)]. Typical true CMOS output-voltage swings are 5 V, while typical TTL output-voltage swings may be as low as 3 V or even less depending upon loading (worst-case minimum TTL voltage swing is 2.5 V). Since dynamic power is a function of the signal swing squared, on a typical basis, CMOS-level dynamic power is greater than TTL-level dynamic power by a factor of

$$\frac{5^2}{3^2} = \frac{25}{9} \approx 2.8$$

The difference in CMOS- and TTL-level dynamic dissipation is illustrated in the following example: The dynamic load power dissipation for a CMOS-level signal driving a 50-pF load at 20 MHz is [using Equation (3.8)]

$$\begin{aligned} P_d &= (50\text{ pF})\,(5\text{ V})\,(20\text{ MHz}) \\ &= 25\ mW \end{aligned}$$

and for the same load and signal-switching frequency, TTL-level signal dynamic power is

$$\begin{aligned} P_d &= (50\text{ pF})\,(3\text{ V})\,(20\text{ MHz}) \\ &= 9\ mW \end{aligned}$$

From the above example, it follows that BiCMOS and second-generation advanced CMOS devices with TTL output levels offer significant dynamic-power savings. At high signal toggle rates, individ-

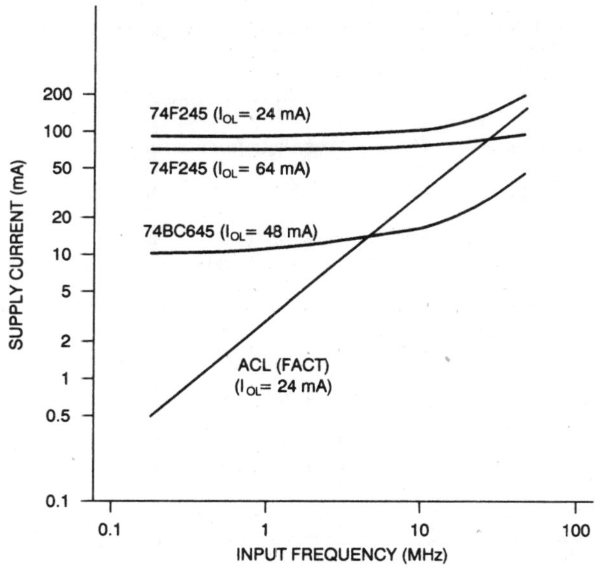

Figure 3.18 Supply current versus frequency for BiCMOS, CMOS, and TTL drivers—no load, 50 percent duty cycle with all buffers switching, V_{cc} equal to 5 V, and T_A equal to 25°C. (*Copyright of Motorola, Inc.; Used by permission.*)

ual first-generation advanced CMOS devices may provide little or no power advantage over bipolar TTL devices, but, in most cases, on an overall basis, high-signal-level CMOS systems dissipate less power than equivalent TTL systems because most control signals toggle at relatively low rates; it is usually only a few signals that toggle at high rates and have high dynamic-power dissipation. However, if all signals toggle at very high rates, for example 40 MHz or greater, the dissipation of a true CMOS-level system may exceed the dissipation of an equivalent bipolar TTL system. The BiCMOS BC logic family and the new FCT TTL-level CMOS logic families offer the best of both bipolar and CMOS technologies—low static and dynamic-power dissipation. Figure 3.18 shows that the total dissipation of a BC645 that has TTL output levels is less than that of an equivalent AC part with true CMOS levels above approximately 5 MHz.

3.7 References

1. Millman, Jacob: *Microelectronics, Digital and Analog Circuits and Systems,* McGraw-Hill, New York, 1979.
2. Mano, M. Morris: *Digital Design,* Prentice-Hall, Englewood Cliffs, N.J., 1984.
3. Lin, Liang-Tsai, and Richard Spehn: "Fast, Low-Powered Logic Array Unites CMOS and Bipolar," *Electronic Design,* April 16, 1987, pp. 82–88.

4. *Toshiba Bi-CMOS Logic TD 74BC Series Data Book,* Toshiba Corp., Tokyo, Japan, 1989.
5. *Bi-CMOS Logic Data,* Motorola Inc., Phoenix, Ariz., 1989.
6. Buchanan, James E.: *CMOS/TTL Digital Systems Design,* McGraw-Hill, New York, 1990.
7. *Advanced CMOS Logic Designer's Handbook,* Texas Instruments Inc., Dallas, Tex., 1987.
8. *FACT—Advanced CMOS Logic Databook,* National Semiconductor Corp., Santa Clara, Calif., 1989.
9. *RCA Advanced CMOS Logic ICs,* GE Corp., Somerville, N.J., 1988.
10. Cox, Gerald C.: "Impedance Matching Tweaks Advanced CMOS IC Testing," *Electronic Design,* April 1987, pp. 71–74.
11. Sokal, Nathan O.: "Check Lists Help You Avoid Trouble with MOS and Memory ICs," *EDN,* November 27, 1986, pp. 229–235.
12. Wakeman, Larry: "Closing in on CMOS Latch-Up," *Integrated Circuits Magazine,* April 1985, pp. 38–44.
13. Troutman, Ronald R.: *Latchup in CMOS Technology,* Kluwer Academic Publishers, Hingham, Mass., 1986.
14. Troutman, R. R.: "Epitaxial Layer Enhancement of n-Well Guard Rings for CMOS Circuits," *IEEE Electronic Device Letters,* Vol. ELD-4, No. 12, December 1983, pp. 438–440.
15. Schroder, J. E., A. Ochoa, Jr., and P. V. Dressendorfer: "Latch-Up Elimination in Bulk CMOS LSI Circuits," *IEEE Trans. on Nuclear Science,* Vol. NS-27, No. 6, December 1980, pp. 1735–1738.
16. Dawes, W. R., Jr., and G. F. Derbenwick: "Prevention of CMOS Latch-up by Gold-Doping," *IEEE Trans. on Nuclear Science,* Vol. NS-23, No. 6, December 1976, pp. 2027–2030.
17. MIL-M-38510/606A, *Microcircuits, General Specification for,* Navy Publications Center, Philadelphia, Pa., March 8, 1988.
18. Oxner, Ed, *Designing with Field-Effect Transistors,* 2d ed., Siliconix Inc., McGraw-Hill, New York, 1989.
19. *Electrostatic Discharge Control Handbook for Protection of Electrical and Electronic Parts, Assemblies and Equipment (Excluding Electrically Initiated Explosive Devices),* DOD-HDBK-263, Department of Defense, Washington, D.C., May 1980.
20. Frederiksen, Thomas M.: *Intuitive CMOS Electronics,* McGraw-Hill, New York, 1989.
21. *Electrostatic Discharge and Electronic Equipment: A Practical Guide for Designing to Prevent ESD Problems,* IEEE Press, New York, 1989.
22. Walsh, M. J.: *Choosing and Using CMOS,* McGraw-Hill, New York, 1985.
23. Yates, Warren: "Department of Defense Orders ESD Protection Guarantees for ICs," *Electronic Products,* October 1989, pp. 13–14.
24. Baker, Stan: "National Guarantees FAST ESD Protection," *Electronic Engineering Times,* September 4, 1989, pp. 87, 94.
25. McAteer, Owen J.: *Electrostatic Discharge Control,* McGraw-Hill, New York, 1990.
26. Lin, H. C.: *Integrated Electronics,* Holden-Day, San Francisco, Calif., 1967.

Chapter

4

Beware of Inductance and Transient Switching Currents

4.1 Inductance

Signal and power distribution system inductance is responsible for many of the difficulties encountered in high-speed systems, but many designers continue to ignore interconnection inductance. The general impression seems to be that inductance and inductive effects are not significant in digital applications. Perhaps that was the case with the older, slower, logic families, but that is no longer the case. When advanced BiCMOS/CMOS logic devices are used, interconnection inductance is a major source of problems. Very short connections have significant inductance and voltage loss when devices have switching transient currents with frequency components greater than 100 MHz which is the case when advanced BiCMOS/CMOS devices are used (see Chapter 1).

Inductance is a measure of the ability of a circuit to convert electromagnetic energy into a magnetic field. In certain applications, it is desirable to create a strong magnetic field, but this is not of importance in digital applications. Inductance must be minimized in digital systems to minimize signal energy loss (generating magnetic fields) and signal degradation. Yet, most texts that deal with the subject of inductance describe means for increasing inductance rather than means of decreasing inductance. They describe inductors that consist of numerous turns of wire and leave the reader with the general impression that a number of turns of wire are needed to achieve significant inductance. Even the symbol for an inductor implies that a number of turns of wire are needed for a useful inductor, and that may be the case at 60 Hz, but that is not the case at 100 MHz. Since voltage loss across an inductor is proportional to frequency, at 100 MHz a very small inductor can cause significant loss. Thus, for the successful application of high-speed logic devices, power and signal interconnection inductance must be minimized.

Inductance can be minimized by package design, pc board design, layout techniques (the shorter and closer to ground the better), and by employing special techniques for interconnection wiring. However, even when the best possible techniques are applied, transient voltage spikes due to inductive effects will be significant with respect to BiCMOS signal levels (the same as TTL) or CMOS signal levels and noise margins. Thus, system design approaches, such as synchronous design practices, that minimize system susceptibility to transient voltage spikes must be followed (see Chapter 10).

4.1.1 Physical and electrical factors that influence inductance

Self-inductance L, most often simply called inductance, of a current-carrying circuit is defined as the flux linkage per unit current.[1,2] That is,

$$L = \frac{\phi}{i} \tag{4.1}$$

where ϕ is magnetic flux and i is current.

Equation (4.1) is generally shown as

$$L = \frac{N\phi}{i} \tag{4.2}$$

where N is the number of turns of wire in an inductor. The more common equation for self-inductance, Equation (4.2), reinforces the idea that an inductor must be composed of numerous turns of a conductor, but it should be noted that 1 is an acceptable number for N in Equation (4.2).

Equations (4.1) and (4.2) show that inductance L is directly proportional to flux linkage. Since low inductance is desired for digital interconnections, it is important to understand what flux linkage is and how to minimize it. Flux density (flux linkage per unit area) is a measure of the ease with which flux can link with itself. The more concentrated the flux is, the greater the magnetic field and the more the inductance. Flux does not need to link with other sources of flux, but flux, like current, must have a complete path. To decrease inductance, *the ease with which flux can link with itself* must be impeded. The two variables that determine the *ease of linkage,* and thus the inductance, are the geometry of the circuit and the permeability of the magnetic medium. An iron core or other high permeability μ material near a coil of wire increases the ease at which flux can link with itself and thus increases the inductance. Likewise, nearby low-permeability material reduces inductance. The permeability of the insulating material of pc boards and of most insulating material is low; it is near that of air. Thus, when dealing with conventional interconnection systems, little reduction in inductance can be accomplished by changing the material adjacent to signals.

However, digital designers have some control over circuit geometry. Circuit inductance can be decreased by decreasing the length of the circuit (coil) or by arranging the circuit so that flux linkage is not enhanced (i.e., by arranging circuit topology so that flux cancels rather than enhances—coiling enhances flux linkage, thus coiling is to be avoided).

Equations for calculating the inductance of various physical arrangements of conductive elements can be found in numerous references. For many configurations the equations are quite complex. However, in all cases, flux linkage and inductance are proportional to the area enclosed by the circuit (current) path.[3,4] That is,

$$L = K \times \text{area enclosed by current path} \tag{4.3}$$

$$L \propto \text{area enclosed by current path}$$

where K is a constant related to conductor geometry and the permeability of the surrounding material. Thus, even though the two current paths shown in Figure 4.1 are the same length, the current path in Figure 4.1(*a*) has less inductance than the current path in Figure 4.1(*b*), because the enclosed area of 4.1(*a*) is less than 4.1(*b*). Since there is often little flexibility in the choice of insulating material permeability or conductor size, minimum inductance is achieved by minimizing the area enclosed by the current path. To minimize area, signal lines and current return paths must be in close proximity. That is, signal lines need to be run close to a ground plane (which serves as a

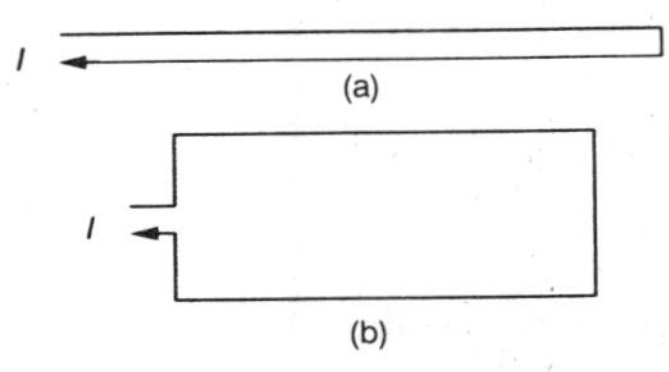

Figure 4.1 Inductance is minimized by keeping lines short and by minimizing the area enclosed by the current path. (*a*) Low-inductance path; (*b*) high-inductance path; even though the lines are the same length.

return path), and signal lines in cables need to be twisted with a return line.

4.1.2 Transient voltage drop across inductors

The effect of a changing magnetic field $d\phi$ on an electric circuit is an induced voltage. The effect is described by Faraday's law, which states that the induced voltage v is

$$v = \frac{d\phi}{dt} \tag{4.4}$$

for a single-turn inductor, which is the case for most digital circuits.

However, the change in voltage across an inductor due to a changing current is of more interest to digital designers. Using Equations (4.1) and (4.4), it can be shown that a change of current i with time causes a corresponding change in the magnetic flux ϕ and induces a voltage v in the circuit.

First rearranging Equation (4.1) to

$$Li = \phi \tag{4.5}$$

and differentiating both sides with respect to time t gives

$$L\frac{di}{dt} = \frac{d\phi}{dt} \tag{4.6}$$

From Faraday's law the induced voltage is

$$v(t) = \frac{d\phi}{dt} \tag{4.7}$$

and it follows that

$$v(t) = \frac{d\phi}{dt} = L\frac{di}{dt} \tag{4.8}$$

and

$$v(t) = L\frac{di}{dt} \tag{4.9}$$

is the familiar equation for the time-varying voltage across an inductor. Equation (4.9) shows that to lessen undesirable transient voltages in signal and power connections, either or both L and di/dt must be minimized. Inductance is minimized as described above in Section 4.1.1. Transient switching currents di/dt are controlled by keeping loads as small as possible and by the use of components that inherently generate less transient currents (see Section 4.2).

4.2 Transient Switching Currents

The fast edges of advanced BiCMOS/CMOS devices rapidly charge and discharge signal line and load capacitances and in the process cause large transient currents. First-generation advanced CMOS logic devices with rail-to-rail voltage swings and no control of output slew rates are notorious noise generators and are very difficult devices to use successfully in systems. For that reason, some of the newer advanced CMOS logic families have returned to TTL-level output swings (for example the FCT-T family), and most of the second generation of advanced CMOS devices have incorporated some control of output slew rates. However, output slew rates can only be reduced a certain amount without reducing potential system operating speeds. Signals must switch relatively fast to support 20-MHz or greater operation, and fast edges inherently cause large load switching transients. Most BiCMOS devices have TTL-level output swings, which means they have lower output transient currents when compared to first-generation advanced CMOS devices.

Load switching transient currents are not the only source of transients in BiCMOS/CMOS systems. BiCMOS/CMOS logic devices are built with complementary circuit structures that have transient internal feedthrough currents when switched. Unless proper high-frequency interconnection and power distribution techniques are used, these load and internal transient switching currents will cause significant shifts in reference levels, large drops in local V_{cc} levels, and crosstalk or coupling into other signals, any one of which can lead to intermittent or total system failure.

4.2.1 Transient internal switching currents

Most advanced CMOS logic devices and the core of most BiCMOS logic devices are built with complementary circuit structures that use *p*-channel MOSFETs for pull-up switches and *n*-channel MOSFETs for pull-down switches.[5,6] Complementary circuits have no dc current paths, other than leakage paths, between V_{cc} and ground, but under dynamic conditions, complementary circuits have the potential for significant internal transient feedthrough switching currents; both pull-up and pull-down MOSFETs may be ON for short overlapping times during switching.[7,8] BiCMOS devices also have bipolar totem-pole output stages[9] (similar to those used in TTL devices) that have some feedthrough current flow during switching.

Most BiCMOS and CMOS memory devices [e.g., dynamic RAM (DRAM), static RAM (SRAM), ROM, and first-in first-out (FIFO) memory devices] have large internal current demands when accessed. To save power, most such devices power down the peripheral circuitry except when they are being accessed. However, to minimize the access time, large currents are required to quickly power up the peripheral circuitry. The various circuit mechanizations used to power down memory devices is not covered since they are very device-specific, but system designers must be aware that most memory devices have large transient-current demands when accessed.

Transient internal switching currents in BiCMOS devices. Each time a BiCMOS device switches, internal transient currents flow between V_{cc} and ground in the complementary stages in the core of the device and in the bipolar totem-pole output stages. In SSI and MSI devices, most of this internal device transient switching current is the result of feedthrough in the totem-pole output stages.[10,11] Feedthrough occurs because of both pull-up and pull-down totem-pole transistors being ON for a short period of time whenever outputs switch. Device designers attempt to minimize totem-pole feedthrough currents, but it is difficult to match the turn ON and OFF characteristics of transistors or the timing of the separate drive signals. Most totem-pole outputs have added circuitry to help match turn ON and OFF of output-stage transistors, but a perfect match is never achieved. Figure 4.2 shows a BiCMOS totem-pole output stage and the equivalent circuit for a totem-pole output when both output transistors are ON at the same

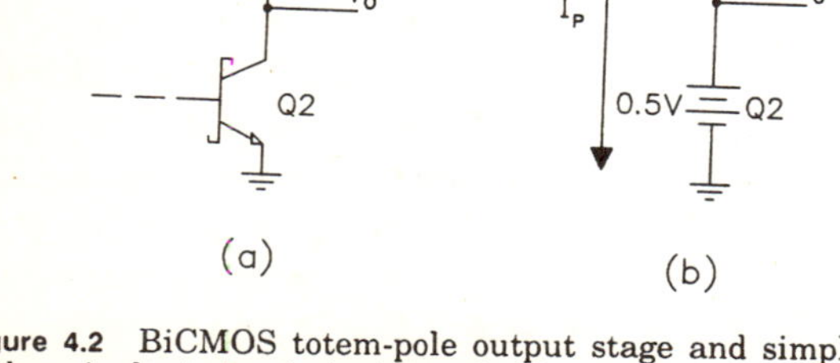

Figure 4.2 BiCMOS totem-pole output stage and simplified equivalent circuit during switching.

time. In the equivalent circuit [Figure 4.2(*b*)], the 0.5-V voltage source represents the lower output transistor $Q2$ (ON V_{CE} for Schottky clamped transistors is near 0.5 V), and the 1.0-V voltage source represents the forward drop across the pull-up Darlington transistor $Q1$.

The upper limit for internal peak transient feedthrough current I_p is a function of the output-stage current-limiting pull-up resistor ($R1$ in Figure 4.2) and the collector-to-emitter voltage drop V_{CE} of the two output transistors. That is,

$$I_p = \frac{V_{cc} - V_{\mathrm{CE(sat)},Q1} - V_{\mathrm{CE(sat)},Q2}}{R1_{\text{current-limiting}}} \tag{4.10}$$

Data sheets never list a value for $R1$, but from output short-circuit current ratings, the value of $R1$ can be estimated. For standard logic elements (NANDs, NORs, etc.) in Motorola's 74BC logic family, the short-circuit rating (to ground or 0 V) is typically listed as 60 to 180 mA, and for drivers (such as 74BC240s, etc.) as 100 to 225 mA at a V_{cc} of 5.5 V.[12]

When the output of a device is shorted to ground, the output short-circuit current I_{OS} is

$$I_{\mathrm{OS}} = \frac{V_{cc} - V_{\mathrm{CE(sat)},Q1}}{R1} \tag{4.11}$$

If I_{OS} is specified, $R1$ can be found by rearranging Equation (4.11) to

$$R1 = \frac{V_{cc} - V_{\mathrm{CE(sat),},Q1}}{I_{\mathrm{OS}}} \tag{4.12}$$

Using the worst-case output short-circuit current rating for drivers, 225 mA, $R1$ is

$$R1 = \frac{5.5\ \mathrm{V} - 1\ \mathrm{V}}{225\ \mathrm{mA}} = 20\ \Omega$$

Using a value of 20 Ω for the current-limiting resistor and with V_{cc} at 5 V, the peak feedthrough current I_p could be as large as [using Equation (4.10)]

$$\begin{aligned} I_p &= \frac{5\ \mathrm{V} - 1.0\ \mathrm{V} - 0.5\ \mathrm{V}}{20\ \Omega} \\ &= \frac{5\ V - 1.5\ V}{20\ \Omega} \\ &= 175\ mA \end{aligned}$$

which is near the value of the short-circuit current used to derive $R1$ as would be expected.

Actual peak output-stage feedthrough currents for BiCMOS parts will be much lower than 175 mA for several reasons. It is unlikely that both transistors will be fully ON during a significant portion of the switching period. Furthermore, simple static transistor models, as used, are not sufficient for ac or transient analysis. In addition, inductance, which is not addressed in the equivalent circuit, tends to have a significant limiting effect at BiCMOS switching speeds. The purpose of the above discussion is to illustrate the basic mechanism that causes internal feedthrough switching currents and to show that they can have significant magnitude, not to determine exact values.

The duration of feedthrough currents is short. Fast logic devices must have close matching of the ON-OFF characteristics of the totem-pole output transistors to prevent power dissipation from being excessive at high data rates. The overlap time in BiCMOS/CMOS devices is very short since they typically switch in 1 to 3 ns.

Output-stage transient feedthrough switching currents will vary with device, logic family, load, and operating temperature. Thus, systems with poor power and ground distribution networks that operate correctly at one given temperature and set of conditions may not operate at another temperature or with another set of conditions. Changing conditions may change transient feedthrough currents, which may cause different noise or reference-level disturbances. Changing parts may also change feedthrough currents and operating margin.

Transient internal switching currents in CMOS devices. Figure 4.3(a) shows a simple CMOS inverter with one complementary stage. During switching, both the pull-up MOSFET $Q1$ and the pull-down MOSFET $Q2$ may be ON for a short time. Hence, the switching transition equivalent

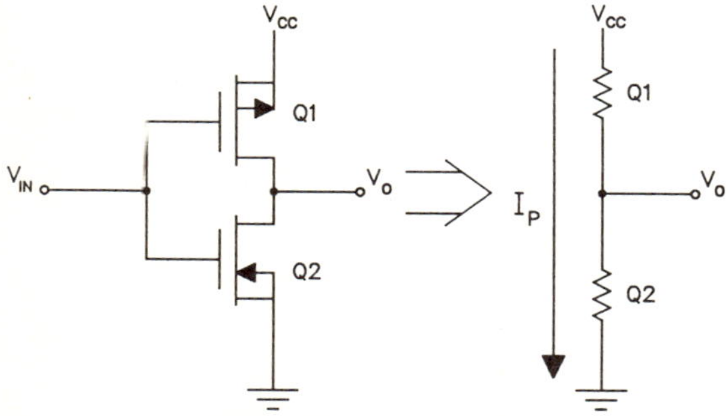

Figure 4.3 Inverting CMOS buffer and simplified equivalent circuit during switching.

circuit is two resistors between V_{cc} and ground as shown in Figure 4.3(*b*). Gates and other CMOS logic devices have more complicated equivalent circuits than the inverter shown in Figure 4.3, but they have analogous dynamic current paths. Gate internal structures have more than one pair of MOSFETs in series between V_{cc} and ground.

One approach for estimating feedthrough current for a simple inverting buffer is to divide the supply voltage V_{cc} level by the sum of the ON impedances (resistances) of the two MOSFETs. Typical ON resistance R_{ON} of advanced CMOS output-stage MOSFETs is near 10 Ω; thus the peak internal transient switching current I_p could approach

$$I_p = \frac{V_{cc}}{2R_{\text{ON}}} = \frac{5\ V}{2(10\ \Omega)} = 0.25\ A \tag{4.13}$$

Actual transient feedthrough currents do not reach levels of 0.25 A. Both MOSFETs are not fully ON at the same time, and most output-stage MOSFETs go into current limit below 0.25 A. Advanced CMOS outputs with 24-mA worst-case static output drive ratings typically have a dynamic current limit of approximately 150 mA (typical value). Thus, a more exact CMOS complementary stage equivalent circuit (during switching) consists of two current sources in series as shown in Figure 4.4. The pull-up current source I_{up} represents the pull-up MOSFET in current limit, and the pull-down current source I_{dn} represents the pull-down MOSFET in current limit.

The more exact equivalent circuit is conceptually correct, but it is of limited use for predicting peak transient feedthrough currents since current-source limits are not specified on data sheets or other manufacturers' literature. Device-output short-circuit current rating may be appropriate in certain cases, but in general, feedthrough current will not reach short-circuit current magnitudes. Both devices will not be fully ON at the same time. Peak transient feedthrough current magnitude is difficult to establish with simple models, but they do show the basic cause of feedthrough currents. Curves of supply cur-

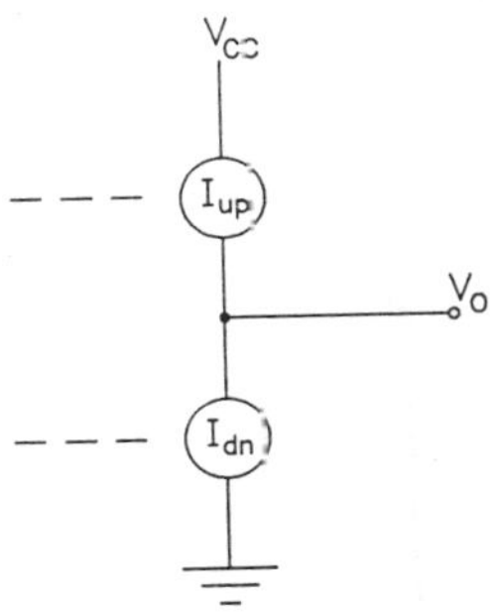

Figure 4.4 More exact switching transition equivalent circuit for complementary CMOS stage.

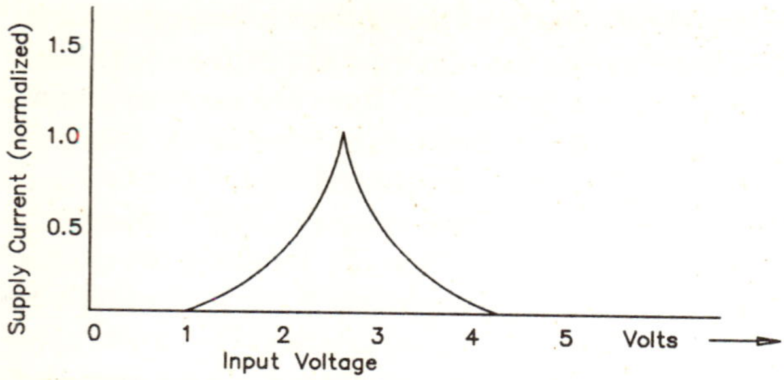

Figure 4.5 Typical profile of supply current versus input voltage for advanced CMOS logic devices.

rent versus input voltage clearly show feedthrough current occurring when input voltage is in the intermediate region (see Figure 4.5). Plots, such as Figure 4.5, showing supply current versus input voltage can be found in most CMOS logic device data books.[7,8] Figure 4.5 is normalized since the plot is not intended to represent a particular device. Its purpose is to show the typical profile of feedthrough current versus input voltage.

4.2.2 Transient load currents

Advanced BiCMOS and CMOS devices, with fast edges and large voltage swings, produce large transient load currents. Each time signals change levels, line and load capacitances must be charged or discharged as shown in Figure 4.6, and the faster the rise or fall time, the larger the charging or discharging current. Large charging currents increase the possibility of system upset due to noise. For example, large charging currents in high-impedance signal, power, or ground connections may cause voltage transients in excess of logic device noise margins. Also, large switching currents increase the possibility

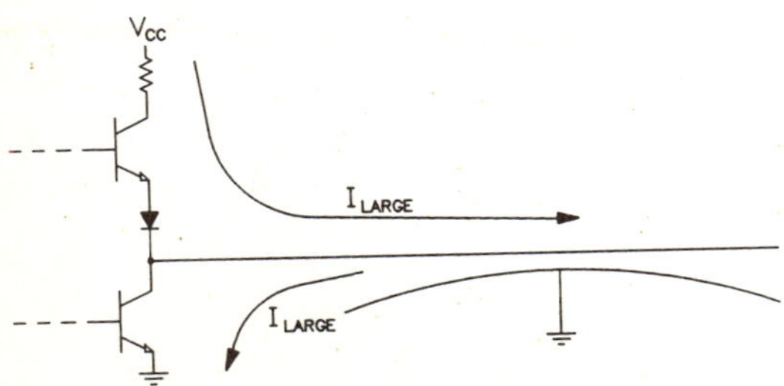

Figure 4.6 Signals with fast rise and fall times cause large charging and discharging currents when driving large capacitance loads.

of cross coupling to nearby signal lines. The potential for system upset is greatest when advanced CMOS is used; CMOS output voltage swings are larger than those of BiCMOS, and as a result, CMOS transient switching currents are larger.

The system designer needs to determine the magnitude of transient load currents so that the signal and power distribution system can be designed to keep noise down to an acceptable level. To aid in that task, the following discussion briefly describes how to calculate transient load currents under several typical system conditions.

Switching currents when driving lumped capacitance loads. When a load can be treated as a lumped load, i.e., the load is very near the source, simple *RC*-network calculations can be used to determine peak charging currents if the output resistance R_o of the driving device is known.[5,13] The charging current i for the lumped capacitance load in Figure 4.7 is

$$i = \frac{V_{cc}}{R_o}\epsilon^{-t/R_oC} \tag{4.14}$$

and the peak transient current I_L at time t equal to zero is

$$I_L = \frac{V_{cc}}{R_o} \quad \text{at } t = 0 \tag{4.15}$$

However, output dynamic resistance R_o in Figure 4.7 and in Equations (4.14) and (4.15) is not specified, so Equations (4.14) and (4.15) are of limited practical value for load current calculations. Static (dc) output resistance can be derived from the worst-case output voltage levels V_{OL} or V_{OH}, which are specified at a given output current on most data sheets. In most cases, R_o is different for *high*- and *low*-level outputs. Even if R_o is known, the charging current does not build to its full value instantaneously as implied by Equation (4.15) because of circuit inductance and other effects.

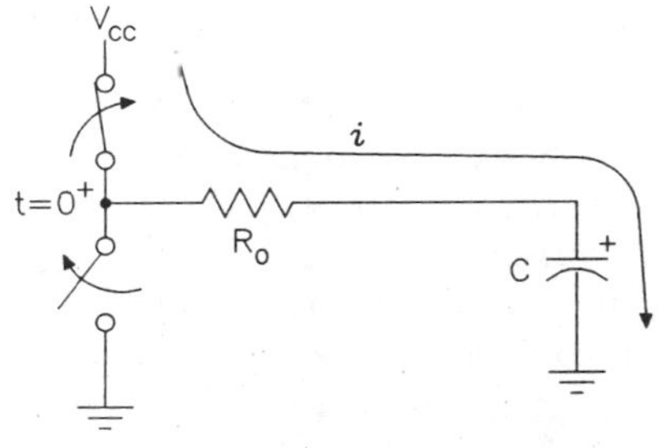

Figure 4.7 Equivalent circuit of a logic device charging a lumped capacitance with the output stage modeled as two ideal switches and a fixed resistance.

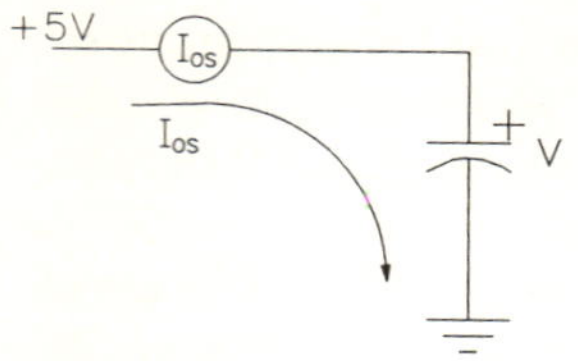

Figure 4.8 Equivalent circuit of a logic device charging a lumped capacitance with the output stage modeled as a current source.

In most applications, BiCMOS or CMOS driver output transistors are in current limit during most of each switching transition interval. Thus, during signal transitions, both BiCMOS and CMOS outputs should be treated as current sources as shown in Figure 4.8, and device output short-circuit current I_{OS} should be used to estimate peak output current. Output short-circuit current is listed on most data sheets. If short-circuit current is not listed on the data sheet of interest, it sometimes may be listed in the general logic family specifications typically found in one of the front sections of data books.

Current when lumped loads are driven with a constant rate of change of voltage. The rate of change of voltage with time when a capacitor is charged with a constant current I is

$$\frac{dv}{dt} = \frac{I}{C} \tag{4.16}$$

When Equation (4.16) is rearranged,

$$I = C\frac{dv}{dt} \approx C\frac{\Delta V}{\Delta t} \tag{4.17}$$

For a typical BiCMOS level change of 3.5 V in 2 ns and a 50-pF load, assuming a constant current output and substituting I_L for I in Equation (4.17), the resulting peak transient load current I_L is

$$I_L = (50 \text{ pF})\left(\frac{3.5 \text{ V}}{2 \text{ ns}}\right) = 87.5 \text{ mA}$$

which is a significant transient current. Large digital systems will have a large number of similar signals switching at any given clock time. A large number of signals demanding 87.5 mA at one time obviously places a burden on the power source and distribution system. However, much larger transient load currents are to be expected when CMOS devices are used. Many advanced CMOS devices have signal transitions of 5 V in 2 ns. When a signal transitions 5 V in 2 ns while driving a 50-pF load, the peak transient load current I_L is

$$I_L = (50\ \text{pF})\left(\frac{5\ \text{V}}{2\ \text{ns}}\right) = 125\ \text{mA}$$

A system using CMOS devices will have many signals with similar transient load currents at any given time. Some signals will cause even larger switching currents (than above) since some first-generation advanced CMOS devices switch in 1 ns. Switching currents increase at cold temperatures since MOSFET ON impedance decreases as temperature is decreased. Lower ON resistance increases drive capability and reduces the rise time. The larger the transient load currents, the greater the demands placed on the power and signal interconnections.

BiCMOS devices with lower nominal switching currents than CMOS are easier to apply in systems. Less switching currents mean less noise and less system disruptions. Furthermore, device switching characteristics and system timing tend to be more constant over temperature when BiCMOS devices are used. The output impedance of BiCMOS devices does not change with temperature as much as that of CMOS devices.

In all high-speed systems, the magnitude of transient feedthrough and load capacitance charging currents must be factored into the design of the power and reference distribution system and into the sizing and placement of local decoupling capacitors. When using either BiCMOS or CMOS technology, designing power distribution and signal interconnection systems that can tolerate such high levels of transient currents is a challenge. Great care must be taken to ensure low dynamic impedances.

Load capacitance. The dynamic performance of most SSI and MSI devices is specified with a 50-pF load, but in many applications, the load capacitance will exceed the standard 50-pF test load. Thus, transient load currents will exceed the values calculated above.

To aid in the calculation of load capacitance, typical BiCMOS or CMOS device input-output and typical interconnection capacitances are listed in Table 4.1. The data in Table 4.1 are useful for quick es-

TABLE 4.1 Typical Device and Interconnection Capacitance

SSI or MSI inputs, 5 pF
SSI or MSI outputs, 7 pF
SSI or MSI bidirectional ports, 15 pF
LSI devices in large packages, 10 to 15 pF
Printed circuit board traces, 2 to 4 pF/in
Wire-wrap wires, 1 to 2 pF/in
Welded-wire wires, 1 to 2 pF/in

timates of load capacitance or as a guide when actual data are unavailable. However, actual specified device capacitance and interconnection capacitance, when available, should be used in transient load current calculations. Most digital interconnections consist of several inches of "wiring" that connect to a number of devices, so it is not uncommon for loads to exceed 50 pF. Buses in particular connect to a large number of loads that are typically bidirectional ports. Thus, it follows from Table 4.1 that loads of 50 pF or more are to be expected on most bused lines.

Switching currents when driving distributed loads. Distributed loads differ from lumped loads in that a finite time is required for the signal to propagate along the signal path. The driving source does not instantaneously see the entire load. Transmission lines are distributed loads. When devices are driving transmission lines (see Chapter 7 for the definition of a transmission line), the transient load current I_L is equal to the magnitude of the signal change ΔV divided by the transmission-line characteristic impedance Z_o [see Equation (4.18) and Figure 4.9].

$$I_L = \frac{\Delta V}{Z_o} \tag{4.18}$$

Transient load currents, as defined by Equation (4.18), exist until steady-state conditions are established on the transmission line, i.e., until all ringing and reflections have subsided.

In most applications where advanced BiCMOS/CMOS devices are used, the dynamic load should be viewed as a transmission line. Most lines have propagation delays that are long relative to the rise time of the signal. When lines are long, it is incorrect to treat the total line capacitance as a lumped load. The line impedance isolates the driver from the more remote capacitance of the line and loads. Thus, in *long* line cases, the Z_o of the line should be used to calculate the transient

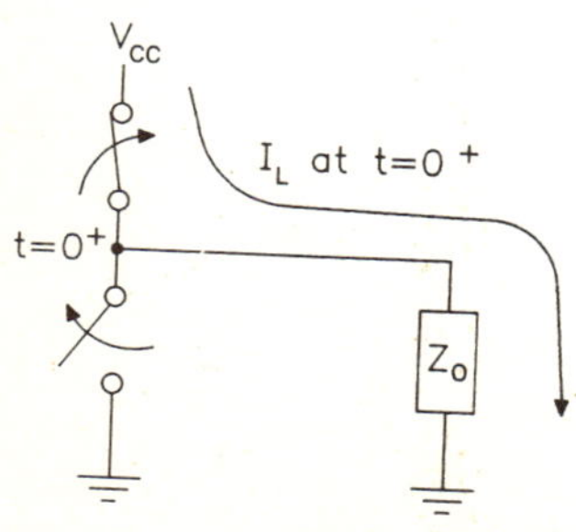

Figure 4.9 The transient load current when a transmission line is driven is equal to the voltage change divided by the transmission-line impedance.

driver and load currents, and transmission-line analysis must be used to determine the transient signal response to determine ΔI_L (see Chapter 7).

It is desirable to have interconnection impedance as high as possible to reduce transient switching currents, but high line impedance increases the chance for crosstalk. Thus, no ideal circuit board or motherboard interconnection impedance exists. When the conflicting requirements are weighed and balanced, line impedance near 60 Ω provides the optimal balance between switching current levels and crosstalk. However, most manufacturers of high-density multilayer pc boards have difficulty achieving line impedance of 60 Ω or greater. Thus, when pc boards are used, the main concern is one of ensuring that the line impedance is high enough so that signal line impedance does not cause excessive dynamic loading. To limit signal degradation, unloaded signal line impedance should be greater than 40 Ω (see Chapters 6 and 7).

4.3 Ground Bounce

Ground bounce and power-supply droop are of great concern when advanced BiCMOS/CMOS devices are used. Their fast edges cause large, high-frequency, transient load currents which also must flow in the ground and power connections of the switching device. The inductance of package leads, bond wires, and chip metalization is significant at the frequencies contained in the switching edges of advanced BiCMOS/CMOS. The combination of large transient currents and significant inductance causes large shifts in chip reference and power-supply levels.[14] If the level shifts are large enough, they may cause

1. Logic errors in the switching device if it contains storage elements or in devices connected to the outputs of the switching device.
2. Nonmonotonic transitions on outputs which can cause double clocking if the outputs are used as clock signals.
3. Degradation of propagation delays due to the supply voltage across the device being reduced which reduces drive.

Many first-generation advanced CMOS devices have severe ground bounce, particularly at cold temperatures, where output impedance is lowest. Ground-bounce problems led some manufacturers to change to center ground pins to reduce the problem.[7] Center power and ground pins lower average and worst-case inductance between internal points on a chip and external ground or supply-voltage levels since average and maximum distances to external connections are reduced. Center ground (or power) pins reduce the ground connection inductance to near 5 nH where conventional dual in-line package (DIP) end pin ground (or power) inductance is from 10 to 15 nH. One of the advantages of BiCMOS devices is that they have less ground bounce or power-supply droop. Thus, most BiCMOS parts have the ground pin in the conventional location.

Table 4.2 shows the range of typical pin inductances for several common packages.[15,16] The two values listed in Table 4.2 are representative of the inductance of the longer and shorter signal paths (pins) into or out of a given package style. Twenty-pin plastic leaded chip carriers (PLCCs) are nearly symmetrical. Hence, there is little difference in the length, or inductance, of the pins. Actual worst-case pin inductance for the various listed package styles may vary considerably from that shown in Table 4.2, so actual package specifications should always be consulted. Note that surface-mount packages such as LCCs and PLCCs have less lead inductance than DIPs and offer an advantage where they can be used.

Ground bounce (or power-supply droop) problems are compounded when several heavily loaded drivers in a common package switch simultaneously.[17] One observable manifestation of ground bounce or power-supply droop is the appearance of transient voltage spikes on stable outputs when other heavily loaded devices in the (same) package switch.[18]

Figure 4.10 shows a common situation where ground-bounce spikes will be observable on a stable output of an octal driver (power-supply droop will occur for the opposite switching conditions). A ground-bounce spike will occur on the stable output when the other drivers switch from a *high* to a *low* level. *High*-to-*low* switching causes a large transient current to flow from the loads to ground through the package ground pin. The inductance of the pin and the large transient current causes a transient-voltage drop (spike) across the ground pin. The transient-voltage spike causes the reference level of the internal chip to shift in the positive direction, causing a positive voltage spike on the stable output as shown in Figure 4.10. The magnitude of the spike

TABLE 4.2 Package-Pin Inductance

Package	Self-inductance, nH	
	Upper value	Lower value
14-pin DIP	10	3
20-pin DIP	15	3
20-pin PLCC	5	4
28-pin DIP	17	5
44-pin PLCC	6	5
100-pin PLCC	15	12

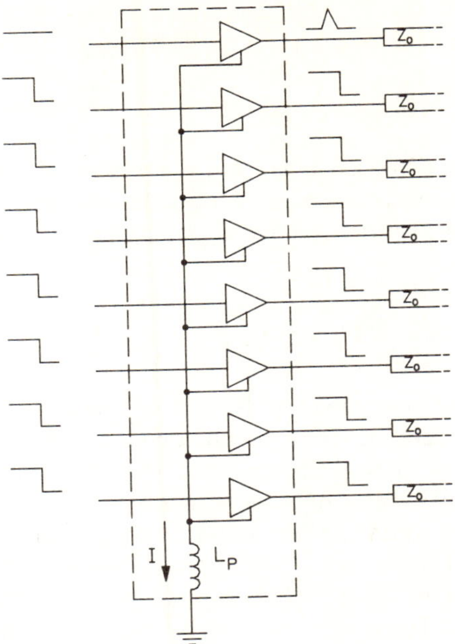

Figure 4.10 Ground-bounce spike on a static output when other buffers in the package switch.

depends on the total load of the switched outputs. In many common situations, the magnitude of the spike will exceed the noise margin of the receiving device.

To find the magnitude of a transient ground-bounce voltage spike for a given application, first the change in current *di/dt* in the package ground pin must be determined. In the octal driver application shown in Figure 4.10, seven buffers driving lines with a characteristic impedance Z_o of 50 Ω simultaneously switch. Assuming that all outputs transition 3.5 V, which is typical for BiCMOS outputs, the change in current ΔI [using Equation (4.18) times 7] is

$$\Delta I = 7\left(\frac{\Delta V}{Z_o}\right) \tag{4.19}$$

$$= 7\left(\frac{3.5\text{ V}}{50\ \Omega}\right)$$

$$= 490\ mA$$

Once ΔI is known, transient ground-bounce voltage is calculated using Equation (4.20),

$$v(t) = L\frac{di}{dt} \tag{4.20}$$

which is approximated by

$$\Delta V = L\frac{\Delta I}{\Delta t} \tag{4.21}$$

For a ΔI of 490 mA, a signal transition of 3 ns, and a ground pin inductance of 15 nH (15 nH is representative of the inductance of a typical 20-pin DIP with an end ground pin—see Table 4.2), the ground-bounce spike ΔV is

$$\Delta V = (15\text{ nH})\left(\frac{490\text{ mA}}{3\text{ ns}}\right) = 2.45\text{ V}$$

but a spike of 2.45 V will not occur because of self-limiting. As ground bounce lifts the device's reference, the device's output drive capability is reduced, and it slows down, reducing the current that produces ground bounce. If a ground-bounce spike of 2.45 V were to occur, it would greatly exceed the static or dynamic *low*-level noise margin of BiCMOS/CMOS devices with TTL-level inputs and would exceed the static noise margin of CMOS devices with CMOS inputs (see Chapter 2 for noise margin specifications). However, 50-Ω dynamic loads and the other conditions are not unusual. In many cases, much lower board impedances must be driven; pc boards may have effective impedances Z_o as low as 20 or 30 Ω. Large lumped capacitance loads have very low dynamic impedances and when switched, may cause very large currents which in turn cause very large ground-bounce spikes.

Repeating the above ground-bounce calculation for a CMOS octal driver with 5 V output swings (with the other conditions the same), the change in current ΔI when seven CMOS drivers switch is

$$\Delta I = 7\left(\frac{5.0\text{ V}}{50\ \Omega}\right) = 700\text{ mA}$$

and the ground-bounce spike ΔV is

$$\Delta V = (15\text{ nH})\left(\frac{700\text{ mA}}{3\text{ ns}}\right) = 3.5\text{ V}$$

which exceeds CMOS *low*-level static noise margin (1.5 V in this case where V_{cc} is 5 V) and greatly exceeds the static *low* noise margin of BiCMOS or CMOS with TTL input levels. Ground-bounce problems are particularly severe when CMOS devices with TTL input levels are used. Most CMOS devices with TTL inputs still have full rail-to-rail output voltage swings and the accompanying large ground-bounce spikes. Some of the latest generation of CMOS logic families do not have rail-to-rail swings[19] and have a number of other design features, such as slew rate control,[20] to reduce ground bounce.

Techniques for minimizing ground-bounce effects. All devices have some ground bounce, even the new improved second-generation advanced CMOS and BiCMOS devices. Even when all external connections are ideal, i.e., zero impedance, some package pin inductance remains which may cause significant ground bounce depending upon load and output slew rate. Ground bounce can be controlled on critical clock and strobe signals by carefully following the design practices that limit ground bounce, but it is impractical to follow those design practices for all signals. The only practical recourse for the system designer is to use logic techniques that negate the possible detrimental effects of ground-bounce voltage spikes on most data and control signals.[21] Synchronous design practices (Chapter 10) are one means of limiting the possible detrimental effects of ground bounce. Synchronous design prohibits the use of asynchronous inputs on storage elements (flip-flops) or cross-coupled gates for performing operational logic functions. Thus, short spikes, such as those generated by ground bounce, cannot directly upset synchronous systems.

To further limit the possible detrimental effects of ground bounce or power-supply droop, the buffers used to drive critical signals (e.g., clocks and signals that go to asynchronous inputs, such as resets and presets) must be segregated into separate packages. The purpose is to prevent ground bounce due to heavily loaded buffers from upsetting other buffers (in a common package). For example, different-frequency clock signals should not use buffers in a common package. The number of buffers used in an octal package should be limited if all the buffers are driving heavily loaded lines and all the signals could switch at once; clock signal fan-out buffers are an example. When the number of buffers used in an octal package is limited because of ground-bounce considerations, those used should be the buffers nearest the ground pin.

4.4 Guidelines for Reducing Inductance and Transient-Current Effects

> Internal device and load-related transient switching currents are a major source of noise and problems in BiCMOS/CMOS systems. Their presence and nature must be understood and the power distribution system designed to minimize them, or else there is little chance of success. It is essential that high-speed systems have low-impedance power distribution systems and local decoupling capacitors to replenish local transient-current demands.

The following guidelines will help reduce the adverse effects of inductance and transient switching currents when using high-speed advanced BiCMOS/CMOS devices in high-performance systems.

1. Use low-impedance planes to distribute power and ground.
2. Use a high-frequency decoupling capacitor located as directly as possible between the power and ground pin of each advanced BiCMOS/CMOS device.
3. Use packages with lower pin inductance such as leaded or leadless chip carriers in critical applications where ground bounce or power-supply droop must be minimized.
4. Avoid the use of sockets or device carriers (or evaluate their use carefully) since they add inductance and increase ground bounce and power-supply droop.
5. Run signal lines near ground (reference) planes to minimize inductance. When signals cannot be routed near a ground plane (such as between separate units), send signals via twisted-pair lines.
6. Keep power and ground planes continuous through package- and connector-pin fields and connect package (or socket) ground and power pins directly to the planes using solder washers or clips when using prototyping boards (such as wire-wrap and welded-wire).
7. Limit loading to reduce transient load currents.
8. Limit loading of signals originating from a common package to reduce total package power and ground pin currents.
9. Limit the number of devices (drivers) used in a package when driving signals with critical waveshape requirements such as clock signals.
10. Use only those drivers nearest the package ground pin in critical applications (particularly where signals have TTL level).
11. Do not mix asynchronous control signals, such as clears and presets, in packages with other heavily loaded signals.
12. Limit the loading and the number of outputs that can switch simultaneously on devices that drive sensitive inputs such as clocks, clock or latch enables, and asynchronous sets and resets to reduce ground bounce and power-supply droop.
13. Segregate critical signals such as clocks of different frequency or phase, clock or latch enables, and asynchronous sets and resets into separate packages to prevent ground bounce or power-supply droop interference.

14. Do not mix critical signals, such as clocks and clock enables, in PLDs that have other signals that are heavily loaded.
15. Be careful to limit the load on byte-wide (or wider) devices to limit ground bounce.
16. Follow synchronous design practices, which prohibit the use of asynchronous inputs on storage elements (flip-flops) or cross-coupled gates for performing operational logic functions, to prevent short spikes, such as might be generated by ground bounce, from upsetting the system.
17. Use devices with TTL-level output swings where possible.
18. Use second-generation advanced CMOS devices with slew rate control and other improvements that reduce switching transients where devices with CMOS output levels must be used.
19. Never use devices or logic families that have more drive or are faster than required, and where possible add series damping resistors (or use devices with them built in) to slow down transitions and reduce transient-current effects.

4.5 References

1. Fitzgerald, A. H., and D. E. Higginbotham: *Basic Electrical Engineering,* 5th ed., McGraw-Hill, New York, 1981.
2. Corcoran, George F., and Henry R. Reed: *Introductory Electrical Engineering,* Wiley, New York, 1957.
3. Langford-Smith, F.: *Radiotron Designer's Handbook,* Radio Corporation of America, Harrison, N.J., 1960.
4. Lee, Reuben, Leo Wilson, and Charles E. Carter: *Electronic Transformers and Circuits,* Wiley, New York, 1988.
5. Millman, Jacob: *Microelectronics, Digital and Analog Circuits and Systems,* McGraw-Hill, New York, 1979.
6. Lin, Liang-Tsai, and Richard Spehn: "Fast, Low-Power Logic Array Unites CMOS and Bipolar," *Electronic Design,* April 16, 1987, pp. 82–88.
7. *Advanced CMOS Logic Designer's Handbook,* Texas Instruments Inc., Dallas, Tex., 1988.
8. *FACT Advanced CMOS Logic Databook,* National Semiconductor Corp., Santa Clara, Calif., 1989.
9. Tuck, Barbara: "TI's BiCMOS Bus Interface ICs Slash Standby Current," *Electronic Products,* June 1987, pp. 17–19.
10. Mano, Morris M.: *Digital Design,* Prentice-Hall, Englewood Cliffs, N.J., 1984.
11. Holt, Charles A.: *Electronic Circuits Digital and Analog,* Wiley, New York, 1978.
12. *Motorola Bi-CMOS Logic Data,* Motorola Inc., Phoenix, Ariz., 1989.
13. Millman, Jacob, and Herbert Taub: *Pulse, Digital and Switching Waveforms,* McGraw-Hill, New York, 1965.
14. Shear, David: "EDN's Advanced CMOS Logic Ground-Bounce Tests," *EDN,* March 2, 1989, pp. 88–97.
15. *Advanced CMOS Logic Design Considerations,* SCLA004, Texas Instruments Inc., Dallas, Tex., 1986.
16. *FAST Applications Handbook 1987,* National Semiconductor Corp., South Portland, Me., 1988.

17. *Simultaneous Switching Evaluation and Testing,* Texas Instruments Inc., Dallas, Tex., 1987.
18. *FCT—Fast, CMOS, TTL—Compatible Logic Tech Note,* Integrated Device Technology Inc., Santa Clara, Calif., December 1986.
19. *Integrated Device Technology Data Book,* Integrated Device Technology Inc., Santa Clara, Calif., 1989.
20. Gunn, Lisa: "Output Control Quiets Noise Usually Found in Advanced CMOS Logic," *Electronic Design,* October 12, 1989, pp. 30–32.
21. Funk, Richard, and James Nadolski: "Advanced CMOS—-Pinouts Are Not the Crucial Factor," *Electronic Engineering Times,* Monday, August 4, 1986, p. 33.

Chapter

5

Power Distribution

The design of power-distribution systems for equipment using high-speed BiCMOS/CMOS devices is a difficult electrical and mechanical design task. High-speed BiCMOS/CMOS devices have severe transient-current demands. Unless the best design practices are followed, power distribution systems will be plagued with inductance that will degrade the quality of the delivered power. The power interconnection system designer's task is to ensure that the inductance and overall impedance of the power distribution system do not degrade the distributed voltage below the operating limits of the logic devices being used. Of primary importance in high-speed BiCMOS/CMOS systems is the use of very low impedance planes throughout the distribution systems and the use of decoupling capacitors located very close to each device.[1]

In the typical large digital system, the power distribution system designer must address possible voltage loss at each of the following power system interfaces:

1. Component power and ground connections to board power and ground planes
2. Circuit board power and ground planes
3. Circuit board-to-motherboard power and ground connections
4. Motherboard power and ground planes
5. Motherboard power and ground plane connections to power and ground feeder lines or buses
6. Power and ground lines or buses between motherboard and power supplies

Both ac and dc losses are of concern at each interface, but power distribution problems are most often caused by ac losses when high-speed BiCMOS/CMOS devices are used. Their large transient switching currents will cause severe voltage loss and degrade or prevent system operation unless power system inductance is minimized. Guidance for dealing with losses 1 to 4 listed above are described in the following sections. Techniques for limiting losses 5 and 6 (above) are covered in Reference 2.

5.1 Component Power and Ground Pin Connections to Power and Ground Planes

The connections between component power and ground pins to power and ground planes must be direct to minimize ground bounce or power-supply droop—no wiring or long pc board tracks. When components that have leads that project through pc boards are used, for example DIPs, power and ground leads should connect directly to the respective plane. When surface-mount packages are used, power and ground connections to the respective planes must be as short as possible. Long tracts to feedthrough vias must be avoided. Wired power and ground connections must not be used when wire-wrap or welded-wire universal boards are used.

Often during the developmental phase of a project, there is a temptation to wire ground and power connections. However, simple calculations of voltage loss as a function of transient switching currents (that are to be expected when BiCMOS/CMOS devices are used) show the inappropriateness of wired power or ground connections. On wire-wrap boards, it is difficult to achieve connections that are much less than 1 in. Various references show the inductance of a 1-in length of number 30 wire to be in the range of 15 to 50 nH, depending upon how close the wire is to a ground plane or other return path.[3,4] Using Equation (4.21), the transient voltage drop ΔV across a 1-in wire with an inductance of 20 nH, and the same load and switching conditions as used for the BiCMOS package pin ground-bounce calculations in Chapter 4 (seven outputs switching in 3 ns with level changes of 3.5 V and 50-Ω loads), is

$$\Delta V = L \frac{\Delta I}{\Delta t}$$

$$= (20 \text{ nH}) \left(\frac{490 \text{ mA}}{3 \text{ ns}} \right)$$

$$= 3.26 \, V$$

Even with a 50 percent allowance for self-regulation, ground bounce due to ground wire and package ground pin inductance (which must be added to the above) is in excess of BiCMOS or CMOS static or dynamic noise margins.

The inductance of wired power or ground connections cannot be tolerated when high-speed logic components are used. Power and ground connections must be made directly to power and ground planes. When using prototyping boards (such as wire-wrap or welded-wire), package (or socket) power and ground pins must be connected directly to the power and ground planes using solder washers or clips.[5] Even sockets must be avoided in most applications since they add inductance and increase ground bounce and power-supply droop.

5.2 Voltage Loss across Planes

Low-impedance planes must be used to distribute power and ground on circuit boards and motherboards in high-performance BiCMOS/CMOS systems. Planes are needed to keep the dc resistance and the ac impedance as low as possible to supply high dc and pulse currents with minimum loss. There should be no interruptions in the planes except clearance holes for vias (feedthroughs).

5.2.1 DC loss

In BiCMOS/CMOS systems, dc loss in the ground distribution system should be kept to negligible levels on circuit boards and motherboards. Slightly more loss can usually be tolerated in the V_{cc} side of the power system especially if all devices have TTL input levels. A typical design goal for circuit boards or motherboards is to keep the total dc loss on both V_{cc} and ground to less than 1 percent (50 mV) of the nominal supply level (5 V).

The dc drop across a circuit board or motherboard plane is difficult to calculate since currents in planes are not confined to known paths. One useful approach for estimating dc voltage drops across a plane is to segment the plane into squares, as shown in Figure 5.1, and sum the more manageable and more easily determined drops across the individual squares. First, determine the current requirements of each square (i.e., the supply current requirements of components or circuit board connector pins that are located in that portion of the plane). Second, average the current for each row of squares and make the assumption that the current density is uniform in each square in a given row across the plane (i.e., from left to right in Figure 5.1). Third, determine the cumulative current in each square in a column of the

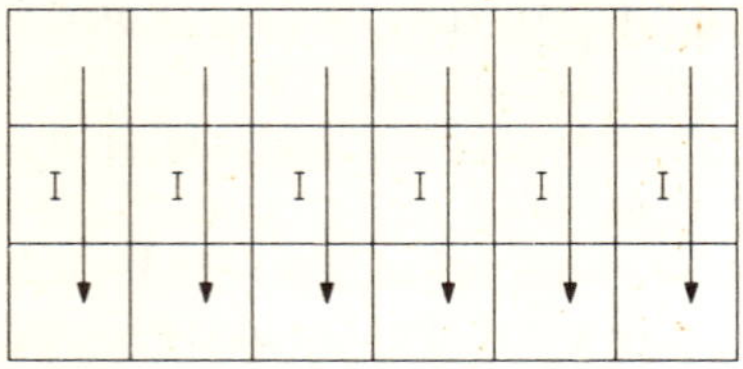

Figure 5.1 Ground or voltage planes divided into squares for IR loss calculations.

asquares working from the top of the plane to the bottom. Use the sheet resistance of the plane and the current in a given square to calculate the *IR* drop across each square. Sum the drops across the squares to arrive at the drop across the plane or to points of interest.

Resistance of a section of a plane. For a section of a plane, with dimensions as shown in Figure 5.2, the resistance is[6]

$$R = \rho\left(\frac{l}{wt}\right) = \rho\,\frac{l}{\text{area of cross section}} \tag{5.1}$$

where ρ is the resistivity of the plane. For a square section ($l = w$), the resistance is

$$R = \frac{\rho}{t} = \frac{\rho}{\text{thickness}} = \text{sheet resistance} = \rho_s \tag{5.2}$$

The approximate sheet resistance for three common weights of copper pc board conductive layers is[7]

0.5-*oz copper plane* ≈ 1.0 *m*Ω *per square*

1.0-*oz copper plane* ≈ 0.5 *m*Ω *per square*

2.0-*oz copper plane* ≈ 0.25 *m*Ω *per square*

Today most multilayer printed backplane and circuit boards are built with 1-oz copper planes. One-half-ounce material is seldom used for backpanel or motherboard power and ground planes. One-half-ounce material is used for signal layers and in a few cases for low-

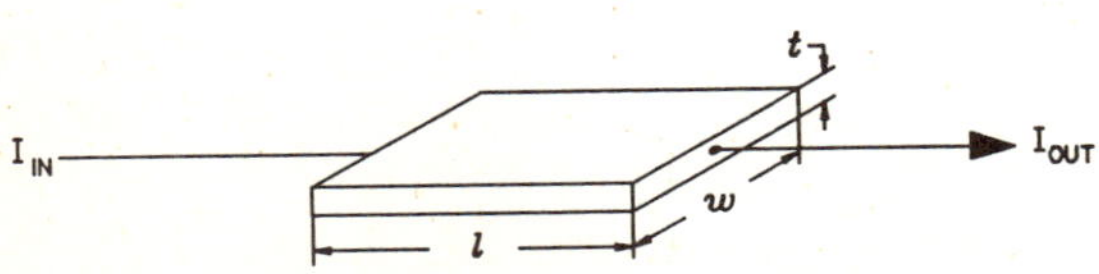

Figure 5.2 Square section of a plane showing dimensions used to define sheet resistance.

current circuit board planes. Two-ounce planes are used in very high current applications.

The sheet resistance used to calculate the drop across a plane should be derated from that of a solid plane based on the number and location of voids. In many applications, voids due to clearance holes for vias and package or connector pins will result in cross-sectional area losses in excess of 50 percent, and in some cases as much as 70 percent. Cross-sectional loss must be determined and sheet resistance adjusted as appropriate. In most BiCMOS/CMOS applications, dc loss in power or ground planes is not significant when 1-oz or heavier copper planes are used.

5.2.2 AC loss

The ac voltage loss in power and ground planes due to inductance is typically of much more concern than dc loss in high-speed BiCMOS/CMOS systems.[3] The inductance of power and ground planes is kept low by the use of continuous planes with no cutouts except for feed-throughs (vias) and by keeping the power and ground planes as close together as possible. The basic equation for the inductance of two parallel plates or planes of equal width w where the width is much greater than the separation b is given by Matick as[8]

$$L = (4\pi\mu_r \times 10^{-7})\left(\frac{b}{w}\right) \ \frac{\mathrm{H}}{\mathrm{m}} \tag{5.3}$$

The relative permeability μ_r is approximately 1 for nonconducting materials[9] such as the epoxy glass used to separate planes in multilayer pc boards. Thus, using a parallel-plate model, the inductance of closely spaced power and ground plane on pc boards is given by (with the units changed to nH/cm)

$$L = 4\pi\left(\frac{b}{w}\right) \ \frac{\mathrm{nH}}{\mathrm{cm}} \tag{5.4}$$

The parallel-plate model assumes uniform current density which is not the actual case in most local situations on a pc board. On a pc board the current is typically flowing to or from a point (the power or ground pin of a package). However, it seems reasonable to assume that most of the transient-current flow, when a device switches, follows a relatively direct path between the nearest decoupling capacitor and the power and ground pins of the device. If that is the case, then the question is, How wide is the path in which the current flows relative to the separation of the power and ground planes? From Equation (5.4) it can be seen that if the width to separation is on the order of 10 to 1, the inductance per centimeter is on the order of 1 to 2 nH.

Figure 5.3 High-speed Schottky circuit board showing ground plane connections between each socket. *(Courtesy of Stitch Wire Systems Corp. Reprinted with permission of Stitch Wire Systems Corp.)*

However, if the power and ground path is very long, 1 or 2 nH/cm quickly becomes significant at BiCMOS/CMOS edge speeds and transient load currents (see ground-bounce calculations in Chapter 4). To minimize the transient-current path in power distribution networks, each high-speed device needs a nearby decoupling capacitor to supply the nearly instantaneous change in current that occurs when high-speed devices switch. Decoupling capacitors are needed because it is generally impractical to get the inductance of planes much below 1 or 2 nH/cm, and an inductance of 1 or 2 nH/cm is significant at the high pulse currents demanded by BiCMOS/CMOS devices. The voltage loss in planes due to inductance is calculated using the basic equation for the voltage drop across an inductor, which is

$$v(t) = L\frac{di}{dt} \approx L\frac{\Delta I}{\Delta t} \tag{5.5}$$

where ΔI is the change in current when a device switches from one level to the other and Δt is the rise or fall time of the signal.

It is important when high-speed BiCMOS/CMOS devices are used that board power and ground planes be continuous throughout all device and connector pin fields as shown in Figure 5.3. Boards with ground cutouts for pin rows should not be used in BiCMOS/CMOS applications. Some board manufacturers' catalogs describe universal boards with the required continuous planes as "Schottky" boards.

5.3 Circuit Board-to-Motherboard Connections

Device to pc board interfaces are not the only system interfaces where ground bounce or power-supply droop is of concern (see Chapter 4). Any circuitry not connected to the ground or supply voltage by a continuous plane can experience ground bounce or power-supply droop. For example, circuit board-to-motherboard or backplane connections by necessity (so that boards can be removed) are not continuous. To keep ground bounce and power-supply droop within limits that allow reliable transmission of single-ended signals, circuit board-to-motherboard connectors must have a sufficient and well-distributed number of ground and power connections. Circuit board-to-motherboard ground integrity is essential. The circuit board reference must be very close to that of the motherboard and the other boards in the system. If a good ground level is maintained and circuit boards have sufficient decoupling (see Section 5.4), the number of circuit board-to-motherboard power connections is not as critical as the number of ground connections.

The number of ground (or voltage) pins needed in a circuit board-to-motherboard connector (or in any connector) should be calculated based on an overall system noise budget[10] that defines the allowable ground bounce at the connector (see Figure 5.4). The parallel inductance L_p of the ground or power pins

$$\frac{l}{L_p} = \frac{l}{L_1} + \frac{l}{L_2} + \cdots + \frac{1}{L_N} \tag{5.6}$$

must be small enough that the transient voltage drop $v(t)$ across the connector

$$v(t) = L_p \frac{di}{dt} \tag{5.7}$$

for the worst-case number of simultaneous switching outputs does not cause a transient-voltage shift greater than that allowed in the error budget. As a rule of thumb there should be a ground pin for every eight outputs, and they must be evenly distributed across the connec-

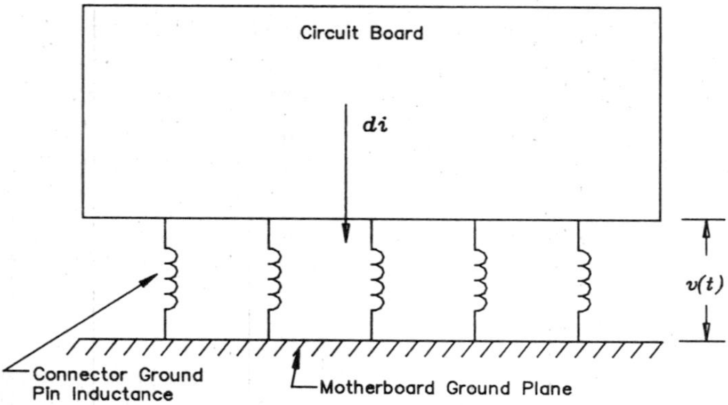

Figure 5.4 Circuit board-to-motherboard connector inductance must be low so that transient interface signal currents do not shift the board reference level a significant amount.

tor. As a minimum there must be a ground pin for every 0.5 to 1 in of connector depending upon connector signal density.

Static dc voltage loss across connectors is generally not a problem in BiCMOS/CMOS systems, but dc loss must be carefully evaluated. Some high-density connectors have relatively high resistance and are not rated for very high currents. As a general guideline in high-reliability applications, connector pin currents should be limited to 50 percent of the manufacturer's rating.

5.4 Decoupling Capacitors

Advanced BiCMOS/CMOS devices require decoupling (also called bypass) capacitors to provide a local source of load switching current to compensate for power and ground system inductance. *Local* decoupling capacitors are needed to supply a nearby source for the large, high-frequency pulse switching currents, and *bulk* capacitors are needed to provide circuit boards with enough capacitance to support longer-term, lower-frequency board transient-current demands.

5.4.1 Local decoupling

All high-speed logic devices need local decoupling capacitors to supply the current demanded during switching intervals (i.e., when changing states). The fast edge rates of advanced BiCMOS/CMOS devices exacerbate the requirement for local decoupling. Local decoupling capacitors compensate for power distribution system inductance and prevent large transient shifts in local power or reference levels when large loads are quickly charged. Even when low-impedance power and ground planes are used, they have significant inductance at BiCMOS/

CMOS edge rates. Planes, just as discrete wires, require a finite time for current to flow from one point to another. When several devices in a given area switch at once, significant amounts of instantaneous current must be available to prevent large shifts in local power or ground levels. Planes alone, even when closely spaced, do not provide sufficient local storage of energy charge (they are not good capacitors) to supply the instantaneous current (charge) demanded when a large number of devices in a given area switch at the same time.[11] Thus, even well-designed power and ground distribution systems may have local transient excursions of V_{cc} and ground that violate device power-supply levels unless local decoupling is present. With low supply voltage, devices slow down, signal transitions slow down, and potential system operating speed is reduced. Decoupling capacitors located near the power terminals of devices limit local power level transient excursions to harmless levels.

Guidelines for sizing local decoupling capacitors. When advanced BiCMOS/CMOS devices are used, it is extremely important to provide an adequate amount of local decoupling for octal drivers or for parts driving large amounts of capacitance. A general guideline for the amount and the placement of decoupling capacitors for BiCMOS/CMOS devices mounted on conventional pc boards or high-speed wire-wrap or welded-wire boards is

Guideline for Sizing Local Decoupling Capacitors
One 0.1-μF high-frequency decoupling capacitor mounted as close as possible to the V_{cc} pin of each SSI or MSI advanced BiCMOS/CMOS device.[11,12]

In all applications, the manufacturer's decoupling recommendation must be followed. In many cases, larger values, such as 0.22 μF are recommended for LSI devices such as dynamic memories and programmable devices. Devices in large packages with multiple power and ground pins have special decoupling requirements; again, the manufacturer's recommendations must be closely followed.

A typical design goal is to provide enough local decoupling capacitance to keep (support) the local dc supply voltage level within 1 percent of the nominal dc supply level under worst-case switching current demands. To meet that goal, the local decoupling capacitance C_d must be at least 100 times the maximum possible simultaneously switched load capacitance C_{load}. That is,

$$C_d \geq 100\ C_{\text{load}}$$

where C_{load} is the signal track capacitance plus the signal line device loads for the worst-case combination of simultaneously switched signals within a package.

The benefit of local decoupling capacitors can be illustrated using the principle of equality of charge. For example, Figure 5.5 represents the output stage of a gate or buffer along with the power source and load circuitry that typically surround a logic device. When $S1$ closes ($S1$ closing represents a gate switching from a *low* level to a *high* level), the nearby 0.1-μF decoupling capacitor $C1$ supports the local supply V_{cc} level and provides the transient current demanded by the 100-pF load $C2$. Without the nearby decoupling capacitor, the V_{cc} level at the gate is reduced by the voltage drop across the power source inductance L_s. A reduction in V_{cc} may upset other gates in the package. For example, other outputs may shift to incorrect levels. In addition, the gate responsible for the change in V_{cc} (i.e., the gate that switched) may not respond as expected because of an out-of-specification V_{cc} level.

For the circuit shown in Figure 5.5, the final steady-state charge Q_f on $C1$ and $C2$, reached sometime after switch $S1$ is closed, is equal to the initial charge Q_i on $C1$ before $S1$ is closed (see Figure 5.6). That is,

$$Q_i = Q_f \tag{5.8}$$

$$Q_i = (C1)V_{cc} \tag{5.9}$$

$$Q_f = (C1)V_f + (C2)V_f \tag{5.10}$$

$$C1V_{cc} = (C1 + C2)V_f \tag{5.11}$$

$$V_f = \left(\frac{C1}{C1 + C2}\right)(V_{cc}) \tag{5.12}$$

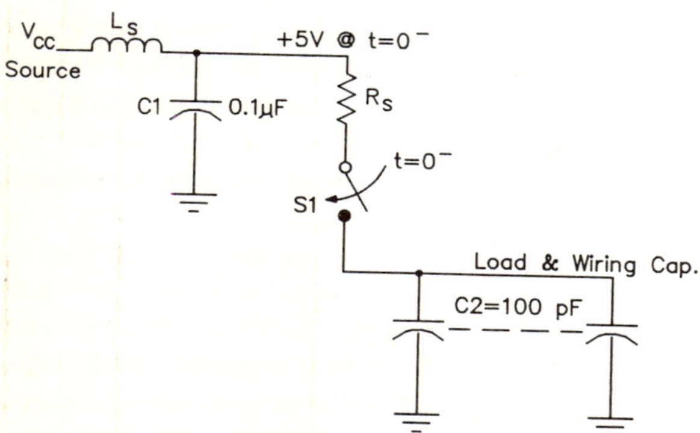

Figure 5.5 Idealized equivalent circuit when a logic device with a nearby decoupling capacitor switches from a *low* to a *high* level (capacitor and logic device lead inductance are not shown).

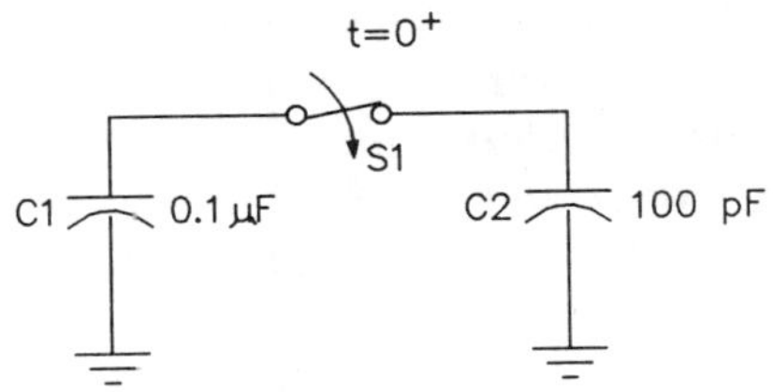

Figure 5.6 Decoupling capacitor $C1$ serves as a reservoir of charge for quickly charging the load capacitance $C2$.

For the circuit shown in Figures 5.5 and 5.6, the transient-voltage excursion is

$$V_f = \left(\frac{0.1\ \mu\text{F}}{0.1\ \mu\text{F} + 100\ \text{pF}}\right)(5\ \text{V})$$

$$= 4.995\ V$$

$$\Delta V_{cc} = 5\ \text{mV}$$

However, the above example is a very benign and ideal case. In most situations, more than one device in a given package will change at the same time. All eight outputs of an octal driver or register may switch at once, resulting in transient switching current levels 8 times that of the above example. In addition, in all real applications there is significant inductance between the decoupling capacitor and the device that is switching that will cause the transient-voltage excursion to be significantly more than the 5 mV calculated above. Even the best capacitors have significant effective series inductance (ESL) and effective series resistance (ESR) at BiCMOS/CMOS edge speeds (Figure 5.7). Thus, it is important to select capacitors with good high-frequency performance. In most applications, ceramic capacitors are the best choice, when size, cost, and frequency response are all considered. Aluminum electrolytic and the various tantalum capacitors do not have adequate frequency response for local decoupling. Where very high frequency response is required, 0.01-µF ceramic capacitors are generally superior to larger-value ceramic capacitors. Because capacitors behave as *RLC* networks, they have a resonant frequency above which they look more like an inductor than a capacitor. Typically, as the capacitance goes down, the inductance goes down also,

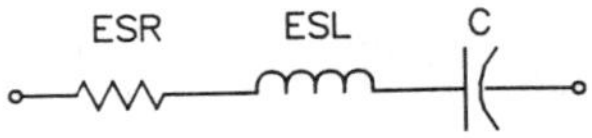

Figure 5.7 High-frequency equivalent circuit for capacitors

which means the resonant frequency goes up. Unfortunately, the series resistance typically goes up as the capacitance goes down.[13] Capacitor manufacturers' catalogs often list ESR but typically do not list ESL. However, ESR is frequency-dependent and the ESR data given are normally low-frequency data (they decrease at higher frequencies). Sometimes graphs of impedance versus frequency are provided or can be obtained from the manufacturer from which high-frequency ESR and ESL can be extracted.

Decoupling capacitor leads need to be as short as possible to minimize the inductance between the capacitor and the device being decoupled.[14] Either surface-mount or radial-leaded devices are best, but DIP profile packaged capacitors made especially for decoupling have reasonable high-frequency characteristics. Axially leaded capacitors should not be used since in most mounting arrangements their leads must be relatively long which means they will have higher inductance.[15]

The actual transient supply voltage excursion seen by a switching device is a function of the total inductance and resistance between the decoupling capacitor and the device plus the decrease in local supply level due to charge transfer to the load (see above). The inductance includes the ESL and lead inductance of the decoupling capacitor and the inductance of the power and ground plane and the package power and ground pins. In the typical application, the only significant resistance in the path is the ESR of the decoupling capacitor. The transient-voltage excursion due to inductance in the current path is determined by the basic equation for the voltage loss across an inductor

$$v(t) = L\frac{di}{dt} \approx L\frac{\Delta I}{\Delta t} \tag{5.13}$$

where ΔI = change in current when device switches from one level to another

Δt = rise or fall time of signal

L = sum of capacitor ESL and lead inductance and power and ground plane and switching device power lead inductance

The loss due to ESR is simply

$$v(t) = \Delta I \times \text{ESR} \tag{5.14}$$

A reasonable design goal is to keep the transient supply voltage excursions seen by the switching device due to inductance, resistance, and charge transfer to below 0.2 V.

Local decoupling summary. Local decoupling capacitors supply local transient-current needs and prevent local degradation of V_{cc}. To serve that function, they must have good high-frequency response and be located as close as possible to the power pins of the package being decoupled to minimize the interconnection inductance.

5.4.2 Bulk decoupling

Bulk decoupling capacitors help compensate for inadequate power or ground connections between circuit boards and motherboards or between motherboards and power supplies. Large bulk capacitors provide low-frequency replenishment of charge to local decoupling capacitors to help maintain V_{cc} at the proper level.[16] In addition,

1. Circuit board bulk capacitors reduce the transmission of board-generated noise to the motherboard.
2. Motherboard bulk capacitors help eliminate low-frequency ripple and ringing due to power-supply conductor inductance.
3. Circuit board and motherboard bulk capacitors help reduce the transmission of digital switching noise back to the power source; thus bulk capacitors help in meeting emission requirements.

Bulk decoupling capacitors must be sized for worst-case low-frequency charge replenishment.[17] A good rule of thumb to follow is: The value should be approximately 50 to 100 times the total worst-case simultaneous switched load capacitance. The simultaneous switched load capacitance is the total signal track capacitance, plus the total device load for the worst-case combination of signals, within a unit, that could switch at one time. In most bulk decoupling applications, solid-tantalum capacitors in the 10-μF or greater range are used.

A note of caution. Tantalum capacitors, or other large-value capacitors, tend to be less reliable than most other devices used in digital systems. Hence, the use of tantalum capacitors should be minimized. It is much more preferable to reduce the inductance of the power distribution system, rather than try to compensate for a poor distribution system with bulk capacitors. If the impedance of the power distribution system can be kept low enough, high-value (0.1-μF or larger) ceramic capacitors, which are more rugged and have a superior high-frequency response, can suffice for bulk decoupling as well as local decoupling.

In cases where bulk tantalum capacitors are required, careful attention must be given to the application so as to limit ripple currents, transient-voltage spikes, in-rush currents, and polarity reversals. Tantalum capacitors can be destroyed or degraded unless precautions are taken to ensure that voltage ratings are not exceeded. Care must be taken to keep tantalum capacitors as cool as possible since voltage ratings decrease at high temperature. When tantalum capacitors are used for decoupling, they should be selected so that the maximum applied voltage during normal operation is close to the derated maximum rated voltage. Tantalum capacitors should not be excessively derated. However, worst-case operating voltage (including transient conditions) must never exceed the maximum voltage rating at the maximum operating temperature (tantalum capacitors are rated for less voltage at high temperature).

Tantalum capacitors are polarized, so care must be taken to not reverse leads during installation. Tantalum capacitors, as well as the location where they are to be installed, should be clearly marked to reduce the opportunity for incorrect installation. Extra care must be taken to ensure that power connections are not reversed. Small power-supply reversals at turn ON or turn OFF, which often occur, will damage certain types of tantalum capacitors.[18] Many of the capacitor styles available in the size range needed for bulk decoupling are only rated for 0.5 V of reverse voltage. Ensuring that power-supply undershoots will not exceed 0.5 V is a difficult task. The use of a normally reverse-biased silicon diode in parallel with a bulk capacitor is not a solution since high-current silicon diodes clamp at 0.8 to 1.2 V, depending upon the temperature and the current. However, there are large-value tantalum-cased tantalum capacitors available (mil type CLR-79) with reverse-voltage tolerance up to 3 V, but they are expensive.

Meeting surge current limits is usually less of a problem than meeting reverse-voltage requirements. Slow turn ON of supplies, plus power conductor impedance, is often sufficient to limit surge currents to acceptable levels.

The application of large bulk capacitors presents mechanical as well as electrical problems. The coefficients of expansion of the case, lead frame, and tantalum slug of solid tantalum capacitors are all radically different. Thus, wide operating temperature ranges and temperature excursions during installation (soldering) can degrade devices.

For maximum decoupling capacitor reliability, the largest-value capacitor for a given case size and voltage should not be used. Compromises may have been made to pack a little more into a little less.

5.5 Power Distribution Requirements Summary

Internal device and load-related transient switching currents offer the potential for serious noise problems in advanced BiCMOS/CMOS systems. The nature and presence of high-frequency pulse currents in advanced BiCMOS/CMOS systems must be understood, and the power distribution system designed to accommodate them, or else there is little chance of success. It is essential that high-speed systems have low-impedance power distribution systems and local decoupling capacitors to replenish local transient-current demands. For minimum inductance, power and its return current paths must be in close proximity and as short as possible. Low-inductance power and ground interconnections are needed to ensure that switching transients generated by active devices driving large capacitance loads do not upset other nearby devices. The inductance of "real world" power connections cannot be ignored when systems are built with advanced BiCMOS/CMOS devices.

5.6 Summary of Techniques for Minimizing Inductance and Transient-Current Effects in Power Distribution Networks

> For minimum inductance, power and ground planes must be continuous and in close proximity.

Techniques for minimizing the inductance and the detrimental transient-current effects in power distribution systems include:

1. Power and ground planes must be used to distribute power and ground. Continuous ground reference planes are essential for high-speed systems. Power may be distributed with a ladder or grid configuration if there is adequate decoupling but never with point-to-point wiring.
2. Continuous ground planes with no avoidance areas in IC package or connector pin fields are required for circuit boards and motherboards.
3. A decoupling capacitor per device is required when advanced BiCMOS/CMOS devices are used. A good rule of thumb is one adjacent 0.1-μF decoupling capacitor per SSI or MSI device and one 0.22-μF capacitor per memory device or other LSI device located as close as possible to the V_{cc} pin of the decoupled device.
4. Power supplies should be located as close as possible to the powered circuitry, and twisted-pair lines (to minimize the area between the lines and hence the inductance) should be used to bring power to local power planes when the power supply cannot be connected directly to the power and ground planes. When the power supply can-

not be directly connected to the motherboard planes, local bulk decoupling may be needed at the power entry point.

5. Ground pins must be evenly distributed across all connector pin fields, including custom chip or multiple-chip packages. The number of connector ground pins required to minimize the effects of transient load currents must be determined based on a system ground upset error budget. A rule of thumb for the number of connector ground pins is: A minimum of one ground pin per inch of connector length or no less than one ground pin per eight outputs.
6. When using prototyping boards (such as wire wrap, welded wire, etc.), the power and ground planes must be continuous through the package, and connector pin fields and package (or socket) ground and power pins must be soldered directly to the respective planes using solder washers or clips on universal boards. Wiring of power and ground connections with discrete wires is not feasible because they have excessive inductance.
7. The use of sockets or device carriers must be avoided (or their use carefully evaluated) since they add inductance and increase ground bounce and power-supply droop. Sockets with built-in decoupling capacitors are a better choice where sockets must be used.

5.7 References

1. *FACT Advanced CMOS Logic Databook,* National Semiconductor Corp., Santa Clara, Calif., 1989.
2. Buchanan, James E.: *CMOS/TTL Digital Systems Design,* McGraw-Hill, New York, 1990.
3. Ott, Henry W.: *Noise Reduction Techniques in Electronic Systems,* 2d ed., Wiley, New York, 1988.
4. Morrison, Ralph: *Grounding and Shielding Techniques in Instrumentation,* 3d ed., Wiley, New York, 1986.
5. Visco, Anthony P.: "Coaxing Top Bipolar Speeds from Prototyping Boards," *Electronic Products,* September 1, 1987, pp. 55–58.
6. Lin, H. C.: *Integrated Electronics,* Holden-Day, San Francisco, Calif., 1967.
7. Blood, William R., Jr.: *MECL System Design Handbook,* 4th ed., Motorola Semiconductor Products Inc., Phoenix, Ariz., 1988.
8. Matick, Richard E.: *Transmission Lines for Digital and Communication Networks,* McGraw-Hill, New York, 1969.
9. Truxal, John C.: *Control Engineers' Handbook,* McGraw-Hill, New York, 1958, p. 7-4.
10. *FAST Applications Handbook 1987,* National Semiconductor Corp., South Portland, Maine, 1988.
11. Tomlinson, Jock: "Avoid the Pitfalls of High-Speed Logic Design," *Electronic Design,* November 9, 1989, pp. 75–84.
12. Hall, William, "Advanced CMOS Design Guidelines: Understanding and Addressing Noise," *National Anthem* (a publication of National Semiconductor Corp.), Santa Clara, Calif., November/December 1989, pp. A6–A8.
13. IEEE Standard P1194.0/D2, *A Guide to Backplane Electrical Performance Measurements,* IEEE, New York, 1989.

14. Cowdell, Robert B.: "Bypass and Feedthrough Filters," *Electronic Design,* No. 17, August 16, 1975, pp. 62–67.
15. *ATMEL Data Book 1989,* ATMEL Corp., San Jose, Calif., 1989.
16. Martin, Arch G., and R. Kenneth Keenan: "Neater Decoupling on Surface-Mount Boards," *Electronic Products,* August 15, 1987, pp. 47–49.
17. Doubrava, Laudie: "Bypass Supply Loads with Care for Optimum Transient Response," *EDN,* September 20, 1979, pp. 113–117.
18. Upham, Arthur F.: "Failure Analyses and Testing Yield Reliable Products," *EDN,* August 8, 1985, pp. 165–174.

Chapter

6

Signal Interconnections

Each distinct class of digital system or subsystem has specific interconnection requirements, some being much more critical than others. However, when advanced BiCMOS/CMOS devices are used to implement a system, the interconnection system and the signal routing, at any of the system levels or boundaries, cannot be left to chance or be based on purely mechanical concerns. Signal-path electrical requirements must be given high priority for system operating speed to approach that of the intrinsic speed of the logic devices used. Until recently, system operating speed was limited by the digital components, but that is no longer the case in high-performance systems. Delays in the interconnections are responsible for a major portion of most signal delays as system operating speeds move into the 20 MHz and higher range.[1]

When interconnecting devices with edge speeds of 1 ns and system clock rates above 20 MHz, the functional design and the signal interconnection system can no longer be treated as separate entities. They must be viewed as an integral portion of the overall functional system design. Time delays through interconnections cannot be ignored since they may be a significant part of most signal delays. Likewise, waveform distortion due to reflections, resonance phenomena, and signal attenuation due to imperfect transmission media cannot be ignored. To further complicate matters, the loading, load placement, conductor topology, and the conducting and insulating media all influence waveform quality. Also, crosstalk between adjacent signals cannot be ignored since it becomes more severe with higher edge speeds, and with higher clock frequencies there is less time for it to dampen out.

6.1 Signal Interconnection Categories

Signal interconnections fall into three basic categories in most large digital systems:

1. Interconnections within a circuit card (component-to-component)
2. Interconnections between circuit cards that are mounted on a common motherboard (board-to-board)
3. Interconnections between separate units such as boxes, chassis, racks, and cabinets (unit-to-unit)

Each category has specific physical and electrical design and implementation requirements. These special requirements must be understood so that the proper interface devices and physical interconnections can be selected. Fundamental issues that must be addressed at each interconnection level include signal dynamic response requirements, interconnection and device propagation delays, transmission-line effects, and crosstalk.[2]

The fast edge transitions of advanced BiCMOS/CMOS devices will cause severe cross coupling (crosstalk) between signals, noise in the power and ground distribution system, and degraded signals due to transmission-line effects (all of which will impact correct system operation) unless the physical interconnection system is designed to limit the side effects of fast edges.[3] When devices have fast edges, the physical routing of all signals is critical.

In the past, interconnection design did not require much effort. Interconnection delays and transmission-line effects are not significant when low-speed devices are used in low-speed systems, but that is no longer the case. Most systems using advanced BiCMOS/CMOS devices are expected to operate near the limits of technology, but this does not come easy. A great deal of design time is necessary to optimize interconnection networks for most signals, particularly clocks and other critical signals, such as memory data and address lines.

6.2 Physical Means of Interconnecting Signals

Large high-speed digital systems typically use the following means of physical interconnection for the three interconnection categories:[4–6]

1. Multilayer pc wiring boards for component-to-component connections
2. Multilayer pc or wire-wrap backpanels for board-to-board connections

3. Shielded twisted-pair cables, twinex, or coaxial cable for unit-to-unit connections

Minimizing signal line length is a fundamental requirement in all three categories. Reducing signal line length reduces the capacitance load, minimizes the chance for crosstalk, and reduces transmission-line effects and signal interconnection propagation delays.

Continuous ground planes are essential and are a fundamental requirement for circuit boards, motherboards, backpanels, etc., when advanced BiCMOS/CMOS devices are used.[7] Ground planes must not be broken or interrupted by device and connector pin fields,[7] and they must be arranged so that they are below (or above) all signal runs, including signal traces going to connector pin fields. Ground planes are needed near each signal trace to provide low-impedance return paths for transient load currents and to reduce cross coupling between signals. A nearby plane provides a shielding effect. Voltage planes are important but not as important as ground planes. Voltage planes must be continuous where boards are populated with circuits, but voltage planes do not have to be continuous in all pin fields.

6.2.1 Component-to-component connections

Multilayer pc boards with multiple voltage and ground planes are required for most BiCMOS/CMOS component-to-component connections because of signal density and electrical requirements.[7] Multilayer pc boards with multiple signal layers provide high signal density, and those with multiple ground and voltage planes provide good signal isolation. It is usually impossible to interconnect high-performance circuits without multiple signal layers. Multiple voltage and ground planes are required for isolation because high-signal density and fast signal edges increase the probability of coupled noise.

6.2.2 Board-to-board connections

Board-to-board connections are typically implemented using one of the following methods:

1. Multilayer pc motherboards with multiple signal layers and ground and voltage planes
2. Wire-wrap signal connections above a multilayer backpanel with ground and voltage planes
3. Wire-wrap signal connections, bus bars for voltage connections, and a backpanel that serves as a ground plane

With attention to the basic concerns of a high-performance signal distribution system, any of the above structures can be used successfully for most advanced BiCMOS/CMOS applications. Very high speed applications require multilayer mother-daughter pc boards with controlled impedance signal layers (instead of standard printed wiring interconnects that do not have controlled impedances) so that the response of signals can be accurately predicted and controlled. In all cases, daughterboard-to-motherboard connections must have sufficient ground pins so that transient load currents do not cause significant shifts in the reference level of daughterboards[3,8] (see Chapter 5).

At the board-to-board level, signal connections tend to be dominated by bus structures that have an increased probability of crosstalk due to long runs across backpanels or motherboards. Care should be taken in the assignment of daughterboard connector signal pins so as to minimize the length of signal lines. Also, care must be taken to isolate signals that can be upset by cross coupling (Section 6.8 discusses cross coupling and its control). In general, buses should be grouped together rather than intermingled.

6.2.3 Unit-to-unit connections

Unit-to-unit connections in high-speed BiCMOS/CMOS systems are typically implemented using some type of twisted-shielded–pair connections. For very high performance applications, controlled impedance twisted-shielded–pair, twinex, or coax interconnections are used so that terminations can be closely matched to cable impedance.[9] In all cable interconnections, cable impedance, propagation delay, dispersion, and shielding factor must be considered.[10] Knowledge of cable impedance is needed so that proper terminations can be selected. Propagation delay must be known to predict transport delay. Dispersion characteristics are important in long cables with high data rate; if dispersion is excessive, data and controls may be skewed sufficiently to cause errors. Sufficient shielding is needed to maintain data integrity and prevent excessive radiation. Cable manufacturers' data books and References 11 and 12 are recommended for guidance on these issues.

All unit-to-unit signal cables must include ground lines for signal return currents. When single-ended high-level interunit communication is used (low-level single-ended signals should never be used—see Chapter 9), ideally there should be one ground line per signal line. When balanced differential interunit communication is used, one ground line per four to eight signal pairs is a good rule of thumb. If differential signals are perfectly balanced, no return ground lines are needed, but differential signals are never perfectly balanced. Thus,

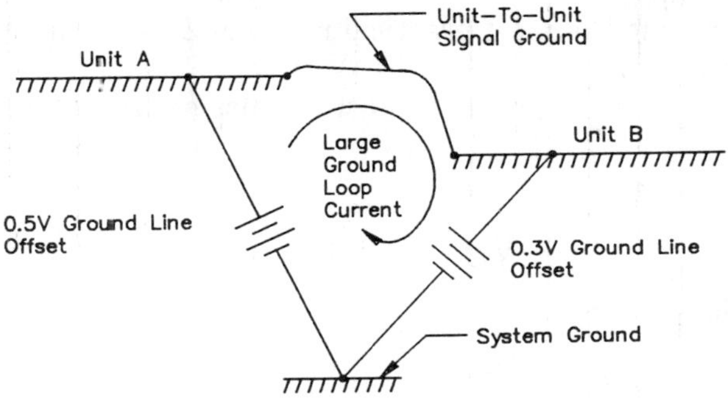

Figure 6.1 Large ground-loop currents may flow in unit-to-unit ground connections because of reference level offsets.

some grounds are needed to provide a direct return path for the unbalanced portion of the signal currents. However, direct unit-to-unit ground connections offer the potential for large ground-loop currents, as shown in Figure 6.1. Thus, when unit-to-unit signal grounds are required to ensure a low-impedance path for signal return currents, care must be taken to ensure that all units have low-impedance ground connections to prevent large interunit reference voltage offsets and large ground-loop currents. If large offset voltages exist between units, ground current in the signal return lines can be of such a magnitude that it will disrupt system operation by introducing noise[13] and in severe cases may overheat and burn out signal return lines.

6.3 Interconnection Impedance

Interconnection characteristic impedance Z_o is the key parameter that interrelates all aspects of high-speed system performance. An optimum range of Z_o exists that is a function of the electrical characteristics of the logic family used and mechanical limitations and requirements. This optimum range of Z_o is the best compromise between noise, delay, crosstalk, and mechanical constraints.[14] Lines with high Z_o are easier to drive, but high Z_o increases crosstalk (coupled signals are attenuated less), *RC* delays, and transmission-line ringing (BiCMOS/CMOS devices tend to have low output impedance, so there is greater mismatch). Mechanical constraints make it difficult to achieve high Z_o in high-density multilayer pc boards. Low Z_o attenuates signals which reduces signal-to-noise margins, increases delay (signals take longer to reach switching thresholds—first incident switching may not occur), and increases transient switching currents which increases system noise. When the cumulative effects are considered, 50 to 60 Ω tends to be the optimum impedance for advanced

BiCMOS/CMOS systems and the most practical range of Z_o to achieve from a mechanical standpoint.

The characteristic impedance Z_o of an ideal transmission line is defined as[15,16]

$$Z_o = \sqrt{\frac{L}{C}} \tag{6.1}$$

and the propagation delay t_{pd} as

$$t_{pd} = \sqrt{LC} \tag{6.2}$$

where L and C are as shown in Figure 6.2. At a sufficiently high frequency, which is exceeded by the frequency components in the switching edges of advanced BiCMOS/CMOS devices, the inductive and capacitive effects cancel and a transmission line appears to the source and load as a pure resistance.[15]

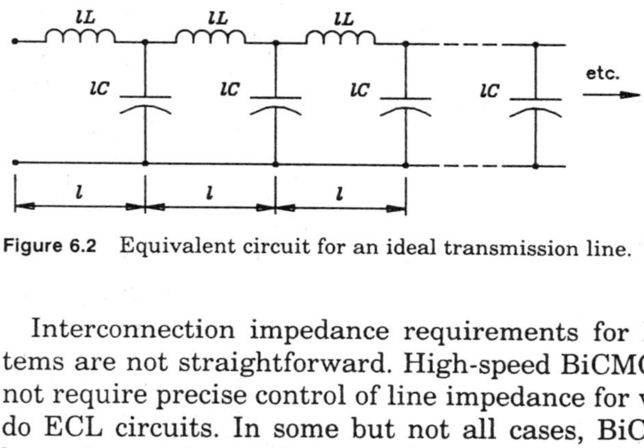

Figure 6.2 Equivalent circuit for an ideal transmission line.

Interconnection impedance requirements for BiCMOS/CMOS systems are not straightforward. High-speed BiCMOS/CMOS circuits do not require precise control of line impedance for waveshape control as do ECL circuits. In some but not all cases, BiCMOS/CMOS circuits have input clamps for waveshape control (see Chapter 7). Input clamps are less sensitive to variations in line impedance, but input dynamic clamping characteristics are seldom specified and many parts do not have effective clamps, which means predicting signal response in BiCMOS/CMOS systems is difficult if not impossible. Practical experience has shown that a two-to-one line impedance range can be tolerated on signal lines that require waveshape control, such as clocks, and even more on lines that do not require waveshape control (assuming settling time is available). Thus, precise control of the impedance of signal lines is not a major issue relative to waveshape control. The main issues relative to pc board characteristics are minimum line impedance and control of crosstalk.

The impedance of signal lines must be high enough so that BiCMOS/CMOS outputs can drive the signal lines in a timely manner. If the line impedance is too low, signals require too long a time to reach final values and system operating speed is reduced. It is essential that certain signals, such as clocks, reach a level that exceeds the switching threshold of the receiving devices on the first trip down the line (this is called first incident switching).[17]

Real BiCMOS and CMOS devices have some finite amount of output impedance; when they switch, the output impedance forms a divider with the line impedance until the line is charged to the new state.[18] Hence, signals tend to step to new levels as shown in Figure 6.3. If line impedance is too low, it may take a number of steps for the signal to reach its final value.[19] In low-speed applications, stepping waveforms may not cause problems. For example, data and control signals with steps are not disruptive in synchronous systems if time is available for signals to reach valid logic levels (see Chapter 10). However, the extra time a stepping signal requires to reach a valid logic level reduces potential system operating speed. In a few cases, a waveform with a step can cause secondary level problems. If a step happened to be near the threshold level of a receiver, the receiver input circuitry may oscillate during the time the signal dwells in the threshold region. In some cases, the oscillation may generate enough noise to upset nearby circuits, but in most cases, the time that signals dwell in the intermediate region is relatively short, and as a result there are no ill effects. However, signals with steps must not be used to clock devices; multiple clocking may occur. For example, signals with steps are inherent near the source of TTL- or CMOS-driven line. To prevent steps from causing multiple clocking, clock lines must be routed so that the de-

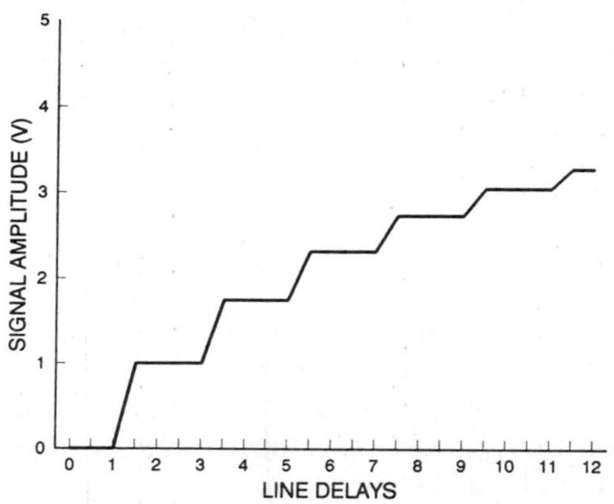

Figure 6.3 Interconnection impedance forms a divider with device output impedance that causes signal transitions to step up or down.

vices being clocked are located near the end of the lines (see Chapter 8 for clock distribution guidelines). Alternately, line impedance must be high enough so that any steps are well above the threshold of the clocked devices.

In BiCMOS applications, minimum line impedance requirements are established by dynamic *high*-state output impedance. Dynamic *high*-state output impedance is higher than *low*-state output impedance because BiCMOS totem-pole output stages have a resistor in the pull-up circuit but not in the pull-down (see Chapter 3, Figures 3.10 and 3.11). BiCMOS data books do not list *high*-state output impedance, but it is reasonable to assume that BiCMOS devices have output impedance in the same range as advanced Schottky TTL devices. Output impedance for advanced Schottky devices is typically near 20 Ω for drivers and near 45 Ω for standard logic function. For first incident wave switching and to prevent multiple clocking from steps dwelling in the threshold region, clock lines or other lines where wave edge shape is important must have an effective line impedance on the same order as device output impedance. The voltage divider formed by the device output impedance and the line impedance must not attenuate the signal so much that it does not exceed the switching threshold of the receiving devices (the switching threshold of BiCMOS devices with TTL levels is near 1.5 V). If the interconnection system minimum line impedance meets the needs of *high*-going signals, it will be more than adequate for *high*-to-*low* BiCMOS transitions since *low* BiCMOS device output impedance tends to be much less than *high*-state output impedance (typically near 10 Ω).

Forty ohms is also near the minimum usable line impedance for CMOS-level signals that must switch on the first incident wave. Advanced CMOS input thresholds are at 30 and 70 percent of V_{cc}. Thus, both *high* and *low* initial signal transitions must equal or exceed two-thirds of V_{cc}. Under nominal +25°C conditions the output impedance of advanced CMOS devices is near 10 Ω. Worst-case output impedance over the extremes of military operating conditions is near 20 Ω. Thus, initial signal transitions on lines with 40 Ω or greater impedance will equal or exceed two-thirds of V_{cc}.

6.3.1 Loaded transmission-line impedance

Distributed lumped loading on lines lowers the effective impedance and increases the propagation delay. The effective characteristic impedance Z'_o and effective propagation time t'_{pd} of a line is a function of the total lumped load capacitance C_{LOAD} and the total intrinsic line capacitance C_{LINE}. The expressions for effective characteristic impedance and for effective propagation delay of loaded lines are[4]

$$Z'_o = \frac{Z_o}{\sqrt{1 + C_{\text{LOAD}}/C_{\text{LINE}}}} \tag{6.3}$$

$$t'_{\text{pd}} = t_{\text{pd}}\sqrt{1 + C_{\text{LOAD}}/C_{\text{LINE}}} \tag{6.4}$$

where C_{LOAD} is the total lumped capacitance (inputs and outputs) of each device connected to the line and C_{LINE} is the total line capacitance. Determining C_{LOAD} is straightforward; C_{LOAD} is the sum of the input or output, capacitance, as the case may be, of all the devices connected to the line. Determining C_{LINE} requires that the line capacitance C per unit length be known ($C_{\text{LINE}} = C$ per unit length × line length). However, deriving C (per unit length) is not a straightforward process. One approach is to use the basic equation for Z_o [Equation (6.1)]

$$Z_o = \sqrt{\frac{L}{C}}$$

and rearrange it to the following form:

$$C = \frac{L}{(Z_o)^2} \tag{6.5}$$

to calculate C per unit length of line. The characteristic impedance Z_o used in Equation (6.5) can be calculated from equations such as (6.12), (6.14), and (6.16) or may be available from the manufacturer's specifications, but L in Equation (6.5) is not readily available and is difficult to calculate.

The most practical means of determining C per unit length of line C is to measure the propagation delay of an actual unloaded pc board track of the configuration of interest and use the measured time $t_{\text{pd(measured)}}$ and the calculated Z_o [from Equations (6.12), (6.14), or (6.16)] in Equation (6.8). This equation is derived from Equations (6.6) and (6.7), which are specific cases of Equations (6.1) and (6.2).

$$t_{\text{pd(measured)}} = \sqrt{LC} \tag{6.6}$$

$$Z_{o\text{(calculated)}} = \sqrt{\frac{L}{C}} \tag{6.7}$$

$$C = \frac{t_{\text{pd(measured)}}}{Z_{o\text{(calculated)}}} \tag{6.8}$$

and the total capacitance of the line C_{LINE} is

$$C_{LINE} = C \times \text{line length} \tag{6.9}$$

where the *line length* is in the same units as those used with C.

It is important to note that lines with a number of loads will have an effective Z'_o much less than the unloaded Z_o, and that the effective t'_{pd} will be much greater. It is not unusual for the effective impedance of bused lines, or other lines that connect to numerous locations, to be reduced to one-half the intrinsic unloaded impedance. It is not uncommon for the effective Z'_o of a loaded pc track to be as low as 20 Ω, and the actual propagation delay as large as 3.5 ns/ft.

The use of uncontrolled-impedance pc boards or motherboards introduces the risk of very low effective impedance lines. However, high-impedance multilayer boards are difficult and expensive to build. To keep the impedance up, lines must be very narrow or the separation between lines and reference planes needs to be large which may increase the thickness of the board beyond practical limits. Also, higher board impedance increases the possibility of crosstalk. Thus, when all issues are considered, the optimal unloaded line impedance tends to be in the 50- to 60-Ω range.

6.4 High-Density Multilayer PC Boards

As system performance goes up, so do interconnection requirements. Each succeeding generation of systems requires devices with more pins packed more closely together. More package pin connections in a given area means less room for signal interconnections because feed-through holes for package pins or vias to surface-mount packages take away signal interconnection area. Yet, at the same time interconnection requirements go up. The only solution is finer lines with closer spacing and more signal layers, both of which increase mechanical complexity and cost. To meet the high signal density and electrical requirements of today's high-performance systems, the typical high-density pc board is built with several signal layers with 4- to 8-mil lines with similar line-to-line spacing. Signal layers or pairs of signal layers are normally sandwiched between voltage and ground planes to limit crosstalk. Signal layer to voltage or ground plane spacing is typically 4 to 8 mil.

6.5 Characteristic Impedance of Some Common Interconnection Structures

From a practical standpoint the basic equations for characteristic impedance and propagation time, Equations (6.1) and (6.2), are of limited value since L and C are difficult to determine. However, most texts that deal with transmission lines show empirically derived equations for determining the characteristic impedance and propagation delay time for most common physical interconnection structures.

For wire-wrap or welded-wire boards, the basic physical interconnection structure is a wire over a ground or voltage plane (voltage planes serve as ac references) as shown in Figure 6.4.

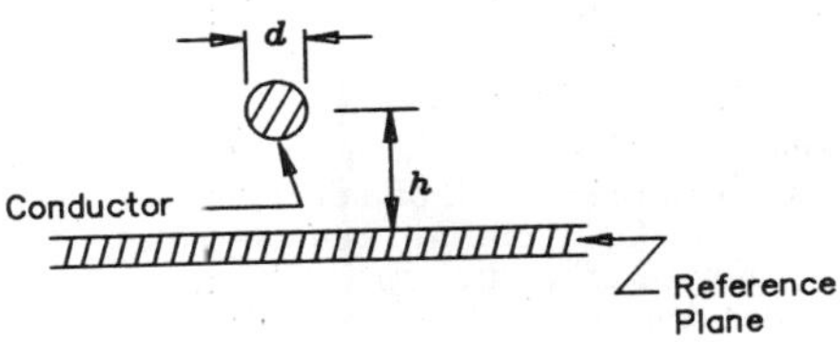

Figure 6.4 Wire over a reference plane with dimensions shown that are used to calculate line impedance and propagation delay.

For a wire above a reference plane,[20]

$$Z_o = \frac{60}{\sqrt{\epsilon_r}} \ln \frac{4h}{d} \tag{6.10}$$

where d is the wire diameter and h is the distance from ground to the center of the wire, and

$$t_{pd} = 1.017\sqrt{\epsilon_r} \tag{6.11}$$

where the constant 1.017 is the reciprocal of the velocity of light in free space.[15] In both Equations (6.10) and (6.11), ϵ_r is the effective dielectric constant of the material in the separation between the reference plane and the wire. For a wire in air the dielectric constant is 1, but the case of a wire in air is of little interest. For an actual wire-wrap or welded-wire board, the separation consists of air, the insulation on the wire, and other conductors. Thus, the effective dielectric constant is difficult to determine, and Z_o and t_{pd} calculations are quite complex. Actual Z_o measurements of wire-wrap and welded-wire boards show Z_o to be in the range of 80 to 150 Ω and t_{pd} to be on the order of 1.5 ns/ft. The value of Z_o is dependent upon the distance between the wires and the reference plane. If the wires are not kept close to the reference plane, Z_o values higher than 150 Ω are possible.

For pc board interconnections two basic physical arrangements of conductors relative to reference planes are possible:

1. Conductors may be located above a reference plane.
2. Conductors may be enclosed between two reference planes.

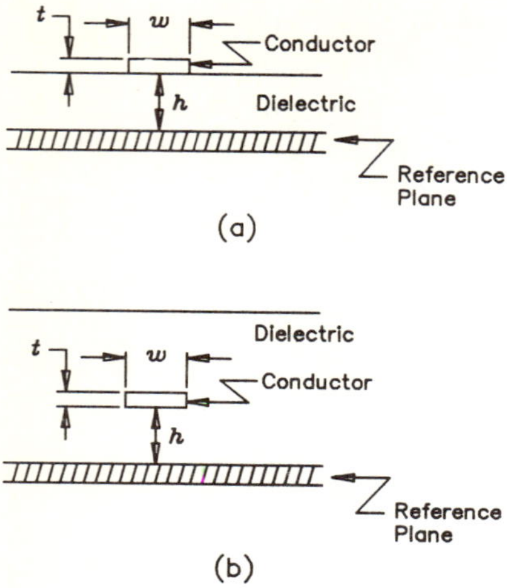

Figure 6.5 (*a*) Microstrip conductor on the surface of a dielectric, and (*b*) microstrip conductor buried in a dielectric with dimensions shown that are used to calculate line impedance and propagation delay.

Printed circuit board connections above a reference plane are called microstrip conductors (see Figure 6.5). A microstrip conductor can be on the surface of a pc board as shown in Figure 6.5(*a*), or buried in a board as shown in Figure 6.5(*b*). Configuration (*a*) corresponds to conductors on uncovered surface layers of pc boards, and configuration (*b*) to conductors on internal layers of multilayer pc boards (that are not enclosed by reference planes). The microstrip conductor of Figure 6.5(*b*) is more common today. Most present-day high-performance systems are built with multilayer pc boards with buried signal layers.

The equation for Z_o (in ohms) for a microstrip transmission line on the surface of a dielectric is

$$Z_o = \frac{60}{\sqrt{0.475\epsilon_r + 0.67}} \ln \frac{5.98h}{0.8w + t} \tag{6.12}$$

which is usually shown as[21,22]

$$Z_o = \frac{87}{\sqrt{\epsilon_r + 1.41}} \ln \frac{5.98h}{0.8w + t}$$

and t_{pd} for a surface microstrip conductor (in nanoseconds per foot) is

$$t_{pd} = 1.017 \sqrt{0.475\epsilon_r + 0.67} \tag{6.13}$$

where h, t, and w are in inches and are as shown in Figure 6.5 and ϵ_r is the dielectric constant of the material between the conductor and the reference plane.

The equation for Z_o (in ohms) for a buried microstrip transmission line is

$$Z_o = \frac{60}{\sqrt{\epsilon_r}} \ln \frac{5.98h}{0.8w + t} \tag{6.14}$$

and t_{pd} for a buried microstrip (in nanoseconds per foot) is

$$t_{pd} = 1.017\sqrt{\epsilon_r} \tag{6.15}$$

where the variables are as defined above.

Printed circuit board connections with a reference plane above and below conductors are called stripline conductors (see Figure 6.6).

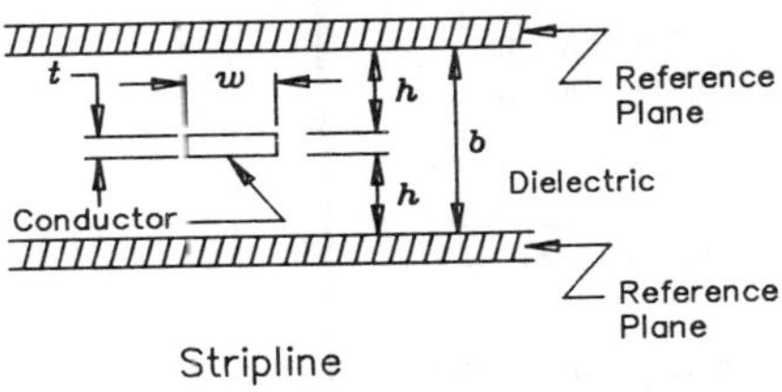

Figure 6.6 Stripline conductor with dimensions shown that are used to calculate line impedance and propagation delay.

The equations for Z_o (in ohms) and t_{pd} (in nanoseconds per foot) for a stripline conductor are[22]

$$Z_o = \frac{60}{\sqrt{\epsilon_r}} \ln \frac{4b}{0.67\pi(0.8w + t)} \tag{6.16}$$

$$t_{pd} = 1.017\sqrt{\epsilon_r} \tag{6.17}$$

where b, t, and w are in inches and are as shown in Figure 6.6.

The dielectric constant ϵ_r for epoxy glass pc board typically ranges from 4 to 5. For interconnection layers on most present-day multilayer pc boards, 1-oz copper plating is used. The thickness t in Equations (6.14) and (6.16) for 1-oz copper is approximately 0.0014 (t of 2-oz copper is 0.0028 in). Today it is not uncommon to see production boards with 4- to 8-mil wide conductors with similar line-to-line spacing and similar line to adjacent voltage or ground plane spacing. Such boards typically have Z_o in the range of 30- to 70-Ω and t_{pd} in the range of 1.8 to 2.2 ns/ft (see Figures 6.7 and 6.8).

The Z_o curves shown in Figures 6.7 and 6.8 were generated using Equations (6.14) and (6.16). They are not exact because they do not take into account effects of other nearby signal lines, but they are accurate enough for most engineering purposes and are useful for quick estimates of Z_o versus line width and dielectric thickness. Figure 6.7

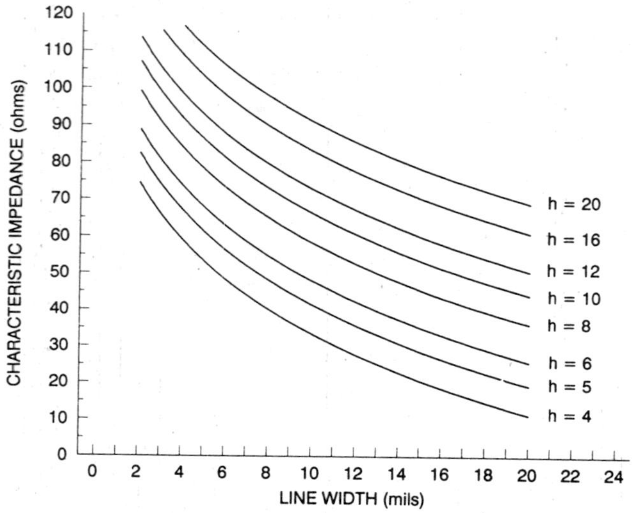

Figure 6.7 Characteristic impedance Z_o of microstrip conductors in epoxy glass pc boards (ϵ_r = 4.5) as a function of conductor width w and height h above a reference plane. (See Figure 6.5 for dimensions.)

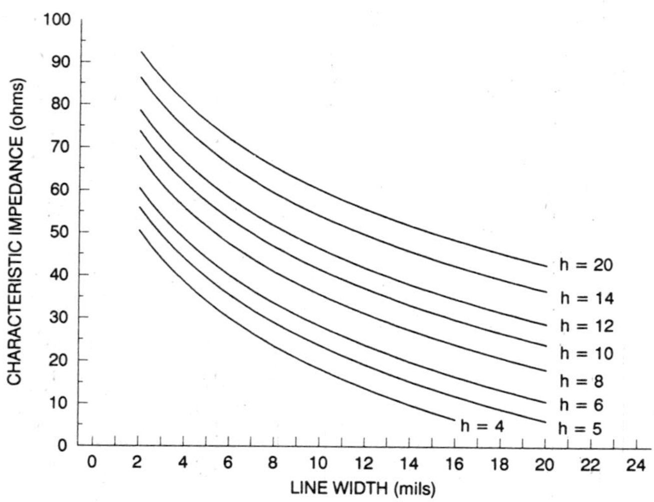

Figure 6.8 Characteristic impedance of stripline conductors in epoxy glass pc boards (ϵ_r = 4.5) as a function of conductor width w and distance h from the two reference planes. (See Figure 6.6 for dimensions.)

shows characteristic impedance Z_o of microstrip conductors in epoxy glass pc boards ($\epsilon_r = 4.5$) as a function of conductor width w and height h above a reference plane. Figure 6.8 shows the characteristic impedance of stripline conductors in epoxy glass pc boards ($\epsilon_r = 4.5$) as a function of conductor width w and distance h from the two reference planes.

In addition to wires over grounds, microstrip, and stripline conductor configurations, digital designers must often deal with coaxial cables and twisted-pair interconnections. Cable parameters, such as Z_o and t_{pd}, are supplied in cable manufacturers' catalogs (which should be consulted when such interconnections are used). Coaxial cables are available with various Z_o values; typical values are in the 50- to 90-Ω range.[23] Twisted-pair lines have Z_o in the neighborhood of 120 to 150 Ω, and shielded-twisted–pair lines are in the 70- to 120-Ω range.[9]

6.6 Breadboard Interconnections

Often during the breadboard phase of a project, welded-wire or wire-wrap interconnections are used at both the component-to-component and the motherboard level to expedite the completion of the initial breadboard unit and so that corrections or modifications can be easily incorporated.

When breadboarding with off-the-shelf welded-wire or wire-wrap boards, only those boards that are described as Schottky boards should be used when high-speed devices are being applied. Schottky boards have ground and voltage planes that are interconnected between all device and connector pins (see Chapter 5 and Figure 5.3) and provide low-impedance return paths for signals.

When advanced BiCMOS/CMOS devices are interconnected with wired boards, the signal response characteristics will not be representative of the final pc board system. Thus, wired breadboards should only be used to establish the functionality of a logic design; they should not be used to forecast the operating speed of the final pc board system. When the limits of the operating speed of a system must be demonstrated, the breadboard system must have the same physical dimensions and electrical characteristics as the final product.

6.7 Routing Guidelines

When advanced BiCMOS/CMOS devices are used, signal routing cannot be left to chance or be based on purely mechanical concerns. Specific routing instructions must be given to drafting departments or routing houses. Generalized statements lead to misunderstandings. Electrical (digital) designers and pc board designers tend to speak different languages.[23] Electrical designers think in terms of impedance, inductance, capacitance, and propagation delays, while pc board designers think in physical implementation terms such as widths, lengths, and layers. Statements such as "keep all clock lines as short as possible" have little meaning to drafting departments or routing houses. Specific physical routing instructions with specific signal names must be called out.

To simplify the task of routing boards, signals should be grouped into categories based on waveshape control requirements, crosstalk limits, or other special requirements. For example, clocks, strobes, buses, memory address, data, chip select, write lines, asynchronous signals, ECL signals, and analog signals all have special routing requirements. ECL and analog signals require special care when mixed with advanced BiCMOS or CMOS devices (see paragraph on ECL and analog signals below).

Clock signals. Clock signals have waveshape, skew, and crosstalk control requirements.

> To meet clock-signal waveshape requirements clock signals must be routed on layers with reasonably controlled impedance. Clock lines must not branch or have long stubs. Termination networks must be properly located. The location of the source and the locations of the loads must be controlled (see Chapter 8).
>
> To control clock skew clock lines may need defined maximum and minimum lengths.
>
> To meet crosstalk limits clock signals must be isolated and confined between reference layers. Other signals must not be mixed with clocks. Clock signals on a given layer must have extra spacing between lines. Clock signals of different frequency must have extra-wide spacing as must clock signals and other signals if they must be mixed.

Strobes. Strobes typically have the same requirements as clocks (see above paragraph).

Buses. Signals in a given bus can generally run next to other signals in that bus (which may save space) but not with other buses or other signals.

Memory address and data lines. High-speed memory address and data signals need to be isolated by reference planes to prevent feedback. Memory address signals can run next to each other, and memory data lines can run next to each other as bused signals (see above paragraph; also see Chapter 13 for more details).

Memory chip select lines. Generally, memory chip select lines have the same routing requirements as address lines (see above paragraph) and in most cases can be grouped with address lines.

Memory write lines. Memory write lines require the highest possible degree of isolation from crosstalk. They must be isolated by reference planes and by extra-wide line-to-line spacing from other signals, particularly other memory chip selects and address and data buses.

ECL and analog. ECL and analog signals require a high degree of isolation from BiCMOS or CMOS signals. They must be physically isolated in separate board areas with separate ground and voltage planes that are isolated from BiCMOS or CMOS switching currents.

Once the electrical designer has established the electrical requirements or limits of each signal category based on system performance requirements and error budgets, the requirements must then be translated to specific mechanical requirements for the pc board designer or routing house.

Routing order. Most pc board routers will route signals from point to point and thus do not cause branches. However, if there are requirements for the location of loads or sources, their locations must be defined and relayed to the routing house. In general, in lower-performance applications, the location of loads and sources on a signal line are not critical. As performance increases, the physical relationship of loads and sources becomes more important. Routing order must be specified in high-speed applications and is of particular importance for clocks and other signals, such as strobes, that have waveshape requirements or settling time requirements. Typically, when waveshape control or minimum settling time is required, the source is located at one end of the line and any parallel termination or clamping is located at the other end with the loads grouped near the termination or clamped end of the line. Alternately, if series termination is used (see Chapter 7), the series-termination network must be located very near the source.

In some applications where minimum settling time is needed, lines are often center-driven or made into loops. Center-driving lines or looping lines (back to the source) reduces the effective line impedance seen by the driver to $\frac{1}{2}Z_o$ and more nearly matches the line impedance to the output impedance of BiCMOS/CMOS drivers which in turn reduces ringing and signal settling times.[24] Looping lines back to the source also has an additional benefit; it typically reduces the worst-case distance between the source and the most remote load and thus reduces the worst-case propagation time. Center driving and looping techniques are often used on clock and memory address lines to improve waveshape and reduce settling time. Loops may also be useful in some bus applications to reduce settling time. Line impedance can be further lowered by connecting additional lines between loops to form grids. However, grids, loops, etc., are only beneficial where the effective line impedance (before being looped or gridded) is much greater than the driver output impedance (which is the typical situation when advanced BiCMOS/CMOS bus drivers are used). If the basic line impedance is much less than the driver worst-case output impedance, grids and loops are not beneficial. Lines may require numerous reflections to step up to the final value, and thus such an arrangement might degrade the response time. If such arrangements are considered, caution must be exercised since CMOS device output impedance may change by a factor of 2 to 1 from nominal with process variations and the extremes of the military operating conditions (see Chapter 3). An additional complication, since most routers will not automatically connect loops or grids, is that manual intervention and layout are required.

6.8 Crosstalk

Crosstalk is the noise voltage developed on signal lines when nearby lines change state. Crosstalk occurs because of capacitive and inductive coupling between adjacent or nearby lines. It is a function of the separation between signal lines, the linear distance that signal lines run parallel with each other, and the height above a ground, or other reference plane.

The faster (relative to older logic families) edge rates of advanced BiCMOS/CMOS devices greatly increase the possibility of coupling or crosstalk between signals.[7] Increased coupling, plus faster device response, greatly increases the possibility that system operation will be degraded by crosstalk. Synchronous design practices reduce the possibility of crosstalk disrupting system operation (see Chapter 10) if time can be allotted for crosstalk to subside, but when advanced BiCMOS/CMOS devices are applied, the goal is usually to achieve high operating speed. To maximize speed, crosstalk must be reduced to levels where no extra time is required for signals to stabilize.

Multilayer pc boards with multiple voltage and ground planes are needed to achieve the isolation required for critical signals in high-performance BiCMOS/CMOS systems.[25] In less critical applications, it is usually possible to use dual-stripline stack-ups to route related signals at right angles on adjacent layers, but in very critical applications voltage and ground planes should be used to isolate all signal layers. For example, when very high speed memory devices are used, address and data signals should be isolated by reference planes. Other techniques of isolating critical signals include using extra-wide spacing between signal lines and running ground traces on each side of critical signals. When signals are not isolated by planes, care must be taken to ensure that noisy signals are not run directly above or below a critical signal for some length.

On welded-wire or wire-wrap boards or backpanels, the wiring should be as direct as possible between points so as to randomize the routing, and the wiring should be kept as close as possible to the ground (or voltage) plane. Care must be taken to ensure that wiring is not channelized and in particular, that the critical signals, such as clocks, do not get channelized with noisy signals.

On welded-wire or wire-wrap boards or backpanels, critical single-ended signals should be run twisted with a ground wire connected to ground at the source and at the load as shown in Figure 6.9. Twisting signal lines with ground lines provides the most direct path possible for return currents and provides some shielding effect. Return currents tend to flow in the nearby twisted ground line and not in the power or ground planes (see Figure 6.10). Thus, transient-current flow and noise are reduced in the power and ground system. In addition, twisted-pair signal lines provide better control of line impedance, which allows more accurate terminations which in turn help control ringing and line reflections. Twisted-pair lines also tend to confine the magnetic fields of the two twisted conductors, which minimizes the chance for coupling into adjacent wiring.

Figure 6.9 Twist signal lines with ground lines to reduce crosstalk.

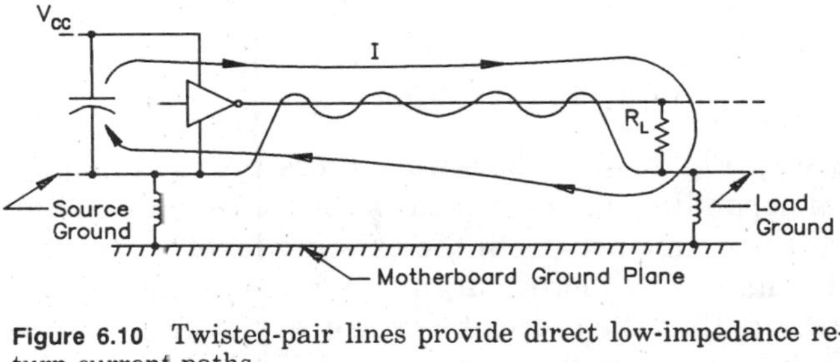

Figure 6.10 Twisted-pair lines provide direct low-impedance return current paths.

In all but the most trivial of systems, some signal line inevitability must run close to and parallel with other signals. Where signals are close and parallel, two forms of crosstalk exist: forward and backward.

Forward crosstalk is present on coupled lines coincident with an active wavefront on a nearby driven line and exists for the duration of the edge transition of the driven line. Backward crosstalk on coupled lines flows away from the wavefront on the nearby active line and exists for twice the propagation delay of the coupled line length.[26] Both forms of crosstalk can cause circuit malfunctions. Backward crosstalk tends to be more detrimental, since it is usually of a higher amplitude and lasts for a longer time, but either form can cause logic errors. The extent of the possible upset of a coupled-to line is dependent upon the polarity and amplitude of the coupled noise relative to the logic level of the signal that is disturbed and the physical topology of the lines. The signal flow may be such that the coupling is of little concern. For example, backward crosstalk is of no concern at output node C in Figure 6.11. However, backward crosstalk could be a major problem for a circuit with the topology shown in Figure 6.12.

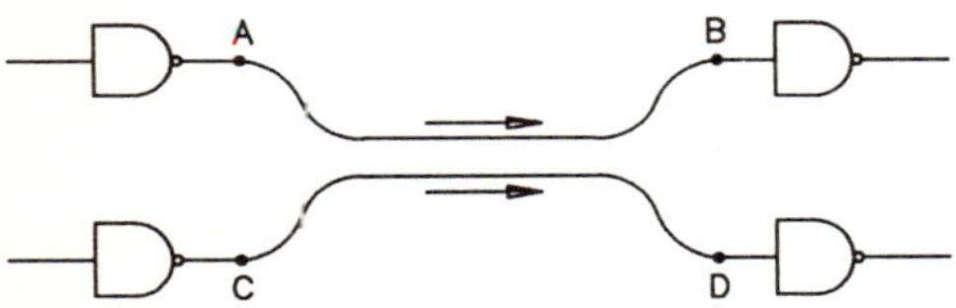

Figure 6.11 Coupled lines with the signal flow in the same direction.

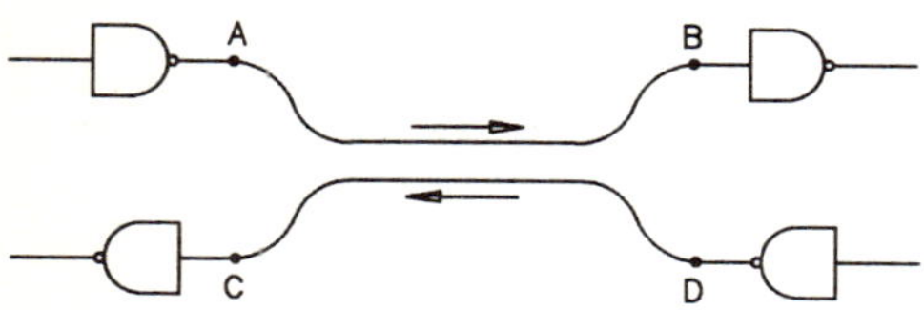

Figure 6.12 Coupled lines with the signal flow in the opposite direction.

In the case shown in Figure 6.12, backward crosstalk may or may not cause a problem depending upon the polarity of the coupled signal. If the coupling adds to or subtracts from the existing level so that the resulting composite signal has more margin, then the coupling is of little concern (assuming that the coupled signal does not cause excessive ringing). Since it may be impractical to analyze all possible combinations, the safest approach is to assume that there will be some combinations of active and inactive signals, such that some inactive signals will be degraded by crosstalk.

In those cases, where the interconnect topology and the polarity of the coupling is in the harmful direction, the crosstalk may or may not be harmful depending on the amplitude and the duration of the coupled voltage. If the amplitude of the coupled voltage is less than the noise margin of the logic components used, then the coupling may not be detrimental. Likewise, if the duration of the coupled pulse is short enough, even though it exceeds the noise margin of the receiving devices, the receiving devices may not react to a narrow pulse. Some device manufacturers are specifying the pulse immunity of their devices to help designers evaluate the possible detrimental effects of ground-bounce transients (see Chapter 2). Such data is also useful for determining the sensitivity of devices to coupled pulses.

The general expressions for crosstalk voltage amplitude between two lines with coupled length l (see Figure 6.13) for the two types of crosstalk are[27,28]

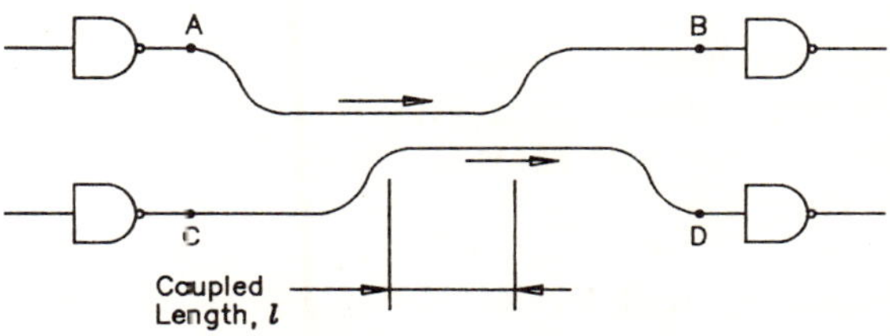

Figure 6.13 Coupling occurs where signal lines are in close proximity which may only be a small portion of the total signal length.

Backward crosstalk V_B

$$V_B = \left(\frac{K_C + K_L}{4}\right)\left(\frac{2t_p}{t_r}\right)\Delta V_S \tag{6.18}$$

for coupled line lengths from $l = 0$ to $l = t_r/2t'_{pd}$, and

$$V_B = \frac{K_C + K_L}{4}\Delta V_S \tag{6.19}$$

for coupled line lengths of $l = t_r/2t'_{pd}$ or greater.

Forward crosstalk V_F

$$V_F = \left(\frac{K_C - K_L}{2}\right)\left(\frac{t_p}{t_r}\right)\Delta V_S \tag{6.20}$$

For Equations (6.18) to (6.20)

ΔV_S = driving signal transition amplitude
K_C = capacitive coupling coefficient
K_L = inductive coupling coefficient
t'_{pd} = effective propagation delay of media
t_p = propagation delay of coupled length ($l \times t'_{pd}$)
t_r = rise time of driving signal
l = coupled length

and

$$K_C = \frac{C_m}{C}$$

$$K_L = \frac{L_m}{L}$$

where C_m = mutual capacitance between lines
C = capacitance between lines and ground
L_m = mutual inductance between lines
L = inductance of each line

In a homogeneous material K_C and K_L are equal and no forward crosstalk exists.[28] However, conductors in typical digital applications are not surrounded by a pure homogeneous material. Welded-wire or wire-wrap interconnections, as well as pc board interconnections, are surrounded by a conglomerate of materials, other conductors, insulation, etc. For practical engineering purposes, however, forward crosstalk is of little concern in welded-wire or wire-wrap circuit board or motherboard interconnections or in embedded conductors in multilayer pc boards. The only exception is when two lines have a long coupled length (relative to the active signal rise time); in this case the t_p/t_r term in the forward crosstalk equation becomes significant. Where forward crosstalk exists, it consists of a pulse with a width equal to the rise time t_r of the driving source. The amplitude is proportional to the coupled length.

Equation (6.18) shows that backward crosstalk magnitude is a function of coupled line length for lengths from $l = 0$ to $l = t_r/2t'_{pd}$. At $l = t_r/2t'_{pd}$ (which is the *critical line length*[4]—see Chapter 1), backward crosstalk reaches a maximum amplitude; for longer coupled lengths, it increases in width but does not increase in amplitude[29] [see Equation (6.19)]. The duration of backward crosstalk is equal to the two-way delay of the coupled length l.

> Backward crosstalk reaches a maximum amplitude where the propagation delay t_p of the coupled length of line is equal to one-half the rise time t_r of the active signal, i.e., when the coupled line length is equal to the critical line length.[4]

It is apparent from the limits of Equation (6.18) that the coupled line length needed for maximum amplitude coupling is reduced as signal rise time decreases. A signal trace adjacent to a trace driven with a signal with a 2-ns edge transition requires one-half the coupled length to achieve the limiting value of crosstalk as is required for a signal with a 4-ns edge transition. It therefore follows that systems with signals with fast rise times will have significantly more crosstalk than systems with signals with slow rise times.

To calculate crosstalk using Equations (6.18) to (6.20), line capacitance C, inductance L, and propagation delay t_p, as well as the mutual capacitance C_m and inductance L_m between lines, are needed. All are difficult to determine manually. Fortunately, today there are programs available, such as Quad Design's XTK crosstalk program,[30] that calculate crosstalk based on conductor topology and driver-receiver input-output characteristics.[31]

When crosstalk calculation tools are not available, the family of crosstalk plots for microstrip, dual-stripline, and stripline conductors (see Figure 6.14) shown in Figures 6.15 to 6.17 allow a quick estimation of the approximate amplitude of backward crosstalk [which is usually of most concern since it tends to be larger than forward crosstalk except where lines are very long—see Equation (6.20)]. The crosstalk curves shown in Figures 6.15 to 6.17 are based on 5-V signal transitions with 2-ns transition times (on the driven line). Faster or slower driving signal rise times move the knee of the curves in or out on the coupled length axis—the knee of the curves corresponds to the critical line length (see discussion above). The curves are directly applicable for signals with CMOS levels since most CMOS signal transitions are in the 4.5- to 5-V range. Crosstalk amplitude estimates for TTL-level signals, which are typically in the 3- to 4-V range, should be proportionally reduced (i.e., three-fifths of the values shown). The curves were generated using data for 8-mil-wide, 2-mil-thick traces in a material with a dielectric constant of 4.5, which is typical for epoxy glass pc boards. Line thickness and width variation, within a two-to-one ratio in either direction, have little effect on the results.

Backward crosstalk estimates based on the curves shown in Figures 6.15 to 6.17 should be accurate enough for most engineering purposes, but sight must not be lost of the fact that crosstalk is usually not a simple phenomenon. In most cases, it is the result of complex interac-

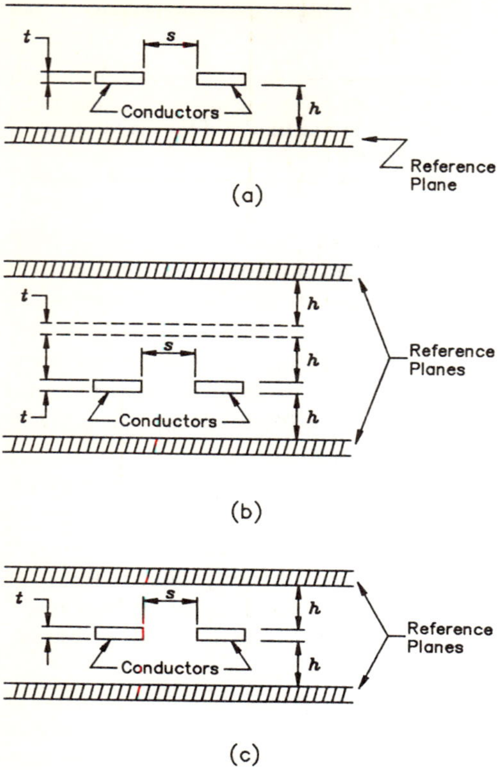

Figure 6.14 Typical pc board interconnection structures. (*a*) Microstrip; (*b*) dual stripline; (*c*) stripline.

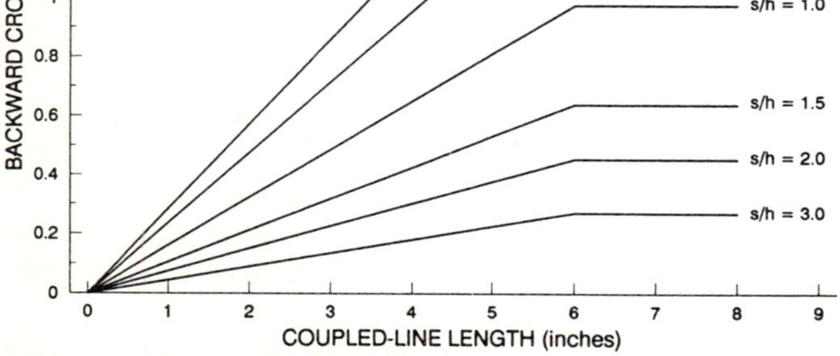

Figure 6.15 Backward crosstalk versus coupled line length *l*, spacing *s*, and height *h* above a reference plane for buried microstrip conductors in epoxy glass pc boards with a dielectric constant ϵ_r of 4.5 for a 5-V signal transition in 2 ns (data calculated using Quad Design XTK crosstalk simulation program). (See Figure 6.14 for dimensions.)

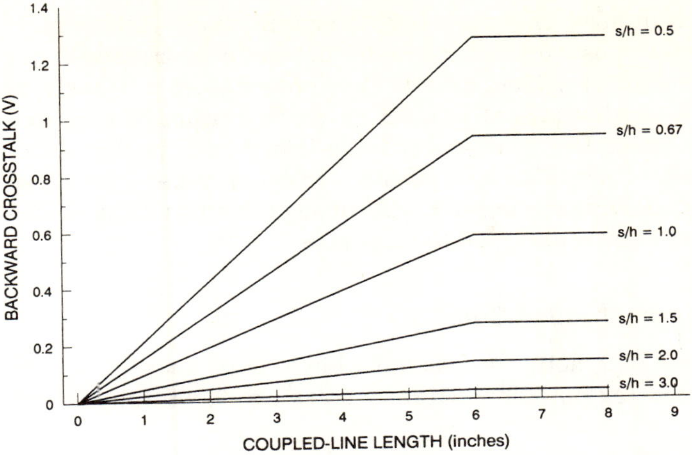

Figure 6.16 Backward crosstalk versus coupled line length l, spacing s, and distance h from reference planes for stripline conductors in epoxy glass pc boards with a dielectric constant ϵ_r of 4.5 for a 5-V signal transition in 2 ns (data calculated using Quad Design XTK crosstalk simulation program). (See Figure 6.14 for dimensions.)

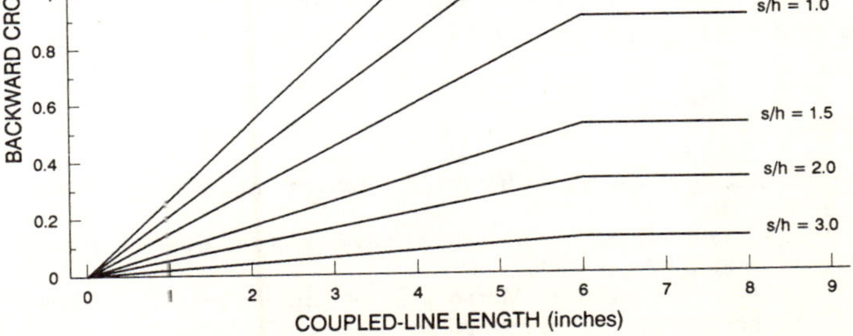

Figure 6.17 Backward crosstalk versus coupled line length l, spacing s, and distance h from reference planes for dual-stripline conductors in epoxy glass pc boards with a dielectric constant ϵ_r of 4.5 for a 5-V signal transition in 2 ns (data calculated using Quad Design XTK crosstalk simulation program). (See Figure 6.14 for dimensions.)

tions between a number of signals and their reflections. Actual measurements are the most reliable source for determining crosstalk, but the practicing engineer seldom has the time or resources to make controlled measurements on final production configuration boards, and measurements "after the fact" are of little benefit except for educational purposes. Crosstalk is a major problem when advanced BiCMOS/CMOS devices are used; it will cause problems unless the interconnection system is designed to minimize it.

6.9 Signal Interconnection Summary

1. Circuit boards and motherboards must have continuous power and ground planes (except for clearance holes for feedthroughs and vias).
2. Signals that are highly susceptible to crosstalk should be run at right angles to one another or be isolated by reference planes (when multilayer pc boards are used). For example, memory address lines must run at right angles to memory data lines as a minimum safeguard. In most advanced BiCMOS/CMOS applications, memory address and data lines should be isolated by reference planes.
3. Signals, such as clocks, that are highly sensitive to crosstalk should be isolated by reference planes from signals on other layers and by extra-wide line-to-line spacing (see Figures 6.15 to 6.17).
4. Interconnection line impedance should be greater than 40 Ω so as to not excessively degrade TTL- or CMOS-level signals.
5. Signal lines must run near reference planes (either power or ground) to minimize cross coupling. When signals cannot be positioned near a reference plane (such as between separate units), signals should be sent via differential twisted-pair lines (the need for differential signals between units not referenced to a common ground plane is described in Chapter 9).

6.10 References

1. Meredith, Mike: "Analyzing Interconnect Timing," *Electronic Engineering Times,* September 11, 1989, p. T6.
2. Cutler, Robert: "Your Logic Simulation Is Only as Good as Your Board Layout," *VLSI System Design,* July 1987, pp. 40–42.
3. Tripp, Tim, and Bill Hall: "Good Design Methods Quiet High-Speed CMOS Noise Problems," *EDN,* October 29, 1987, pp. 229–236.
4. Southard, Robert K.: "High-Speed Signal Pathways from Board to Board," in 1981 WESCON Records, Session 18, September 1981, Paper No. 2.
5. Harper, Charles A.: *Handbook of Electronic Packaging,* McGraw-Hill, New York, 1969.

6. Harper, Charles A.: *Handbook of Wiring, Cabling and Interconnections for Electronics,* McGraw-Hill, New York, 1972.
7. Tomlinson, Jock: "Avoid the Pitfalls of High-Speed Logic Design," *Electronic Design,* November 9, 1989, pp. 75–84.
8. DiCerto, Joseph: "Poor Packaging Produces Problems," *The Electronic Engineer,* September 1970, pp. 91–93.
9. Crouch, Ronald A.: "Choose Cable with Care to Optimize System Design," *EDN,* November 5, 1978, pp. 113–116.
10. Timmons, Frank: "Wire or Cable Has Many Faces, Know Them All Before Choosing," Part II, *EDN,* March 1, 1970, pp. 49–55.
11. Ott, Henry W.: *Noise Reduction Techniques in Electronic Systems,* 2d ed., Wiley, New York, 1988.
12. Morrison, Ralph: *Grounding and Shielding Techniques in Instrumentation,* 3d ed., Wiley, New York, 1986.
13. Brown, H. C.: "Get Rid of Ground-Loop Noise," *Electronic Design,* No. 15, July 19, 1969, pp. 84–87.
14. Arvanitakis, N. C., and J. J. Zara: "Design Considerations of Printed Circuit Transmission Lines for High Performance Circuits," in 1981 WESCON Records, Session 18, September 1981, Paper No. 4.
15. Matick, Richard E.: *Transmission Lines for Digital and Communications Networks,* McGraw-Hill, New York, 1969.
16. Kaupp, H. R.: "Characteristics of Microstrip Transmission Lines," *IEEE Trans. on Electronic Computers,* Vol. EC-16, No. 2, April 1967, pp. 185–193.
17. *FACT Advanced CMOS Logic Databook,* National Semiconductor Corp., Santa Clara, Calif., 1989.
18. Heniford, William: "Muffling Noise in TTL," *The Electronic Engineer,* July 1969, pp. 63–69.
19. DeClue, Joseph L.: "Wiring for High-Speed Circuits," *Electronic Design,* No. 11, May 24, 1976, pp. 84–86.
20. Springfield, William K.: "Designing Transmission Lines into Multilayer Circuit Boards," *Electronics,* November 1, 1965, pp. 90–96.
21. *Applications Handbook,* Cypress Semiconductor Corp., San Jose, Calif., 1989, pp. 1–12.
22. Blood, William R., Jr.: *MECL System Design Handbook,* 4th ed., Motorola Semiconductor Products Inc., Phoenix, Ariz., 1988.
23. Tasker, Shiv C.: "Making the Best Use of On-Board Interconnections," *Electronic Engineering Times,* December 4, 1989, pp. 41, 66.
24. *A Guide to Backpanel Performance Measurements,* IEEE Standard P1194.0/D2, IEEE, New York, June 9, 1989.
25. Catt, Ivor: "Crosstalk (Noise) in Digital Systems," *IEEE Trans. on Electronic Computers,* Vol. EC-16, No. 6, December 1967, pp. 743–768.
26. DeFalco, John A.: "Predicting Crosstalk in Digital Systems," *Computer Design,* June 1973, pp. 69–75.
27. Gheewala, Tushar, and David MacMillan: "High-Speed GaAs Logic Systems Require Special Packaging," *EDN,* May 17, 1984, pp. 8–14.
28. Kozuch, John J.: "A High Speed Approach to Controlled Impedance Packaging," Multiwire Division, Kollmorgen Corp., January 1987.
29. DeFalco, John A.: "Reflection and Crosstalk in Logic Circuit Interconnections," *IEEE Spectrum,* July 1970, pp. 44–50.
30. Nass, Richard: "PC-Board Speeds Skyrocket," *Electronic Design,* September 28, 1989, pp. 31–32, 37–38.
31. Beresford, Roderic: "How to Tame High-Speed Design," *High Performance Systems,* September 1989, pp. 78–82.

Chapter

7

Transmission-Line Effects

Transmission-line effects were of little concern when the older, slower logic families were used. However, when advanced BiCMOS/CMOS devices are used, transmission-line effects are of concern at all interconnection levels. Transmission-line effects begin to appear when signal rise times are near the propagation delays of the interconnecting lines. As rise times decrease relative to line propagation delays, transmission-line effects become more pronounced. For all practical purposes because of their fast rise times, all signal lines are transmission lines when advanced BiCMOS/CMOS logic components are used.

The common criterion for the line length, called the "critical line length," for the onset of significant transmission-line effects is as follows: The *critical line length* is when the effective propagation delay of the line t'_{pd} is equal to one-half the rise time t_r of the signal (measured between 20 and 80 percent).[1–3] That is,

$$\text{Critical line length} = \frac{1}{2}\left(\frac{t_r}{t'_{pd}}\right) \tag{7.1}$$

Effective line propagation delays for typical pc boards range from 2 to 4 ns/ft. For a signal with a rise time of 2 ns on a board with a 2-ns/ft line propagation delay, the critical line length is

$$\text{Critical line length} = \frac{1}{2}\left(\frac{2\text{ ns}}{2\text{ ns/ft}}\right) = 0.5\text{ ft}$$

Since most advanced BiCMOS/CMOS devices have rise times of 2 ns or less, transmission-line effects must be considered in most applications where they are used since most signal runs, even on small pc boards, exceed 0.5 ft.

7.1 Basic Transmission-Line Theory

The following discussion is meant to convey a basic introduction to transmission-line theory and its application as it relates to high-speed advanced BiCMOS/CMOS devices and systems. A thorough discussion of transmission lines is beyond the scope of this book. A number of good references on the subject of transmission lines exist (see the reference list and bibliography at the end of this chapter). All designers involved in the application of advanced BiCMOS/CMOS logic devices should quickly procure (if they do not already have) some of the listed referenced material and undertake a thorough study of transmission lines if they are not currently conversant on the subject. A basic understanding of transmission lines and transmission-line effects is essential to the successful application of high-speed logic devices.

Signals are well behaved when the signal source impedance, signal line impedance, and load impedance are very closely matched (i.e., almost equal) and when the line has only one source and one load with no discontinuities, branches, or stubs. In such cases, one line propagation delay (approximately 2 ns/ft if the line is in a pc board) after a signal is launched into a line at the source, a signal very much like the signal launched at the source can be observed at the load. If the signal source, line, and load are not matched, or the circuit topology does not consist of the ideal case of one source and one load, the signal quality will degrade as a result of transmission-line ringing and other effects. Signal sources, loads, and lines are not matched on a typical pc board or welded-wire or wire-wrap circuit board or motherboard. Thus, when signal runs are greater than the critical line length [see Equation (7.1) for the definition of the critical line length], signal quality will degrade as a result of transmission-line effects, and additional time, beyond the one-way propagation delay of the lines, will be required for signals to settle.

As unterminated signal lines increase in length beyond the critical line length, signal quality progressively degrades and signals take longer to stabilize. Depending upon where along a line a signal is observed, transmission-line effects will be apparent on lines longer than the critical line length under most conditions, even when lines are properly terminated. When signal lines are not significantly longer than the critical length, transmission-line effects may not be severe enough to impact normal operation, but they are present and observable.

Certain logic families incorporate features that help control transmission-line effects. Most BiCMOS components are built with current-limiting resistors in the pull-up section of totem-pole output stages and input clamp diodes (see Chapter 3) both of which help control transmission-line response. Some CMOS components have effective input clamps that help control transmission-line effects, but there is little uniformity in CMOS input-output circuitry or of the response of CMOS input-output circuits. Thus, when CMOS logic components are used to drive lines longer than the critical length, additional time beyond the one-way line propagation time must be allowed for signals to settle unless some form of termination is added. However, it is impractical to terminate all signals in most digital systems since terminations require additional components and may dissipate additional power (depending upon the method of termination used). Thus, in most applications where advanced CMOS devices are used, additional time beyond the one-way line propagation time must be allowed for signals to settle.

7.1.1 Ideal transmission-line response

A long transmission line[4] as shown in Figure 7.1, with a one-way delay greater than the signal rise time, with one source and one load, responds to signal transitions approximately as follows:

1. A signal transition is initiated at the source.
2. The signal travels along the transmission line with a delay time depending on the dielectric constant of the surrounding material.
3. The signal arrives as the first incident wave at the load after a delay that is a function of the line length and the dielectric of the material adjacent to the line.
4. Part of the energy of the first incident wave is absorbed by the load in establishing the initial signal level at the load.
5. The part of the energy of the first incident wave not absorbed as a result of mismatch between the line and the load impedance is reflected back toward the source.
6. When the reflection of the first incident wave arrives at the source, part of it may be absorbed by the source in establishing a new signal level; any mismatch is reflected back toward the load.
7. When the second incident wave arrives at the load, part of the energy is absorbed in establishing a new signal level; any mismatch is reflected back to the source as in item 5 above.
8. The cycles of reflections between load and source are repeated with part of the energy absorbed in each cycle; after some time, depending upon the characteristics of the source, line, and load, the reflected transient levels are so small that they are negligible.

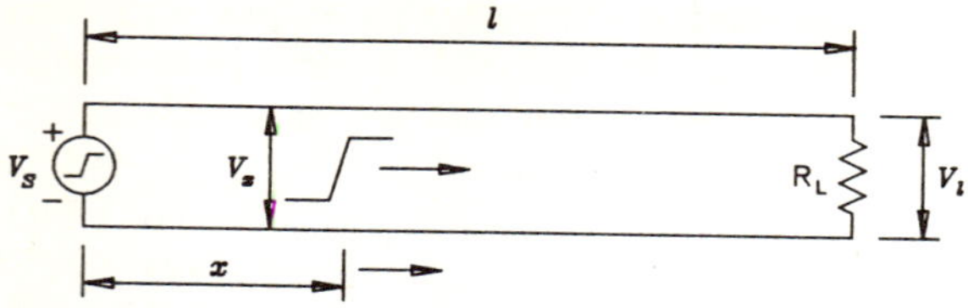

Figure 7.1 Ideal representation of a long transmission line with one source and one load.

Reflections. The voltage reflected when an incident wave arrives at a load is determined by the mismatch between the impedance of the line and the load (or source). If a transmission line is terminated with an impedance that is equal to the characteristic impedance Z_o of the line, there will be no reflection from the end of the line, and the only signal appearing on the line will be the incident wave. If some other value of termination is used, a portion of the incident wave will be reflected, and the signal appearing on the line will be the sum of the incident and reflected waves. The magnitude and polarity of the reflection from a load or a source is quantitatively described by the reflection coefficient ρ.

The equation for the reflection coefficient ρ_L for the load end of a line is[5,6]

$$\rho_L = \frac{R_L - Z_o}{R_L + Z_o} \tag{7.2}$$

and the reflection coefficient ρ_S for the source end of the line is

$$\rho_S = \frac{R_S - Z_o}{R_S + Z_o} \tag{7.3}$$

If either end of the line is exactly matched to Z_o, that is, $R_S = Z_o$ or $R_L = Z_o$, the reflection coefficient is 0 and the incident wave is completely absorbed and no reflection occurs.

The action of a transmission line with nonzero reflection coefficients is shown diagrammatically in Figure 7.2 (the dimensions x and l are as shown in Figure 7.1).

There are a couple of cases of ρ that are of special interest. They are

1. $\rho = +1$ when $R_L = \infty$ or $R_S = \infty$ (an open-circuited line). When $\rho = +1$, the signal doubles when the incident wave arrives at the end of the line.

2. $\rho = -1$ when $R_L = 0$ or $R_S = 0$ (a short-circuited line). When $\rho = -1$, the incident wave reverses its polarity and subtracts an amount equal

to the incident wave from the existing voltage at the load (or source) and the new voltage level is reflected back toward the other end of the line. The energy in an incident wave is *not* absorbed by a short.

Case 1, above, is the typical situation for BiCMOS/CMOS logic circuits; inputs are sensitive to voltage (not power) and require only a small portion of the energy in the incident wave arriving from the transmission line to maintain steady-state operation (i.e., the load impedance is very large with respect to the line impedance). Therefore, the voltage at the load device is increased (up to double) compared to the incident wave, and a significant portion of the energy is reflected back toward the source.

Case 2, above, is approximated by typical CMOS *high* and *low* outputs and BiCMOS *low* outputs; their effective output impedance is not zero, but it is much less than normal line impedance. Thus, *low* BiCMOS outputs, or CMOS *high* or *low* outputs, have negative reflection coefficients. Energy is reflected with a reversal of voltage polarity. A negative reflection coefficient at the source end of a line will cause overshoot that may have occurred at the load end of the line to be converted to undershoot on alternate reflection cycles. The undershoot, when it arrives at the load, may cause the signal level to transition into the threshold region of the receiving device, upsetting the logic sense of the signal. Under such conditions, which are typical of advanced BiCMOS/CMOS devices, a number of round-trips may be required for the excess energy in the signal to be absorbed and for the

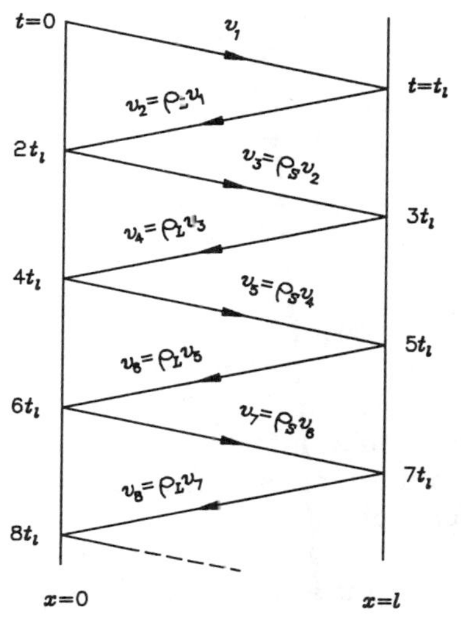

Figure 7.2 Lattice diagram for a long transmission line with one source and one load.

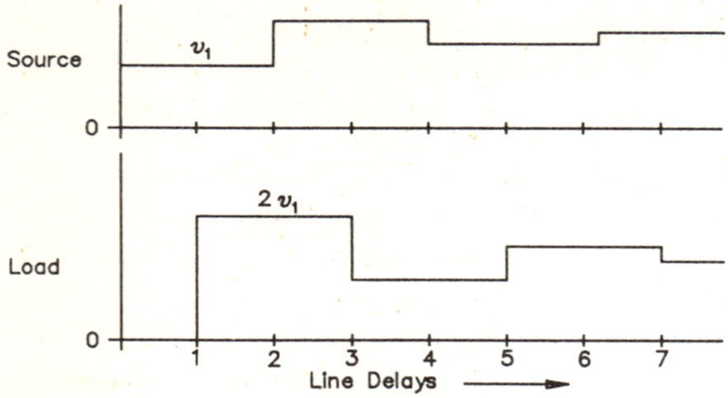

Figure 7.3 Typical response of a long transmission line with one source and one load.

signal to achieve a steady-state value. Figure 7.3 illustrates the classical response of a line with a low-impedance source and a single high-impedance load.[4,6,7] Figure 7.3 is drawn for a load reflection coefficient of +1 and a source reflection coefficient of −0.5 which are typical of BiCMOS/CMOS devices.

In the case illustrated in Figure 7.3, the waveform at the load end does not undershoot (on the second incident wave) below V_1, but in many actual cases the signal will undershoot back into the threshold region of the receiving device and will remain there until the signal travels back to the source and back to the load. Thus, for most lines that are not terminated, five line delays must be allowed for signals to settle. After five line delays, most of the energy in a wavefront will have been absorbed and the undershoot will have effectively dampened out in most cases. However, it is difficult to generalize a very complex phenomenon. Critical situations must be carefully analyzed.

> A rule of thumb that will serve for most timing analysis is: <u>Allow five line delays for unterminated lines to settle.</u>

7.1.2 Transmission-line termination

There are two basic means of terminating transmission lines—series (source) termination or load termination. Either method can be used to achieve a stable signal at the far (load) end of a line after one line delay. However, the response at the source and at intermediate points is different for the two methods of termination. Series-terminated lines require at least two line delays for signals to stabilize at the source end of a line; properly designed load-terminated lines reach steady-state conditions at the most remote point after one line delay.

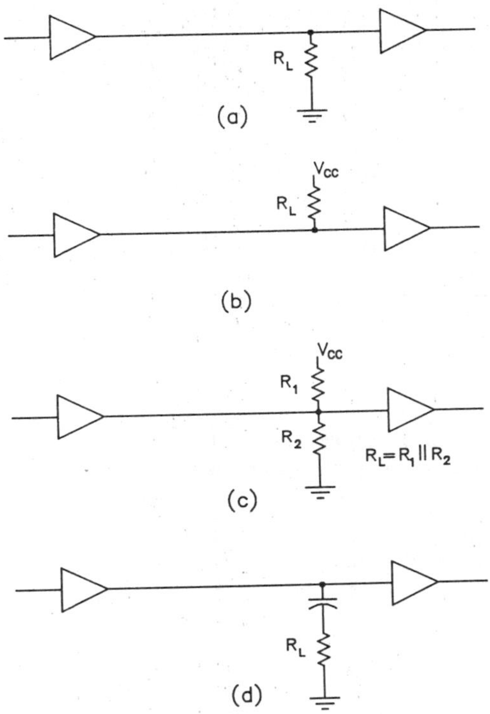

Figure 7.4 Load termination. (*a*) Termination to ground; (*b*) termination to V_{cc}; (*c*) split termination; (*d*) ac termination.

Load termination. A line is defined as load terminated when the load at the end of the line is matched to the impedance of the line. Figure 7.4(*a*), (*b*), (*c*), and (*d*) show various means of load termination.[8,9] The same dynamic results are achieved with each of the configurations.

When a line is load terminated with a matching impedance (R_L of the load = Z_o of the line), regardless of the dc circuit configuration of the terminating network, the signal at the source, at intermediate points, and at the load should appear as a well-behaved signal as shown in Figure 7.5. In such a case, the signal is launched into the line and at some later time (equal to the propagation time of the line t_{LINE}) arrives at the load. The energy in the wavefront is absorbed, and a steady-state level is established upon the arrival of the incident wave at the load (see Figure 7.5). In Figure 7.4 it is implied that the receivers have infinite impedance, and for all practical purposes, that is the case for BiCMOS/CMOS device inputs, as long as the signal levels remain between ground and V_{cc}. Thus, load-termination resistors (networks) do not need to be adjusted to compensate for receiver input impedance.

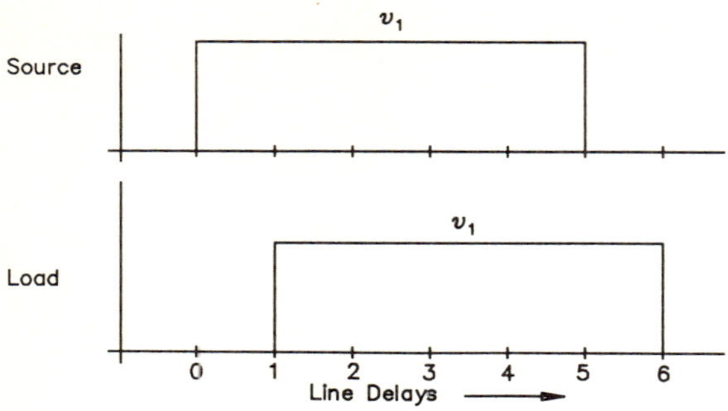

Figure 7.5 Idealized source and load waveforms in a load-terminated line.

Source termination. Source termination, often referred to as series termination, consists of matching the source impedance to the line impedance, as shown in Figure 7.6. The waveforms at the source, at intermediate points, and at the load for a source-terminated circuit that is exactly matched to the line and that has an infinite load are shown in Figure 7.7.

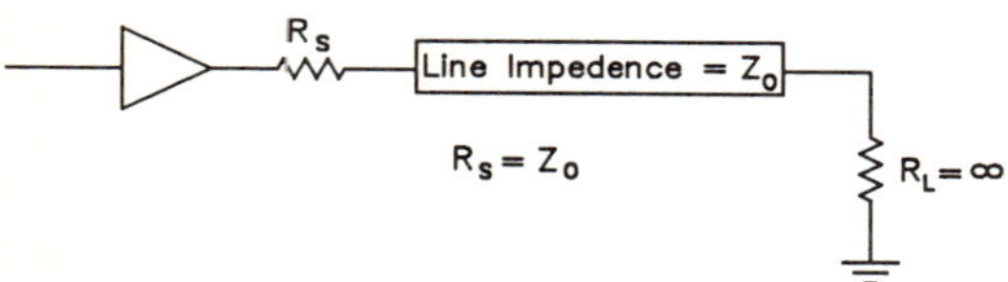

Figure 7.6 Source termination.

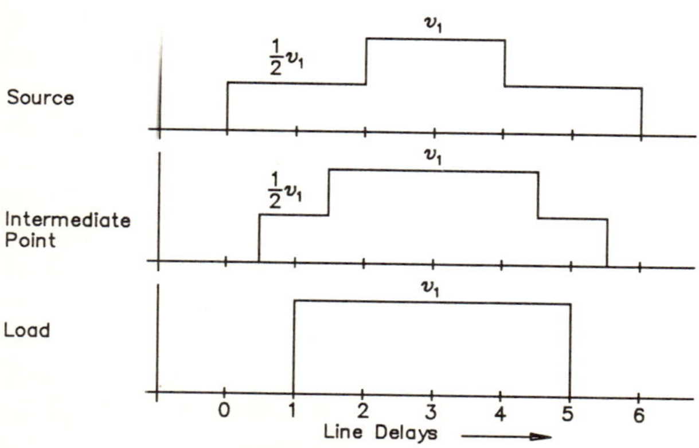

Figure 7.7 Waveforms in a source-terminated line.

Series termination works best where there is one source and one load. The waveform at the load end of a series-terminated line is well behaved, but at the source and at intermediate points near the source, the leading edge of the waveform will step up rather than make a smooth transition (see Figure 7.7). The initial step (or steps) occurs as a result of the divider formed by R_S and Z_o of the line (see Figure 7.8). The amplitude of the initial step is

$$V_{t=0}(\text{at the source}) = \left(\frac{Z_o}{R_S + Z_o}\right) V_{cc} \tag{7.4}$$

When R_S is matched to Z_o, the amplitude of the initial step is

$$V_{t=0}\,(\text{at the source}) = \frac{1}{2} V_{cc} \qquad \text{when } R_S = Z_o \tag{7.5}$$

Driver output impedance must be considered when selecting series-termination resistors. The effective source impedance is the sum of the driver output impedance and the source-termination resistor. Hence, for an exact match to a line, a source-termination resistor must have a value equal to Z_o less the driver output impedance. In BiCMOS applications, it is impossible to exactly match the line impedance for both *high*- and *low*-going signals since the output impedance of totem-pole output stages is different for the two states (see Chapter 3).

One advantage of series termination is that no additional dc power is dissipated as a result of the termination. However, care must be taken to ensure that noise margin is not significantly reduced by voltage dropped across series-termination resistors as a result of dc load currents. In CMOS or BiCMOS applications where devices have CMOS input circuits, dc load currents are very low and dc drop across series-terminating resistors is not significant (CMOS inputs have very high input impedance). As a rule of thumb to minimize noise margin loss and to provide slightly underdamped signals (which reach valid logic levels faster), series-termination resistors should be approxi-

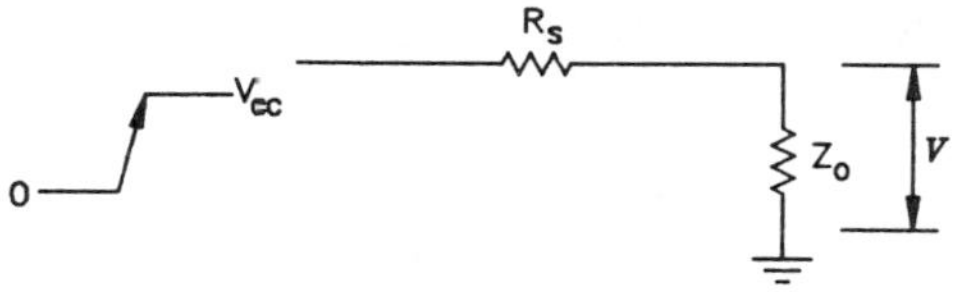

Figure 7.8 Series-termination resistance and line impedance form a divider that causes a step in the initial waveform launched into a series-terminated line.

mately one-third to one-half the line characteristic impedance Z_o. A value in the range of 22 to 33 Ω is appropriate for most applications.

7.2 BiCMOS/CMOS Device Transmission-Line Response

The response of advanced BiCMOS/CMOS logic devices in a transmission-line environment cannot be determined using simple ideal transmission-line theory. Ideal transmission-line theory is based on fixed linear source and load impedances. The input-output current-voltage relationships of BiCMOS/CMOS devices are nonlinear. Graphical techniques or transmission-line analysis programs that incorporate nonlinear models are needed to predict their response.

A graphical technique, sometimes called "Bergeron plots," is useful for predicting the response of BiCMOS/CMOS devices.[10] Bergeron plots predict signal response by combining nonlinear device input-output *I/V* characteristics with interconnection impedance. The succession of changes in voltage and current at the source and load are predicted by intersections of the input-output *I/V* characteristics and the load lines that represent the transmission-line impedance. Bergeron plots assume that quasi-steady-state conditions are achieved at each end of the transmission line before any reflection from the other end alters those conditions (this condition is met if the two-way delay is greater than the rise time). Losses, distortions, and source switching transients are neglected. Common dielectrics and conductors do not result in significant signal losses for typical CMOS rise times (1 ns or greater) and line lengths of less than 1 m.

The basic obstacle to using Bergeron plots to predict signal response is that in most cases no dynamic *V/I* interface specifications are listed or shown on BiCMOS/CMOS device data sheets, or if they are supplied, they are typical values without minimum and maximum limits. Without worst-case dynamic limits, system designers cannot accurately predict worst-case dynamic performance. Parts from different lots or from different manufacturers may behave much differently. If worst-case dynamic response cannot be predicted, potential performance must go unrealized; extra and wasted time must be allocated for signals to settle.

7.2.1 Typical advanced CMOS device transmission-line response

The output impedance of advanced CMOS devices is low relative to typical pc board line impedance, and the input impedance is much greater than the line impedance for normal signal levels.[11,12] For signal levels beyond the static operating range (transient levels greater than the supply or reference), the input impedance of CMOS devices may remain very high or it may change drastically. Inputs may have parasitic diodes, electrostatic protection diodes, and in some cases signal clamping diodes that begin to conduct as signals exceed the power rails.[12] When clamping circuits have an effective high-frequency impedance near or less than the interconnection impedance and the break point is near or within a diode drop of V_{cc} or ground, input clamps control transmission-line ringing in most applications. However, if receiving devices do not have clamps or the clamps are not effective at high frequencies, severe ringing may occur.

Typical clamped input line response. The typical source-load response for CMOS devices where the input is clamped by a low-impedance diode beyond the power rails is*

1. The source launches excess energy into the line.[13,14]
2. The load presents a high impedance to the line in the normal signal range which causes overshoot into the clamping region where most of the signal energy is absorbed, but some is reflected back to the source.
3. The source converts the reflected overshoot to undershoot and reflects the excess energy back to the load. If most of the excess energy in the first incident wave was absorbed at the load, the undershoot is small and does not impact operation.
4. The alternating overshoot-undershoot cycles quickly subside.

The above sequence of events for a FACT advanced CMOS gate driving a 50-Ω line is shown in the reflection diagram in Figure 7.9.[15] The reflection diagram for a *low*-to-*high* transition is constructed by first drawing a 50-Ω load line from the V_{OL} quiescent point to the

*The following nomenclature and definitions are used. Overdamped signals occur when the source impedance is higher or the load impedance is lower than the transmission-line impedance; underdamped signals occur when the source impedance is lower or the load impedance is higher than the transmission-line impedance. Overshoot is a signal transition beyond the steady-state level (i.e., below the normal *low* state or above the normal *high* state). Undershoot is a signal excursion between steady-state *high* and *low* levels (undershoot usually occurs after the nominal signal level has been reached the first time). Rise time is the time required for a signal to switch from its previous steady-state level to a new steady-state level. To reduce ambiguity, rise time is usually measured for only a central portion of the interval between steady-state signal levels, such as the 20 to 80 percent levels.

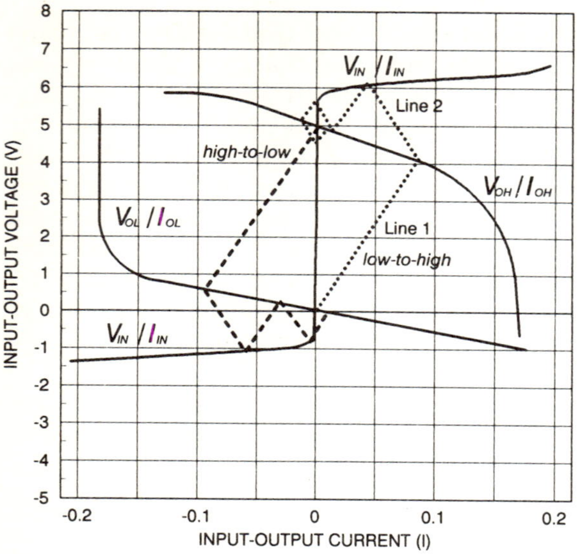

Figure 7.9 Reflection diagram for FACT CMOS gates with a 50-Ω interconnection. (*Reprinted with permission of National Semiconductor Corp.*)

point where it intersects with the V_{OH}/I_{OH} curve as shown by line 1.† The voltage at the intersection, approximately 4 V, is the amplitude of the first step at the source and of the first incident wave launched down the line. Next a line is drawn at a −50-Ω slope from the intersection to the V_{IN}/I_{IN} curve (see line 2). The voltage at this second intersection represents the amplitude of the initial signal seen at the load and the first reflection back from the load. The process of drawing lines is continued until they converge at the V_{OH} quiescent voltage. The reflection diagram for a *high*-to-*low* transition is constructed in a similar manner (see Figure 7.9). Figure 7.10 shows the theoretical waveforms (for the *low*-to-*high* transition) at the source and load based on the results of the reflection diagram of Figure 7.9.

Note that the low-impedance input clamps at the load limit the overshoot for both *low*-to-*high* and *high*-to-*low* transitions to near 1 V. The low source reverse characteristic then converts the reflected 1-V overshoot to about 1-V undershoot. Several reflection cycles are required before steady-state conditions are reached, but the signal never undershoots into the critical region near the threshold.

---

†Load lines terminating on V_{OH}/I_{OH} and V_{OL}/I_{OL} curves have positive slopes, and lines terminating on V_{IN}/I_{IN} curves have negative slopes. The transmission line acts as a load to the source and thus changes at the source have a positive Z_o slope. Likewise, the transmission line acts as a source to the load, and thus changes at the load have a negative Z_o slope.

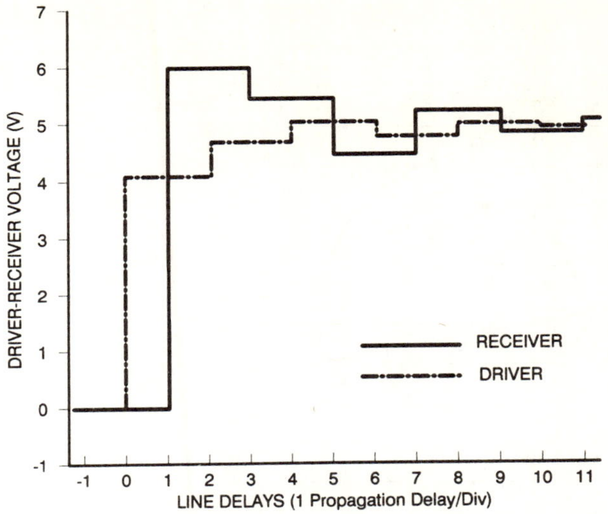

Figure 7.10 *Low*-to-*high* source and load waveforms predicted by the reflection diagram shown in Figure 7.9. (*Reprinted with permission of National Semiconductor Corp.*)

Typical unclamped input line response. The typical source-load response for CMOS devices where the input remains a high impedance beyond the power rails is

1. The source launches excess energy into the line.
2. The load presents a high impedance to the line which causes overshoot and reflection of most of the signal energy back to the source.
3. The source converts the overshoot to undershoot and reflects the excess energy back to the load.
4. The alternating overshoot-undershoot cycles repeat as damped ringing.

The above sequence of events is shown in the reflection diagram in Figure 7.11. The sequence is the same as for Figure 7.9 except that since the load does not have input clamps, line 2 is drawn to the zero-current axis (the V_{IN}/I_{IN} relationship is a vertical line at zero current). The voltage at the intersection of the zero-current axis is approximately 8 V which is the amplitude of the reflection of the first incident wave from the load. Note that the unclamped load results in more than 3-V overshoot. When the 3-V overshoot returns to the source, the intersection of line 3 and the V_{OH}/I_{OH} curve, the low output impedance of the source converts the 3-V overshoot to 2 V of undershoot (line 4). An undershoot of 2 V results in a signal level that is nearly

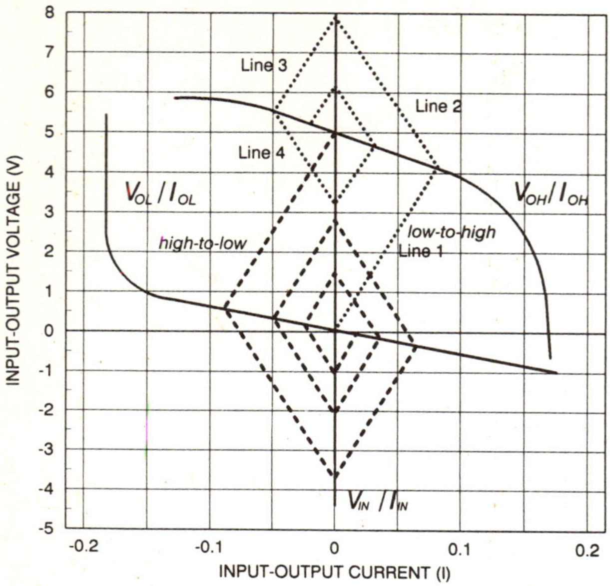

Figure 7.11 Typical reflection diagram for advanced CMOS input-output circuits without input clamps for signals below ground or above V_{cc} when interconnected with a 50-Ω transmission line.

midway between the *low* state and the *high* state. Several reflection cycles will be required before steady-state conditions are reached. Settling times are notably prolonged when devices do not have effective high-frequency input clamps.

7.2.2 Typical BiCMOS device transmission-line response

Reflection diagrams for BiCMOS devices look very much like those of advanced Schottky devices since both device families have similar input and output circuits.[10] Both have totem-pole output stages and inputs with clamps for signal excursions below ground, but neither have clamps for excursions above V_{cc}. Figure 7.12 is a reflection diagram for BiCMOS 74BC645 octal transceivers based on the output *V/I* curves shown in the Motorola book *Bi-CMOS Logic Data*[16] and on an assumed input V_{IN}/I_{IN} curve for signal excursions above and below ground when interconnected with a 50-Ω transmission line. Note that signals quickly stabilize and do not have dangerous undershoot levels. *High*-to-*low* transitions overshoot (in the negative direction) but do not ring back into the dangerous region. *Low*-to-*high* transitions do

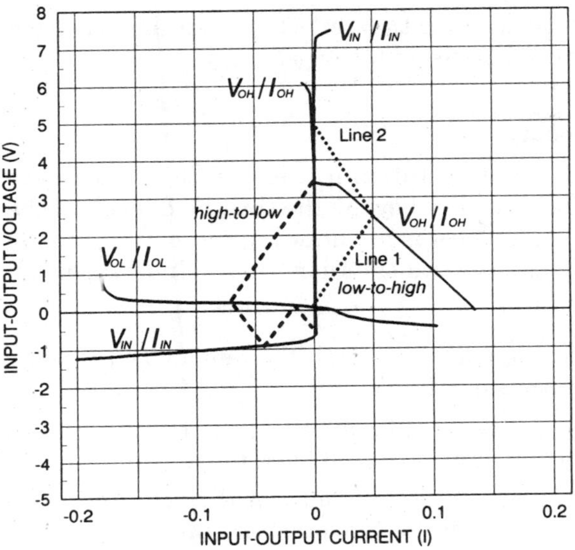

Figure 7.12 Typical reflection diagram for BiCMOS input-output circuits with input clamps for signals below ground but not above V_{cc} when interconnected with 50-Ω transmission line.

not overshoot V_{cc} or ring back. They stabilize at about 5 V where line 2 intersects the vertical (high-impedance) V_{IN}/I_{IN} curve. At that point, both the V_{OH}/I_{OH} and the V_{IN}/I_{IN} are vertical on the zero-current axis which precludes drawing additional reflection diagram lines (i.e., the diagram converges at that point).

The starting point for *high*-to-*low* transitions depends upon load and leakage currents and signal frequency. In low-frequency applications, *high* signals that dynamically stabilize above the level where the output can supply current tend to drift down to the point where the output can supply current—the intersection of the V_{OH}/I_{OH} curve with the zero-current axis. In high-frequency applications, *high* signals in the zero output current region may not have time to drift to the stable static level. Hence, in certain circumstances, signal characteristics are not constant but are a function of recent signal history. In Figure 7.12, the *high*-to-*low* transition is shown starting from the point where the V_{OH}/I_{OH} curve intersects the zero-current axis (i.e., at the point where *high*-level drive current is first available). Depending upon circumstances, a *high*-to-*low* refection diagram starting at the intersection of line 2 with the V_{IN}/I_{IN} curve should also be constructed. Note that more reflection cycles are required to reach a stable *low* state if the *high*-to-*low* transition reflection diagram is started from the point where line 2 intersects the V_{IN}/I_{IN} curve.

7.2.3 Controlling undershoot

Undershoot may impact circuit operation if it is of such a magnitude that signal levels degrade to near or beyond the threshold levels of the receiving devices. Thus, undershoot is the most important condition to avoid in networks that behave as transmission lines.[2] Undershoot can be controlled as follows:[17]

1. The source can be designed to control the amount of energy launched into the transmission line.
2. The load response can be designed to absorb some of the overshoot to control the amount of energy reflected back to the source.
3. The source can be designed to absorb some of the reflected overshoot from the load and control the conversion to undershoot.

The benefits of limiting the amount of energy launched into interconnections include less transient current, less crosstalk, and less overshoot energy to be absorbed or reflected. However, a moderate degree of overshoot is beneficial. The benefits include increased first incident signal amplitude to improve initial noise margin, reduced rise time degradation due to discrete capacitance loads, and compensation for line losses. If the source impedance is greater than the line impedance, not enough energy is sent down the line and the initial signal amplitude is reduced at the load. The signal may not have sufficient amplitude for first incident wave switching which slows the response of the line. Reduced signal amplitude causes loss of signal-to-noise margin at the load. Therefore, BiCMOS/CMOS devices need both maximum and minimum output dynamic characteristics specified to assure their operational characteristics in a transmission-line environment. High-performance ECL specifies both minimum and maximum output values.

High-load impedance minimizes steady-state power consumption and maximizes signal levels and noise margins. However, excess energy in incident waves will cause signals to overshoot, and transient disturbances (such as crosstalk) are reflected with increased amplitude. Thus, it is advantageous for the load to absorb a reasonable portion of any excess energy to minimize overshoot and ringing. Overshoot response should be well-matched to the lowest transmission-line impedance for relatively low currents. For large currents, a dynamic impedance less than the line impedance will reduce the overshoot and overshoot recovery time. A low-impedance clamp diode provides the desired response.[5]

The effect of source impedance on reflected waves is often overlooked. However, source impedance is very critical because it is usually responsible for the conversion of reflected overshoot to undershoot. Source response to overshoot should be either nearly matched or underdamped to prevent conversion of overshoot to undershoot and loss of signal noise margin. However, the underdamping should be limited so that steady-state conditions are approached quickly (if steady-state conditions are not achieved before the next clock interval, the response is a function of the previous signal history). Therefore, minimum and maximum specifications are necessary to characterize source response to reflected overshoot.

7.2.4 Optimum BiCMOS/CMOS source-load impedance characteristics

For optimum high-speed performance of advanced BiCMOS/CMOS devices in a transmission-line environment, with no external terminating resistors, the ideal source-load impedance characteristics are as follows:

1. The source output response at high currents, i.e., for the initial step, must be underdamped (lower impedance) with respect to the minimum transmission-line impedance to assure first incident wave switching. However, the amount of underdamping must be minimized to limit the initial voltage and current and the resultant crosstalk and power source and reference-level transient upset.
2. The load input impedance between the power rails should be very high to minimize power dissipation and maximize steady-state signal levels.
3. The load response to overshoots at high currents must be overdamped (lower impedance) with respect to the minimum transmission-line impedance to minimize the first incident wave overshoot and the amplitude of the reflected wave.
4. The source response to reflected overshoots must be underdamped (higher impedance) with respect to the maximum transmission-line impedance to assure that conversion to undershoot does not occur. However, the amount of underdamping must be minimized to limit the time required to approach steady-state conditions.

7.3 Driver-Receiver Response Test Circuit

Since the dynamic characteristics of advanced BiCMOS/CMOS devices in transmission-line environments are not specified, system designers need some means of establishing how candidate devices may react in a transmission-line environment. A simple test setup that is useful for observing signal response is easily implemented as follows: Connect the output of a driver (for example an AC240) to the input of a device of interest using 2 or 3 ft of 50-Ω coaxial cable (coax); apply a clock to the input of the driver driving the coax and observe the output of the coax at the receiving device (see Figure 7.13). The purpose of the coax is to simulate the 50-Ω impedance of a typical multilayer pc board (75- or 90-Ω coax can be used if more appropriate for a given application). The longer the length of coax used, the more pronounced will be the transmission-line effects. A 2- to 3-ft length is usually sufficient for most observations. A test as described will quickly establish the effectiveness of the clamps and other techniques for controlling transmission-line effects that the manufacturer may have incorporated in the device under test. However, a test of one device does not establish that all devices of that type or of that logic family will behave in a similar manner. They more than likely will not. Not all devices in a logic family have the same input-output circuitry. Manufacturers are constantly updating and changing devices to increase producibility, but that does not mean the transmission-line characteristics are always improved. There are wide variations in the transmission-line response characteristics of the same generic part from different manufacturers. Response characteristics change significantly with temperature, power-supply voltage, and process limits. Thus, device response needs to be checked at the extremes of the operating temperature and power-supply voltage range. Yet, no matter how many tests are performed, there is no guarantee that production lots of a given device or devices will behave in a similar fashion. With

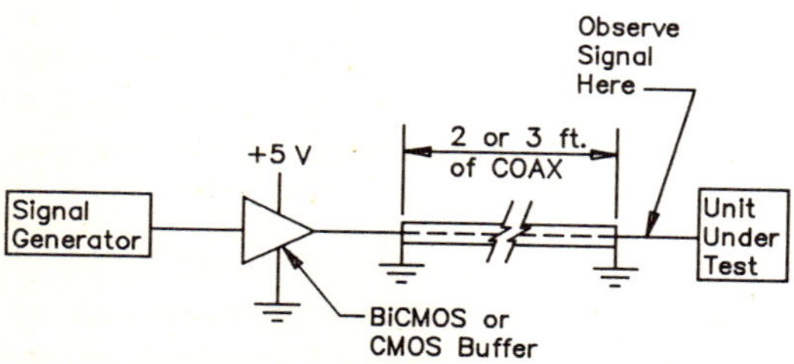

Figure 7.13 Circuit for observing transmission-line response and input clamping circuit efficiency of BiCMOS/CMOS devices.

all the variables, the best that can be hoped for by testing devices is to identify logic families and LSI devices built with a given technology or by a given manufacturer that tend to behave well in a transmission-line environment. Until manufacturers start specifying advanced BiCMOS and CMOS devices in a manner that allows a determination of their worst-case dynamic performance in transmission-line environments, the system designer is left with little solid information to predict performance.

7.4 Summary of Techniques for Dealing with Transmission-Line Effects

1. Signal lines with waveshape requirements (such as clocks) must be terminated.
   a. Series termination is best for most single-source to single-load applications.
   b. Load termination is best for single-source to multiple-load applications.
   c. Terminations that result in slightly underdamped lines provide optimum response.
2. When signals are sent between circuit boards, allowance must be made for possible signal degradation due to transmission-line effects since most board-to-board interconnections will be long. Worst-case board-to-board signal propagation time budgets must allow for possible added signal length due to extender boards, since most digital systems must function with boards extended for troubleshooting. Unterminated board-to-board and unit-to-unit signal interconnections must allow five line delays for ringing and reflections to subside.
3. Interconnection line impedance should be greater than 40 Ω (unloaded) so as to not excessively degrade initial signal levels.
4. Signals routed to test points should be buffered to prevent test equipment and transmission-line effects caused by interconnecting lines from interfering with normal system operation.
5. Signals that are used internally and externally to a board or unit should have external lines buffered to isolate internal signals from transmission-line reflections or other external disturbances such as shorts.
6. Input signals to boards, racks, or systems should be received at one place with a device with effective input clamps so as to minimize load-induced transmission-line effects.
7. CAUTION: Many LSI and VLSI parts, such as PLDs and FIFOs, ei-

ther do not have input clamps or have ineffective input clamps. The result is that signals may ring severely.

7.5 References

1. Blood, William R., Jr.: *MECL System Design Handbook,* 4th ed., Motorola Semiconductor Products Inc., Phoenix, Ariz., 1988.
2. Balph, Thomas: "Implementing High Speed Logic On Printed Circuit Boards," in 1981 *WESCON Conven. Record, Session 18,* September 1981, Paper No. 1.
3. Royle, David: "Transmission Lines and Interconnections," Part Three, *EDN,* June 23, 1988, pp. 155–160.
4. Millman, Jacob, and Herbert Taub: *Pulse, Digital and Switching Waveforms,* McGraw-Hill, New York, 1965.
5. Saenz, R. G., and E. M. Fulcher: "An Approach to Logic Circuit Noise Problems in Computer Design," *Computer Design,* April 1969, pp. 84–91.
6. DeClue, Joseph L.: "Wiring for High-Speed Circuits," *Electronic Design,* No. 11, May 24, 1976, pp. 84–86.
7. Nguyen-huu, Anh: "An Analysis of the Ringing Phenomenon in Digital Systems," *Computer Design,* July 1971, pp. 39–45.
8. *FAST Applications Handbook 1987,* National Semiconductor Corp., South Portland, Maine, 1988.
9. Royle, David: "Transmission Lines and Interconnections," Part Two, *EDN,* June 23, 1988, pp. 143–148.
10. Stehlin, R. A.: "Bergeron Plots Predict Delays in High-Speed TTL Circuits," *EDN,* November 15, 1984, pp. 293–298.
11. Southard, R. K.: "High-Speed Signal Pathways from Board to Board," in 1981 *WESCON Conven. Record, Session 18,* September 1981, Paper No. 2, pp. 1–8.
12. *Advanced CMOS Logic Designer's Handbook,* Texas Instruments Inc., Dallas Tex., 1988.
13. Leung, K.: "Controlled Slew Rate Output Buffer," *IEEE 1988 Custom Integrated Circuits Conference CH2584-1/88,* 1988, pp. 5.5.1–4.
14. "Electronic Design Report," *Electronic Design,* May 12, 1988, pp. 77–79.
15. *FACT Advanced CMOS Logic Databook,* National Semiconductor Corp., Santa Clara, Calif., 1988.
16. *Bi-CMOS Logic Data,* Motorola Inc., Phoenix, Ariz., 1989.
17. Burton, E. A.: "Transmission-Line Methods Aid Memory-Board Design," *Electronic Design,* December 8, 1988, pp. 87–91.

7.6 Bibliography

Applications Handbook, Cypress Semiconductor Corp., San Jose, Calif., 1989.

Harper, Charles A.: *Handbook of Wiring, Cabling and Interconnections for Electronics,* McGraw-Hill, New York, 1972.

Matick, Richard E.: *Transmission Lines for Digital and Communications Networks,* McGraw-Hill, New York, 1969.

Skilling, Hugh H.: *Electric Transmission Lines,* McGraw-Hill, New York, 1951.

Chapter

8

Clock Distribution

Clock-signal quality is a critical matter that must not be neglected when designing high-speed digital systems. Clock signals must be as near perfect as possible which means clock distribution networks and clock generation circuitry must be given a great deal of attention.

8.1 Universal Clock Distribution Guidelines

Since clock signals require an optimal electrical environment, transmission-line effects, crosstalk, and skew must be tightly controlled. In BiCMOS/CMOS systems, clock distribution circuitry must incorporate some form of line termination for waveshape control to limit undesirable transmission-line effects.[1] If not properly terminated, ringing and other transmission-line effects may cause extra clocking. To allow proper termination, clock signals must be routed to their destinations using reasonably well controlled impedance paths.

Clock lines must be physically arranged so as to minimize cross coupling or other adverse interference that could cause extra clocking. Parallel runs adjacent to other signals or other clocks must be kept as short as possible to minimize the opportunities for cross coupling. When multilayer pc boards are used, clock signals should be isolated and shielded from other signals by locating clock traces between reference planes. On welded-wire or wire-wrap circuit boards or backpanels, each clock line should be twisted with a ground line to provide some shielding and control of line impedance. Care must be taken to ensure that clock lines do not get bundled with other signals.

Clock distribution networks (trees) must be designed to minimize skew between clock signals. Skew reduces potential system operating speed by decreasing the time signals have to settle and may cause logic malfunctions if the skew is sufficient to cause hold-time violations. To accomplish the needed alignment, clock distribution circuitry must be standardized. Clock drivers at each level in distribution networks must be identical generic devices of the same logic family so that propagation delays are matched. Where possible, all drivers in a given level should be located in a common package. Drivers and loads must be arranged so that loads are equalized. Clock lines must be very short so that wiring propagation delays are insignificant, or lines must be of near equal length in each distribution level to equalize physical propagation delays.

Clock driver loading must be limited to much less than the specified rating to prevent loading effects from corrupting clock signals. Dynamic (ac) loading is usually of most concern and has the most effect on clock-signal quality. Static (dc) drive limits are seldom approached in advanced BiCMOS/CMOS systems. In most cases, much before dc drive limits are reached, ac capacitance loading will exceed the standard 50-pF test load used to specify most dynamic switching characteristics. Capacitance loading increases signal rise time and driver ground bounce. Clock signals with slow rise times are prone to double clocking because of noise, and slow rise times make it more difficult to control skew. Excessive loading also causes ground bounce to increase. Excessive ground bounce in clock driver packages will corrupt clock signals. To minimize ground bounce and power-supply droop, the capacitance loading of individual drivers and of groups of drivers in a common package must be limited to minimize the switching currents in package power and ground pins. As a rule of thumb, the number of loads driven by any one driver should be limited to 6 and the total package load limited to 36. If octal parts, such as 240s or 244s are used, no more than six of the eight drivers should be used. Those used should be the six nearest the ground pin (or pins) of the package. The two spare drivers should not be used for other purposes. Even with limitations on the loading of the active drivers, six drivers switching six loads at once can generate significant ground bounce (or power-supply droop) which will appear as spikes on the outputs of the other drivers in the package (see Chapter 4).

Different clock frequencies or phases should not be mixed in a common package. If two different frequencies are mixed in a common package, spikes due to ground bounce or power-supply droop may appear on the lower frequency when the higher frequency switches (see Figure 8.1).[2]

To supply the transient-current needs of clock drivers, local decoupling capacitors and circuit boards with continuous low-impedance power and ground planes are a necessity.[3,4] Each clock driver package should be decoupled with a high-frequency capacitor (typically 0.1 μF) located as close as possible to the package (see Chapter 5 for guidelines on selecting the value). Clock driver package

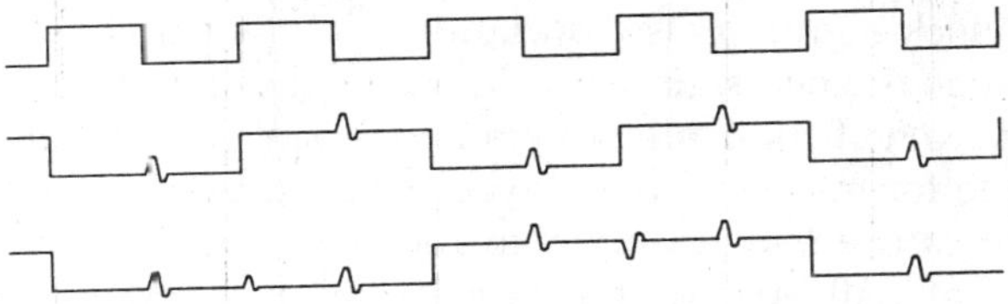

Figure 8.1 Low-frequency clock signals may have noise spikes when higher-frequency clocks switch if different clock frequencies are mixed in a package.

power and ground pins must connect directly to the respective circuit board power and ground planes. When universal welded-wire or wire-wrap prototyping boards are used, solder clips or washers must be used for clock driver package power and ground pin connections to power and ground planes.[5] Actual welded-wire or wire-wrap wiring should never be used to make clock driver power or ground connections; the inductance is too large.

To minimize the effects of cross coupling, all clocked devices with TTL input thresholds should be selected so that they change states on the *low*-to-*high* transition of the clock. The goal is to have all local clock signals in the *high* state, which has maximum noise immunity, when the majority of system signals are switching and generating noise (see Figure 8.2).

When devices with TTL input levels are used for clock distribution, clock-signal phasing at all major distribution levels (i.e., at all major interfaces) should be such that all interface clock signals are in phase with component-level clock signals (i.e., all clock signals should switch from a *low*-to-*high* level at the same time). The design objective is to ensure that long clock lines that have a high probability of being exposed to noise, such as the clock lines that traverse backpanels or motherboards, are in the maximum noise immunity state during the time of peak system noise (Figure 8.2).

When devices with TTL input levels are used, low-frequency clocks (10 MHz or less) should be shaped so that the *low* state is as short as

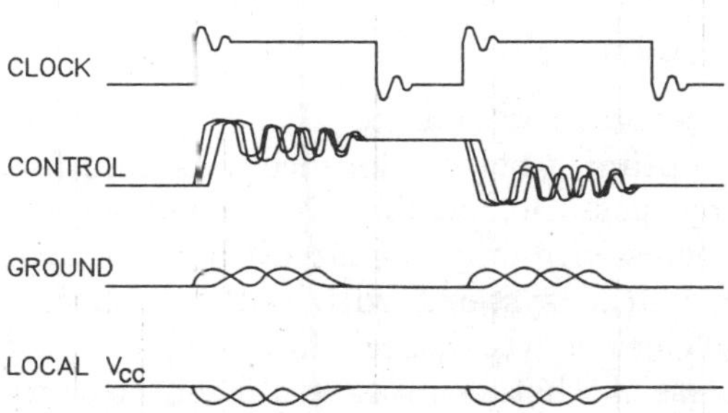

Figure 8.2 Clock phasing for maximum noise margin in a TTL-level system.

possible to optimize clock-signal noise immunity (see Figure 8.2). Such an arrangement keeps clock signals in the maximum noise margin state for a longer period of time, allowing a longer time for system signals to transition and for noise due to delayed transitions to die out before clock signals go to the *low* state which has less noise margin than the *high* state. Above 10 MHz a nonsymmetrical clock signal is usually not practical, but for systems with slower clock rates such a configuration should be considered. If clock pulses are narrowed, care must be taken to ensure that clock pulse width is adequate for the slowest logic family or device that might be used in the system.

When clock sources and loads are not solidly referenced to a common uninterrupted ground plane, clock signals must be transmitted by some means that provides a high degree of noise margin. The noise margin of single-ended BiCMOS or CMOS devices with either TTL or CMOS levels is not adequate for clock distribution when the transmitting and receiving devices are not directly referenced to a low-impedance common ground plane. When an adequate low-noise reference system does not exist, clock signals are typically distributed using some form of balanced differential transmission that has a high level of common-mode noise-rejection capability.

There are three distinct categories of clock distribution in the typical large digital system:

1. Component-to-component connections (i.e., circuit board clock distribution)
2. Board-to-board connections (i.e., motherboard clock distribution)
3. Unit-to-unit connections (i.e., between chassis, racks, and cabinets)

Interconnection categories 1 and 2 are present in most systems. The need for category 3 depends upon system size and the functional relationship of separate units. Small self-contained systems have no need for unit-to-unit connections. Likewise, many large systems are composed of separate autonomous functional units, each separate unit having its own internal clock source.

8.2 Board-Level Clock Distribution

Each circuit board should be designed to operate with only a single clock frequency and phase to optimize noise tolerance. A single frequency and phase are not always possible, but that should be the goal. The clock source for small systems with only one circuit board is typically an on-board oscillator. Large systems with multiple circuit boards typically have a central clock source (board) that provides clock signals to each board. Regardless of whether an external source or an on-board oscillator is the source for the board clock signals, a standard clock tree, such as the one shown in Figure 8.3, should be used to buffer and fan out clock signals. The clock tree should be the same on all boards. If boards have an external source, each board clock input signal (or signals, if there are more than one) should be received with a single receiver located near the input connector of the board and then fanned out to the loads as shown in Figure 8.3. Incoming clock lines to a circuit board should be terminated as close as possible to the receiver (if load termination is used), and the receiver and termination should be located as close as possible to the point where the clock signal enters the board. Note that load termination can only be used with TTL-level signals (see Section 8.3).

A board-level clock tree as shown in Figure 8.3 provides a means of fanning out to multiple loads on a board without presenting a large load to the clock source. It is desirable to use only one receiver to receive input clock signals to minimize load capacitance. However, requiring that board clock inputs be limited to one load may force board clock trees to extra levels on large boards with large fan-out needs. Extra levels are undesirable in clock trees, since extra levels add uncertainty to clock edge phasing. Thus, in applications where boards are very large, additional receiving devices are usually a better solution than increasing clock tree depth. Up to six receivers can be used when 220 and 330-Ω split terminations are used, but the number should be kept as small as possible. When source termination is used, only one receiver should be used.

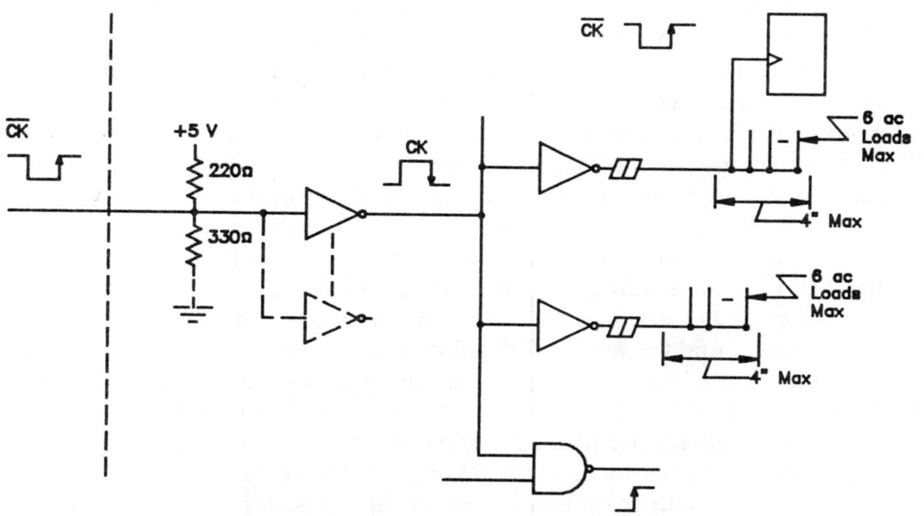

Figure 8.3 Typical circuit board clock tree.

Alignment of board clocks is critical in synchronous systems. Active clock edges must be closely aligned to prevent hold-time violations when register-to-register transfers must be made. To optimize alignment of board clock signals, the following steps should be taken:

1. The final level of the board clock tree buffers should be centrally located.
2. Identical generic devices, in the same package where possible, should be used for buffers in each given level.
3. Loading for each level should be balanced.
4. Line lengths should be minimized and be similar in each level. For very large boards, definite minimum and maximum clock line lengths should be established to limit possible skew.

Ground bounce can be a severe problem in clock buffer packages, since all drivers in clock buffer packages switch at the same time. When all devices in a package switch simultaneously, load and totem-pole feedthrough currents combine to exacerbate ground bounce. To minimize the potential for ground-bounce disturbances, the number of clock buffer loads should be limited to well below dc fan-out ratings. For example, Texas Instruments' BiCMOS 54BCT240s have a dc drive rating of 48 mA and thus can drive forty-eight 54BCT240 inputs (I_{IL} = 1 mA). However, 48 loads at 5 pF per load results in a 240-pF load (with no allowance for the wiring capacitance), which is well above the 50-pF load at which the ac parameters are specified. In addition, if all eight devices in an octal driver package, such as a BCT240 or BCT244, are used, with each buffer driving 48 loads, the total ac load that would have to be charged or discharged at each clock edge transition would be 1920 pF (240 pF $\times$ 8). Such a large ac load would result in severe ground bounce. Ground bounce, when an octal buffer package is used as a clock driver, is further aggravated by the totem-pole transient feedthrough current of eight simultaneously switching buffers (see Chapter 4). Using all eight buffers, with each fully loaded, is somewhat an extreme example, but it illustrates the need to limit ac loading of clock buffer packages. If octal devices are used as clock buffers, it is best to use no more than six of the drivers in the package, and the six used should be the six nearest the ground pin (or pins). A good rule of thumb when using advanced BiCMOS/CMOS devices is: Limit the ac load to no more than 6 loads per driver and no more than 36 loads per package. The extra drivers in octal clock buffer packages should not be used for other purposes since they may exhibit severe ground-bounce spikes during clock transitions. It is important that they not be used for signals that might go to asynchronous inputs, such as flip-flop sets or resets, that are sensitive to spikes. In cases where multiple clock frequencies or phases must be used on a board, they should not be mixed in common buffer packages. Ground bounce from the higher clock frequency transitions may appear as spikes on the lower-frequency clock signals (see Figure 8.1).

When advanced BiCMOS/CMOS devices are used on boards of any significant size, some means of waveshape control must be used for clock lines. When actual routed line lengths are considered, clock-signal lines on most boards will have propagation delays that will exceed one-half the rise time of the clock signal at its source. Thus, on most boards termination must be used to ensure waveshape control. Board clock lines should be terminated so that they are slightly underdamped for optimum edge speed but not so underdamped that excessive ringing occurs. Underdamped signals have faster rise times and thus reach threshold levels faster than critical damped or overdamped signals. A suggested range for termination impedance for load terminated lines is 1.2 to 2 times the characteristic impedance Z_o of the line being terminated.

On circuit boards, it is impractical to limit clock lines to a single source and load which is the optimum configuration for terminating lines; multiple loads must be driven by drivers in all but the most trivial designs. A typical circuit board is populated with numerous clocked devices, such as flip-flops, counters, registers, and shift-registers, all of which require clock signals. When synchronous design practices are followed, as they should be, all clocked devices on a board require the board clock signal. Thus, in most cases, board-level clock distribution needs are such that a driver per load is not practical.

Load termination works best in most multiple-load situations, but in most board clock distribution applications it is impractical to load terminate each clock line on a board. Load termination dissipates a great deal of power, and split load termination increases system parts count and cannot be used with CMOS input threshold levels. There are schemes for ac load terminating lines that reduce power dissipation, but they require a larger number of discrete components.

Series or source termination is the most practical method of terminating board clock lines. Series or source termination works best when there is only one source and one load and for that reason would not seem suitable for board clock termination, but when all the issues are considered, the disadvantages of load termination outweigh the disadvantages of series termination at the board level. Series termination requires care in the physical placement of the loads, but when the placement requirements are understood, series termination works well.

Either ferrite shield beads or small-value resistors located near the driving source can be used to terminate clock lines, but ferrite shield beads work best in most applications. Ferrite shield beads (see Figure 8.4), which are made of a lossy ferrite material, present little impedance to clock signals at low frequencies, but at high frequencies act as small lossy impedances. When a ferrite shield bead is used for source termination, the terminated line must pass through the bead and the bead must be located very near the driver (within 1 in). Beads with impedances in the range from a few ohms to near 50 Ω at 100 MHz are available. Figure 8.5 shows a typical impedance-versus-frequency profile for a ferrite shield bead. The bead impedance needed in a given application is set by the clock driver edge rates and circuit board characteristic impedance. When welded-wire or wire-wrap boards, which tend to have a characteristic impedance on the order of 100 Ω, are used, beads with an impedance near 50 Ω at 100 MHz should be selected. For clocks on pc boards, which typically have a characteristic impedance near 50 Ω, beads with an impedance of 25 Ω at 100 MHz should be used. The desire is to not overdampen clock signals and thus not slow down edges too much. The goal is to remove enough energy to prevent excessive overshooting, which in turn will prevent undershooting of clock signals.

The disadvantage of series resistor termination of board clock lines is reduced dc noise margin when driving BiCMOS devices with bipolar inputs. However, the reduction is generally very minimal for ad-

Figure 8.4 Ferrite shield beads of the type used for series-terminating clock lines.

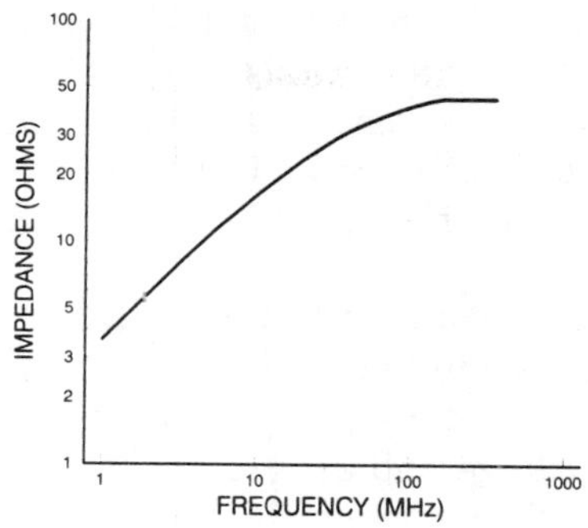

Figure 8.5 Impedance versus frequency curve for a ferrite shield bead of the type used for series-terminating clock lines (based on catalog data for Fair-Rite Products Corp. part number 2743019446). (*Reprinted with permission.*)

vanced CMOS or BiCMOS devices with CMOS inputs since most devices with CMOS inputs have very low dc input current (typically on the order of ±10 μA or less), but caution must be exercised. The noise margin of clock lines can be compromised unless care is taken to evaluate the actual load in each given situation. For example, the *low*-level input current specification for Texas Instruments' BiCMOS bus interface logic devices is 1.0 mA. Assuming a device is driving six loads, the total dc load current is 6 mA. If a 51-Ω series resistor is used, the drop across the resistor is 306 mV (6 mA × 51 Ω). A drop of 306 mV exceeds the 300-mV *low* noise margin and cannot be tolerated. Thus, series resistors cannot be used in such an application or else much lower value resistors must be used.

Series termination, either with ferrite shield beads or resistors, has the advantage of requiring only one discrete component per line and does not dissipate any extra dc power. Series termination does present one problem; when long lines are driven, signal transitions near the source will step up (or down) as shown in Figure 8.6. Steps are inherent in signal (clock) transitions near sources when long lines are source terminated (see Chapter 7). The result is that signals may dwell for some period of time in the threshold region of receiving de-

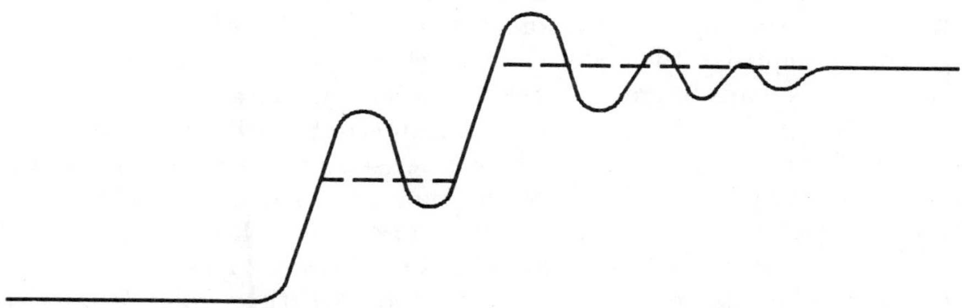

Figure 8.6 Source-terminated lines have an inherent step near the source that may ring and cause extra clocking.

vices. When a step dwells in the threshold region of a device, double clocking of the device may occur because of ringing or noise on the line operating in conjunction with the indeterminate logic level of the step. To avoid extraneous clocking, clock lines must be kept very short or loads must be grouped very close together at the end of lines (within 4 in is a good rule of thumb).

Board clock circuitry must conform to the best possible electrical design standards. Each clock driver package should be decoupled with a high-frequency decoupling capacitor, such as an 0.1-μF ceramic capacitor, located as close as possible to the package and as directly as possible between the power and ground pins of the package. Clock drivers must be connected as directly as possible to the power and ground planes or grids of the host board. When welded-wire or wire-wrap boards are used, clock driver package power and ground connections and decoupling capacitor connections must be made directly to the power and ground planes with solder clips or washers. Wiring should never be used for power or ground connections of clock circuitry; the inductance of wire connections aggravates ground-bounce problems.

When board-to-motherboard connectors do not contain an adequate number of ground pins, or under other noisy reference conditions, some form of balanced differential connection to the clock source must be used to ensure a clean, noise-free, board clock source. If standard TTL- or CMOS-level interfaces are used, when board-to-motherboard connectors do not contain an adequate number of ground pins, noise may be injected into the clock signals and appear as extra clock pulses. Thus, where single-ended clock signals are used, a low-impedance reference path must exist between board reference planes and the motherboard reference plane to prevent reference-level shift in excess of single-ended TTL- or CMOS-level noise margins when large load currents are switched.

Gating of board clocks should be avoided where possible since gating of clocks places severe restrictions on the timing of the gating signals relative to the clock. It is more desirable to gate a clocked element's control signals instead of the clock itself. However, when gating is required, two-level clock trees as shown in Figure 8.3 provide a safe means of gating clocks. The intermediate level (CLK) can be safely gated with signals that originate from devices clocked with the final stage (CLK-) of the clock tree (see Figure 8.7). No spiking occurs on the gated clock line since the control signal safely brackets the intermediate clock signal (CLK in Figure 8.7). If boards are built with only one level of clock-signal buffers, there are no easy means of ensuring that gated signals have the proper time relationship to nongated clock signals (gated signals will lag), nor are there means of

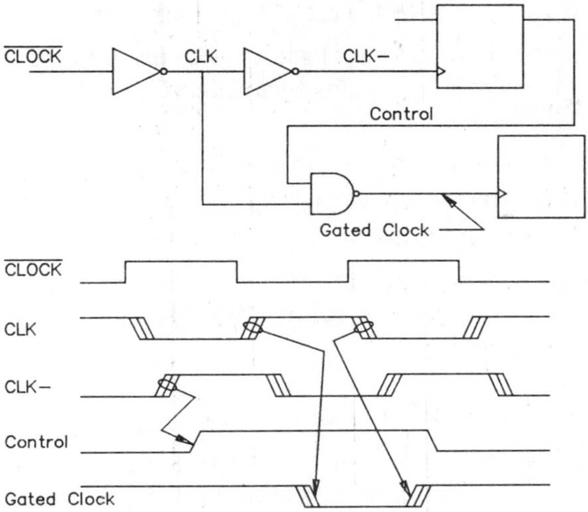

Figure 8.7 Two-level clock trees provide a safe means of gating clocks.

ensuring that gate control signals bracket clock signals. Thus, gated clocks tend to spike.

Summary of board-level clock distribution techniques

1. Use only a single clock frequency and phase.
2. Use at least a two-level clock tree so that clocks can be safely gated where necessary.
3. Use the same type of devices (preferably devices in the same package) in each given level of a clock tree.
4. Balance clock-signal loads.
5. Centralize the distribution point for board clock lines.
6. Keep clock lines short, or balance clock line lengths.
7. Limit the number of ac loads per clock driver package to keep ground bounce within limits.
8. Do not mix clock frequencies, phases, or other signals in clock driver packages.
9. Provide a decoupling capacitor for each clock driver package as near as possible to the package power pins.
10. Directly connect power and ground pins of clock driver packages to board power and ground planes (no wiring).

11. Terminate all clock lines to ensure high-quality clock signals.
12. Closely group all loads near the end of clock lines to prevent false triggering due to edge steps near series-terminated board clock sources.
13. Isolate clock-signal lines to minimize cross coupling.
14. When using devices with TTL input levels, use only those devices that clock on *low*-to-*high* transitions.
15. Avoid gating of clocks where possible, but if it is necessary, be certain the phasing of the clock and the control signal is such that the gated clock will not spike.

8.3 Board-to-Board Clock Distribution

Board-to-board clock-signal interconnections, where the individual circuit boards are referenced to a common backpanel or motherboard ground plane, are generally transmitted and received using single-ended devices. Single-ended interconnections are used for economical and practical reasons. Single-ended drivers are readily available at low cost, and single-ended interconnections use the minimum number of connector pins. However, providing the electrical environment needed for single-ended clock signals requires a great amount of design effort. Single-ended BiCMOS/CMOS devices do not have a great deal of noise margin, and board-to-board clock lines tend to be long and run through noisy areas. When single-ended interconnections are used, it is essential that circuit board-to-motherboard connectors have an adequate number of properly distributed ground pins to ensure low-impedance board-to-motherboard ground connections (see Chapter 5). Single-ended board-to-board clock-signal lines must be properly terminated and care must be given to their routing to reduce the possibility of coupled noise. Noise on clock lines that run between boards is an all too common problem.

Clock signals for boards common to a backpanel or motherboard should all originate from a central source. In the typical application, the central source consists of a clock board that houses a master oscillator and a clock buffer tree to drive clocks to other boards on the common motherboard. In some applications, instead of housing an oscillator, the central clock board receives an externally generated clock and redistributes it to the local boards. In either case, an individual clock signal should be sent to each circuit board on the backpanel using a dedicated clock driver and receiver for each board in the unit as shown in Figure 8.8.

Clock signals should never be run to multiple boards when advanced BiCMOS/CMOS devices are used because of the difficulty of

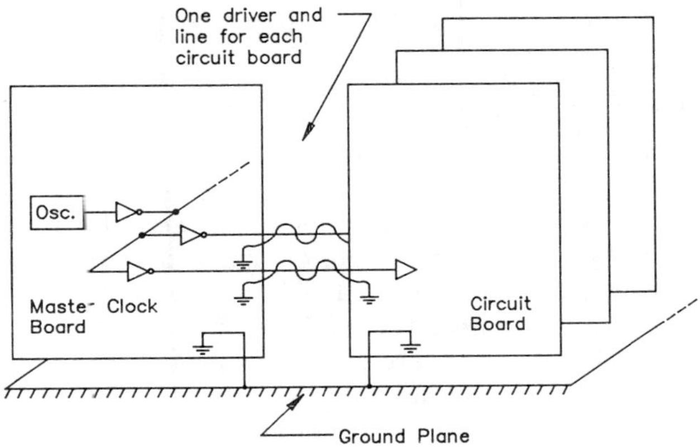

Figure 8.8 When clock signals are distributed to multiple boards, each board should have a dedicated driver and interconnecting line to prevent transmission-line effects from corrupting clock signals.

properly terminating multiple board loads. If a clock signal is sent to multiple boards, when one or more boards (loads) are not present or when a board is extended out of the housing for troubleshooting purposes, the physical arrangement is changed and the termination may no longer be effective. To achieve the clock-signal quality needed for proper system operation, board-to-board clock lines must be properly terminated at all times when advanced BiCMOS/CMOS devices are used. Likewise, all systems must be designed so that they will operate when boards are absent or when extender boards are in use. All systems will have component failures or other faults and therefore must be able to operate with some number of circuit boards extended out of the housing for troubleshooting. The only practical way to achieve such a design is to send an individual clock line to each board. Each board should receive its clock signal with a single receiver that can be used to buffer and redrive the clock within the circuit board. Clock lines from the central source should never be used to directly drive devices on circuit boards; such an arrangement is certain to degrade the clock signal and result in improper system functioning.

Great care must be taken in the routing of clock-signal lines on backpanels or motherboards. Clock interconnections must be implemented with controlled impedance lines so that lines can be properly terminated and crosstalk can be controlled. On wire-wrap backpanels, clock lines should be run twisted with a ground wire as shown in Figure 8.9. The ground line should be connected to ground at both the source and at the load, as close as possible to the actual driver and receiver, to provide a direct return current path. However, in most ap-

Figure 8.9 Twist clock lines with ground lines when wire-wrap backpanels are used to reduce interference and control impedance.

plications it is not practical to run ground lines to the actual drivers and receivers on circuit boards. A ground pin on the source and load circuit board-to-motherboard connectors near where the active clock line exits the source board and enters the load board is close enough for most applications. Twisted-pair interconnections provide a known and controlled characteristic impedance (around 100 to 120 Ω) and some shielding from adjacent signals. Care should be taken to ensure that clock lines do not get bundled with other signals.

Care must also be taken to ensure that board-to-board clock signal interconnections do not have significant discontinuities (in characteristic impedance), or large branches or stubs (which there should not be, since there should only be one source and one load). Large discontinuities, branches, or stubs can cause severe reflections which may degrade clock-signal edges to the extent that false clocking occurs.

To control crosstalk, clock-signal traces on pc motherboards must be physically isolated from other signals. If possible, they should be isolated in the vertical dimension by reference planes. In the horizontal dimension, line-to-line spacing between traces with different-frequency clock signals or between clock signals and other signals must be such that little coupling occurs. As a rule of thumb, the line spacing should be such that the coupling does not exceed 50 percent of the worst-case noise margin of the clock receivers. A clock-signal noise budget should be established for each design that includes an allowance for crosstalk.[6]

In all cases, board-to-board clock lines should be as short as possible, but where clock lines cannot be kept very short, all lines should be made equal in physical length so as to match line propagation delays. To minimize the length of board-to-board clock lines, the clock source board should be located near the center of the motherboard or card cage. All drivers and receivers should be of the same generic type and logic family, and the terminations and the effective load at each receiving board should be the same to minimize skew in the active clock edge.

When advanced BiCMOS/CMOS devices are used, clock lines that transition backpanels must be terminated to ensure clock quality. Such lines generally have two-way propagation delays that are long relative to the rise (or fall) time of the signal and hence must be treated as transmission lines. Either source or destination termination may be used for TTL-level clock signals. Each method has advantages and disadvantages, but destination termination as shown in Figure 8.10 is recommended for most TTL-level applications. When CMOS devices with true CMOS input threshold levels are used, split termination cannot be used; split termination prevents *high* inputs from reaching proper *high* CMOS levels. Thus, CMOS board-to-board clock signals must be source-terminated as shown in Figure 8.11.

Source termination can be implemented with small-value series resistors or ferrite shield beads located very near the driver. One advantage of series termination is that no additional dc power is dissipated as a result of the termination. However, in TTL-level applications, care must be taken to ensure that *low*-level noise margin is not significantly reduced by series-termination resistors. In CMOS or BiCMOS with CMOS-input applications, the dc drop across series-terminating resistors is not significant since CMOS inputs source or sink very little current. To minimize TTL-level noise margin loss and to provide underdamped signals, series-termination resistors should be approximately one-third to one-half the line characteristic impedance Z_o. A value in the range of 22 to 33 Ω is appropriate for most applications. The value should be low enough so that the *low*-state noise margin is not reduced by more than 50 to 100 mV. When determining the response of a source-terminated transmission-line, driver output impedance must be considered. The effective source impedance

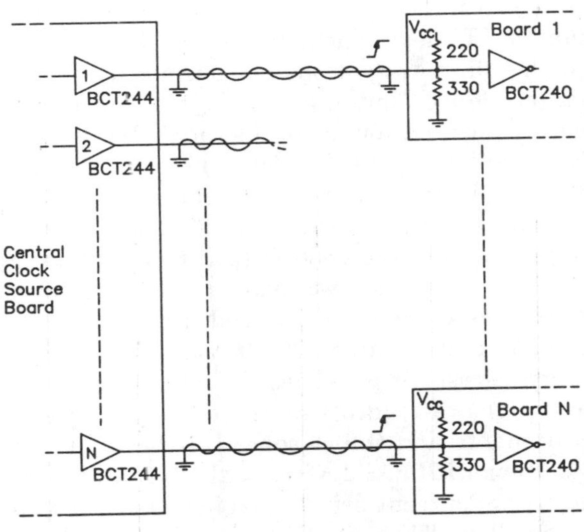

Figure 8.10 Recommended board-to-board clock distribution scheme when clock signals have TTL levels.

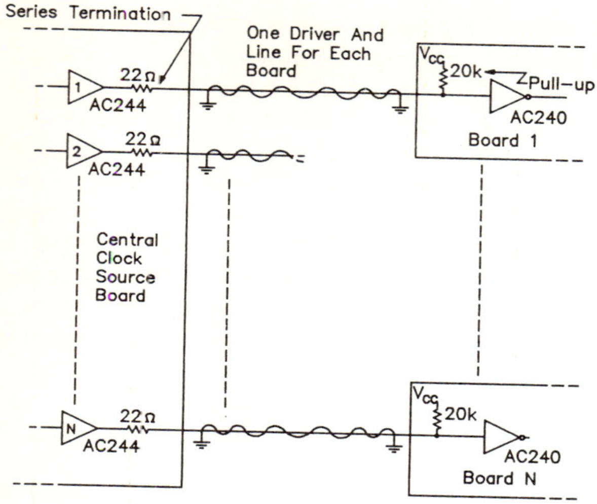

Figure 8.11 Recommended board-to-board clock distribution scheme when clock signals have CMOS levels.

is the sum of the driver output impedance and the external series-terminating resistor.

Ferrite shield beads are recommended when BiCMOS devices with bipolar inputs that have significant *low*-level input current must be series terminated.[7] Ferrite shield beads, with zero static impedance, eliminate possible dc offset problems but provide the needed amount of impedance at high frequencies.

Destination termination of TTL-level backpanel clock lines is customarily implemented with split terminations (sometimes described as a Thevenin's termination). For optimum response, the Thevenin's equivalent impedance of split-termination networks should be slightly higher than line characteristic impedance Z_o. For most pc board and welded-wire or wire-wrap applications, split-termination networks with resistor values of 220 and 330 Ω, as shown in Figure 8.10, are used.[1] The equivalent impedance of a 220-330 Ω split termination is 132 Ω which provides a near match for welded-wire or wire-wrap boards and approximately a two-to-one underdamped termination for pc boards (which is acceptable in most cases); 220-330 Ω networks are available in standard multiple-resistor packages. When selecting the impedance of split terminations, a tradeoff must be made between driver current sink requirements and the closeness of the match to line impedance. Most CMOS or BiCMOS devices with TTL-level outputs do not have a great deal of current drive capability, but they do have high current sink capability. Thus, termination networks must be optimized for minimum loading of the driver in the *high* state. Split termination accomplishes that purpose. No *high*-level static drive current is required when split-termination resistor values are properly selected. The divider effect of the split-termination network establishes the *high* level. In the *low* state the driver must sink a current equal to V_{cc} divided by the value of the termination network resistor connected to V_{cc} (220 Ω in the case of a 220-330 Ω network). When 220-330 Ω networks are used, the sink current is about 22 mA under nominal conditions. Since most standard CMOS or BiCMOS logic devices (NANDs, NORs, etc.) are only rated to sink 20 to 24 mA (depending on the logic family), drivers with high-current capability, such as 240s or 244s, must be used to meet the current sink requirements of 220-330 Ω split terminations.

The advantages of destination termination include:

1. Coupled noise and crosstalk tend to be damped quicker.
2. Edge rates do not degrade as much as they would under similar circumstances (line and load impedance) with source termination.
3. More than one device can be used to receive the signal (on the receiving board), which helps minimize board clock tree depth on large boards.

The disadvantages of destination termination include:

1. Increased power dissipation.
2. Higher current drive requirements placed on the driver.

In most applications, the advantages outweigh the disadvantages. Thus, destination termination is recommended for most board-to-board applications with TTL-level interfaces. Destination termination cannot be used on CMOS interfaces.

With either source or destination termination, clock lines should be terminated so that they are slightly underdamped for optimum edge speed but not so underdamped that excessive ringing occurs. Underdamped signals have faster rise times and thus reach threshold levels faster than critical damped or overdamped signals. A suggested range for termination impedance is 1.2 to 2 times the characteristic impedance Z_o of the line being terminated.

The phasing of TTL-level clock signals that traverse backpanels or motherboards should be such that the *low*-to-*high* transition of the clock signal is the system's active clock edge. Such a phasing will ensure that the clock lines are in the *high* state with maximum noise margin when the majority of system signals are switching. Systems with clock rates that are relatively slow, with respect to the minimum clock pulse width requirements of the devices being clocked, should shape clock signals so that the logic *low* state is as short as possible to further enhance noise margin. Such an arrangement keeps clock signals in the maximum noise margin state for a longer period of time. Thus, a longer time is allowed for system signals to transition and a longer time for noise due to delayed transitions to die out before clock signals go to the *low* state which has less noise margin than the *high* state.

The need for a large number of well-distributed grounds in board-to-motherboard connectors is an issue that is often overlooked. It is important that the clock distribution board be solidly referenced to the motherboard ground plane to ensure that the clock distribution board does not experience excessive ground bounce when all the clock drivers switch simultaneously. An admirable design goal is to provide an adjacent board-to-motherboard ground pin for each clock-signal pin in the connector. In most cases that is not practical; connector pin limitations often force a compromise in the number of ground pins. If a ground pin for each clock signal is not possible, a careful analysis must be performed to determine the number of ground pins required to keep board ground bounce within limits. The design task is to keep clock source board transient reference-level shifts due to ground bounce below the threshold level of the most sensitive clock receiver or receiving device on the clock source board.

When board-to-motherboard connectors do not contain an adequate number of ground pins, or under other noisy reference conditions, some form of balanced differential connection to the clock source (if it is not on the central clock board) and loads must be used to ensure clean, noise-free clock transmission. If standard TTL- or CMOS-level interfaces are used, when board-to-motherboard connectors do not contain an adequate number of ground pins, noise may be injected into the clock signals and appear as extra clock signals. Thus, if single-ended clock inputs are to be used, a low-impedance reference path must exist between board reference planes and the motherboard reference plane to prevent reference-level shift in excess of single-ended TTL or CMOS noise margins.

Summary of board-to-board clock distribution techniques

1. Centralize the source and distribution point for board-to-board clock signals.
2. Use a separate clock driver and clock line for each board.
3. Terminate all clock lines to ensure clock signal quality.

4. Use very short lines or make all lines equal to keep line propagation delays equal.
5. Twist each clock line with a ground line to reduce coupling when wired backpanels are used.
6. Phase TTL-level board-to-board clock signals so that *low*-to-*high* transitions are in phase with the active edge of the clock at the component level to optimize noise margin.
7. Use the same type of device for all drivers to match propagation delays.
8. Use the same type of device for all receivers to match propagation delays.
9. Limit the number of drivers or receivers used in octal packages to keep ground bounce within limits.
10. Prevent system upset due to ground bounce, do not mix clock frequencies, phases, or other signals in clock driver or receiver packages.
11. Provide a decoupling capacitor for each clock driver and receiver.
12. Directly connect power and ground pins of clock drivers and receivers to circuit board power and ground planes (no wiring).

8.4 Unit-to-Unit Clock Distribution

The interconnection of high-frequency clocks between physically separate units (that is, between units that do not have a common backplane, backpanel, or motherboard) is one of the most difficult clock distribution tasks. When clocks must be sent between components that are not directly referenced to a common ground plane, they must be transmitted by some means that provide a sufficient amount of noise tolerance for the application. In Chapter 4, high-frequency transient switching currents are shown to result in the generation of significant voltage transients across the inductance of a 1-in length of wire. It follows that the inductance of several feet of wire, as is generally required to interconnect separate units, will result in very large transient-voltage shifts that will exceed the noise margin of single-ended BiCMOS/CMOS devices with either CMOS or TTL levels. Thus, to prevent noise from corrupting clock signals transmitted between units, some form of transmission is needed that has a high level of noise rejection. Differential line drivers connected to differential line receivers with shielded twisted-pair cable as shown in Figure 8.12 are typically used (see Chapter 9).

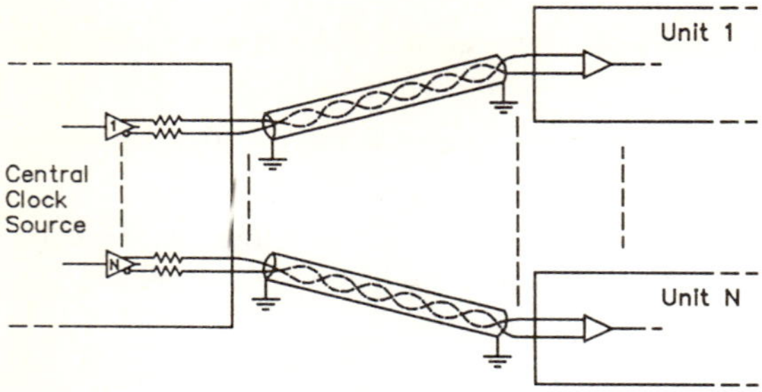

Figure 8.12 Balanced differential driver-receiver circuits are required for unit-to-unit clock distribution.

Most unit-to-unit interconnections are long transmission lines (see Chapter 7) and thus must be properly terminated to ensure clock-signal quality. To allow proper termination, unit-to-unit clock signals must be transmitted using dedicated driver-receiver pairs connected with controlled impedance twisted-pair lines. Unit-to-unit clock lines should be terminated so that they are slightly underdamped for optimum edge speed but not so underdamped that excessive ringing occurs. Underdamped signals have faster rise times and thus reach threshold levels faster than critical damped or overdamped signals. Series termination is recommended for most unit-to-unit clock connections. Series termination provides short-circuit protection for drivers, dissipates less power than load termination, and reduces unit-to-unit current flow. A suggested range for termination resistors is 0.2 to 0.4 times the characteristic impedance Z_o of the line being terminated.

Summary of unit-to-unit clock distribution techniques

1. Centralize the source and distribution point for system clock signals.
2. Use balanced differential driver-receiver interconnections for noise immunity (single-ended interconnections are not adequate).
3. Use a separate driver and shielded twisted-pair line for each unit-to-unit clock.
4. Use a single receiver at the load.
5. Terminate all lines to ensure clock-signal quality.
6. Use equal line lengths to balance propagation delays.

7. Use the same type of device for all drivers to balance propagation delays.
8. Use the same type of device for all receivers to balance propagation delays.

8.5 Test Clock Input Port

High-speed digital systems need some means of varying clock frequency or frequencies during initial design verification and during test. A lower-frequency clock is often needed during initial checkout. A lower-speed clock allows fundamental logic errors to be separated from errors caused by transmission-line effects and other complications caused by full-speed operation. A variable higher-frequency clock signal is needed to check for operating frequency margin once a system is operational.

A circuit for injecting a test clock is shown in Figure 8.13. The test port is activated when the TEST pin is jumpered to ground. Many other configurations are possible. A multiplexer will serve equally well. In all cases, clock switching circuit inputs connected to external test clock sources must be current limited to protect the switching circuit from overvoltage conditions that are likely to occur in test environments. For example, the test input port must be protected so that excess input current does not flow when system V_{cc} is OFF and the test clock source is still ON or when the clock source signal amplitude is not properly adjusted. In the circuit in Figure 8.13, resistor $R1$ limits the input current and diode $D1$ shunts current to V_{cc} in case of excessive signal amplitude or if the clock source is turned ON when system V_{cc} is not present. Test clock input ports must not float when the

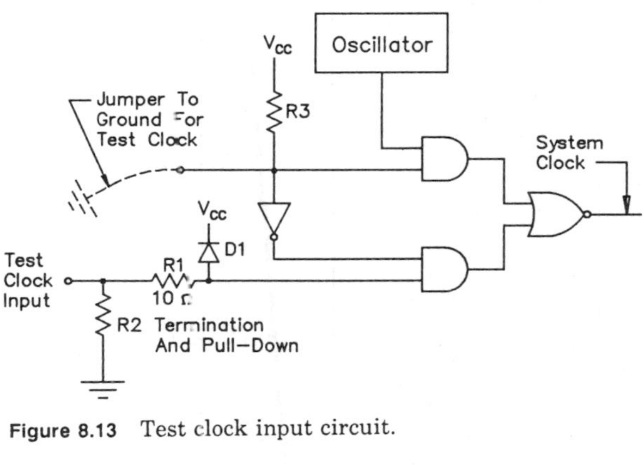

Figure 8.13 Test clock input circuit.

test clock is not present; a pull-up or pull-down resistor is required which can also serve as a termination for the external clock signal (*R*2 in Figure 8.13).

8.6 References

1. Royle, David: "Designer's Guide to Transmission Lines and Interconnections," Part Two, *EDN*, June 23, 1988, pp. 143–148.
2. Abramson, S., C. Hefner, and D. Powers: *Simultaneous Switching Evaluation and Testing Design Considerations,* Texas Instruments Inc., Dallas, Tex., 1987.
3. Cowdell, Robert B.: "Bypass and Feedthrough Filters," *Electronic Design,* No. 17, August 16, 1975, pp. 62–67.
4. Ormond, Tom: "Backplanes Play a Crucial Role in High-Speed Systems," *EDN,* July 10, 1986, pp. 222–228.
5. Visco, P. Anthony: "Coaxing Top Bipolar Speeds from Prototyping Boards," *Electronic Products,* September 1, 1987, pp. 55–58.
6. *FAST Applications Handbook 1987,* National Semiconductor Corp., South Portland, Maine, 1988.
7. Parker, C., B. Tolen, and R. Parker: "Prayer Beads Solve Many of Your EMI Problems," *EMC Technology & Interference Control News,* Vol. 4, No. 2, April–July 1985.

Chapter

9

Device, Board, and Unit Interfaces

Interface signal-level requirements and interface circuit operating characteristics are linked to the stability of the ground reference system. If a low-impedance reference can be maintained between devices, single-ended interfaces can be used. If a low-impedance reference cannot be maintained, differential or high-level signal interfaces are required. In most systems, it is possible to maintain a solid ground reference at the component-to-component and board-to-board level but not at the unit-to-unit level. In most large systems, the physical spacing between units or other practical limitations prevent implementation of a low-impedance reference path between units. Thus, the typical high-performance systems built with advanced BiCMOS/CMOS components use the following electrical interfaces:

1. Single-ended TTL or CMOS levels for component-to-component and board-to-board communication.
2. Balanced differential line drivers and receivers for high-speed unit-to-unit communication.
3. High-level single-ended drivers and receivers for low-speed unit-to-unit communication.

Noise is of particular concern for interconnections between physically separate units. Unit-to-unit interface circuitry must have the capability of operating properly in the presence of high levels of noise and dc reference offsets.

Single-ended interfaces. Most component-to-component and board-to-board connections are implemented with single-ended signal interconnections. Single-ended interconnections require the minimum possible number of interconnections (one), and single-ended driver-receivers are as simple as possible, thus single-ended communication minimizes complexity and cost. However, single-ended TTL- or CMOS-level communication is only possible when the interconnected components are referenced to a common solid ground plane (see Figure 9.1). Thus, circuits and motherboards must have continuous ground planes with no cut-out areas, and circuit board-to-motherboard connectors must have a sufficient number of evenly distributed ground connections so that the circuit board ground plane appears as an extension of the motherboard ground plane.[1,2] Circuit board-to-motherboard ground connection inductance must be low enough so that transient return currents do not shift the reference level of circuit boards beyond single-ended threshold limits.

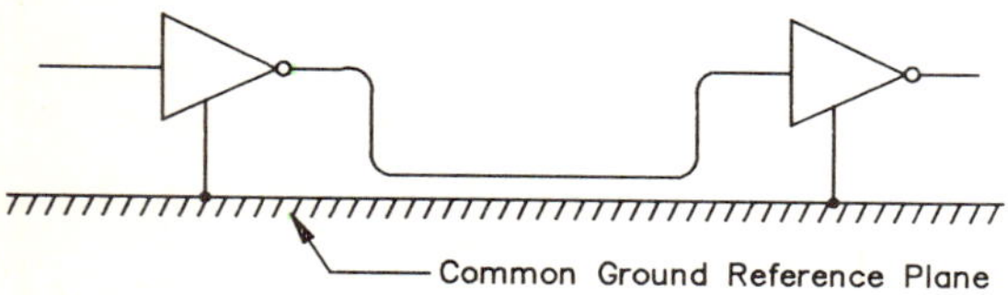

Figure 9.1 Single-ended TTL- or CMOS-level circuits must be referenced to a common ground plane.

> Single-ended BiCMOS/CMOS communication requires a solid common ground reference plane.

The basic interface characteristics of advanced BiCMOS/CMOS devices are bounded by standard TTL and CMOS conventions. System designers have little choice with respect to fundamental interface characteristics, but they do have the choice between TTL and CMOS levels. In general, devices with TTL input and output levels are the best choice; TTL output swings create less noise than CMOS rail-to-rail output swings and dissipate less dynamic power. As operating speed increases, there is less time for noise to die out and dynamic power becomes more and more of a concern since it is a direct function of frequency. Other advantages of TTL levels include the following: Many LSI and VLSI parts, such as memories and PLD, are not available with CMOS input or output levels; much of the standard test equipment is designed for TTL levels; and use of TTL levels allows a fallback to bipolar parts if it is suddenly discovered that a part is not available in the desired technology. Only where maximum noise margin is needed, such as clock signals, should CMOS levels be considered.

9.1 Component-to-Component Interfaces

Once either TTL or CMOS interface levels are selected, the system designer's task is to select devices with adequate drive. In general, the drive characteristic needed depends upon whether lightly loaded control signals or heavily loaded buses are being driven. Standard advanced BiCMOS/CMOS logic functions typically have drive levels of 20 to 24 mA, which are adequate for most lightly loaded applications. However, drive requirements in high-performance systems are a complicated function of dynamic loading and transmission-line driving requirements and whether first incident wave switching is required (see Chapter 7). Static (dc) drive is usually not an issue in advanced BiCMOS/CMOS systems since dc input currents for devices with CMOS input stages are typically very small (less than ± 10 μA), but caution must always be exercised—some BiCMOS bus interface parts have 1 mA input currents.

9.1.1 Input level requirements

Failure to provide proper input voltage levels to BiCMOS/CMOS devices may result in destruction of parts as well as logic malfunctions. Input levels that fall between logic *high* and *low* input limits and inputs that exceed V_{cc} or ground are both classified as improper inputs and both can be destructive. In general, advanced CMOS parts with true CMOS inputs require inputs to have levels between ground and 30 percent of V_{cc} or between 70 percent of V_{cc} and V_{cc} to ensure that one or the other of the MOSFETs in the input complementary structure are OFF. Advanced CMOS devices with TTL-level inputs (those designated ACT, FCT, etc.) and BiCMOS devices with CMOS input stages require inputs to have levels between ground and 0.8 V or between 2.0 V and V_{cc} to ensure that one or the other of the MOSFETs in the input complementary structure are OFF. When input levels fall in the intermediate region, both MOSFETs in the input complementary structure may turn ON and current flow from V_{cc} to ground through the input complementary structure can result in overheating of parts and possibly eventual failure. For example, such a condition may exist when a device with TTL output levels is used to drive a device with CMOS input levels. A worst-case TTL minimum *high* output level, which is 2.5 V, does not fall within the range between V_{cc} and 70 percent of V_{cc}. Hence, pull-ups or other circuitry are required on TTL-to CMOS-level interfaces to ensure that inputs do not dwell in the danger zone. See Chapter 2 for BiCMOS/CMOS interface level specifications.

When input levels exceed V_{cc} or go below ground, even for a very short time or by a very small amount, BiCMOS/CMOS devices can be damaged. When inputs or outputs exceed V_{cc} or ground, parasitic SCRs in electrostatic protection circuits can be triggered ON, which can result in excessive V_{cc} to ground currents and destruction of parts. If input (or output) levels can exceed V_{cc} or ground when devices are powered, some means of current limiting must be provided so that the current levels needed for SCR triggering are not reached (see Chapter 3). It is also important to limit current at interfaces where one side of the interface may be powered before the other side is. If current is allowed to flow into input (or output) protection networks of unpowered devices, those devices may latch up when power is applied. It is a good practice to have some series resistance in circuit board or system interfaces to low-impedance sources or loads to ensure that input or output current ratings are not exceeded.

9.1.2 Unused inputs

Unused inputs on powered BiCMOS or CMOS devices must never be left open or floating; a valid *high* or *low* logic level is required at all times.[3] Valid input logic levels are required to ensure that:

1. Input circuits do not oscillate and upset device operation or increase system noise levels
2. Inputs do not float into the undefined input region between a valid *high* and *low* logic level where both the pull-up *p*-channel and the pull-down *n*-channel complementary MOSFETs in the input stage both turn ON creating a low-impedance path between V_{cc} and ground that causes the device to overheat and be destroyed

Figure 9.2(*a*) shows a CMOS or BiCMOS input stage with an open input, and Figure 9.2(*b*) shows the resulting equivalent circuit. The equivalent circuit consists of two resistors between V_{cc} and ground. The resistors represent the ON impedance of the two MOSFETs. Since the ON impedance of the MOSFETs used in advanced BiCMOS/CMOS devices is relatively low,[4] internal paths initiated by open inputs can cause excessive internal device currents. The typical profile for BiCMOS/CMOS device supply current versus input voltage is shown in Figure 9.3.[5–7] Excessive internal current as a result of open inputs can cause overheating and destruction of parts. Even if the open input condition only lasts for a very short time, device temperature is increased as a result of extra current flow. High device temperature in-

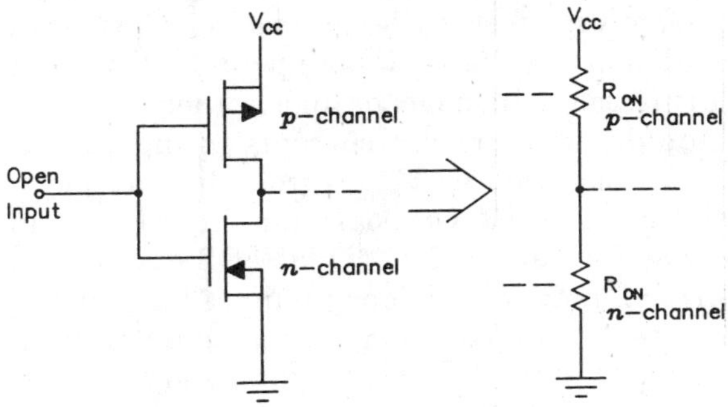

Figure 9.2 CMOS input stage and open input equivalent circuit.

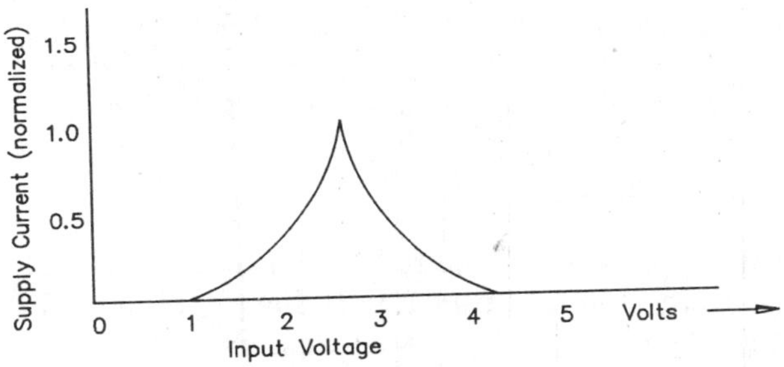

Figure 9.3 CMOS input stage supply current versus input voltage.

creases the susceptibility of BiCMOS/CMOS parts to latch-up which leads to additional high current flow and a greater possibility of device destruction. The time a part can endure with an open input depends on the internal impedance of the device and on other considerations such as how deep into the internal circuitry the intermediate logic level propagates. In general, floating input conditions should not be allowed to last longer than a few microseconds. However, some BiCMOS/CMOS parts have specified minimum rise time requirements that translate into more stringent requirements. In all applications, when BiCMOS/CMOS logic devices are used, minimum rise time requirements must be determined and adhered to in the design.

Unused BiCMOS/CMOS inputs on used devices should be pulled *high* or *low* as required for proper logic device operation. Unused inputs on unused devices can be connected to the most convenient level. Theoretically all advanced BiCMOS/CMOS device inputs can be connected directly to V_{cc} or ground; no current limiting is required from an input circuit operation standpoint. However, it is best to use a pull-up or pull-down resistor to facilitate test (a *high* or *low* input level can be injected) and to prevent the possibility of latch-up due to transient input current injection during power turn ON or under other transient supply conditions. Latch-up due to transient currents in inputs connected to low-impedance sources (such as V_{cc} or ground) is a known possibility with some of the older CMOS logic families. Latch-up should not be a problem when advanced BiCMOS/CMOS devices are used, but it is safer to current limit connections to unused inputs.

Unused inputs that are often overlooked and not pulled up or down include bidirectional ports, unused PAL or PLD inputs, and PAL outputs that serve as feedback paths. CMOS input structures must not float, not even for a short time, no matter what the terminal is called.

BiCMOS/CMOS inputs that connect directly to external sources must have pull-up or pull-down resistors so that they do not float in the absence of the normal source. Often during board or system test not all input sources are present, and without pull-ups or pull-downs those BiCMOS/CMOS devices with open input lines will be damaged.

Techniques for tying off unused inputs

Pull-up or pull-down resistors. A resistor tied to V_{cc} (hence called a pull-up) or a resistor tied to ground (hence called a pull-down) provides the simplest means of tying off unused inputs. A great number of inputs can be connected (tied off) with one pull-up or pull-down resistor without violating minimum *high* or *low* input level requirements since input currents are very low, 1 to 10 μA, for most BiCMOS/CMOS devices. However, it is generally not wise, nor practical, to tie too many points together. Troubleshooting a shorted line tied to a great number of points can be very difficult, and the longer the line the greater the chance for coupled noise. Thus, for practical considerations, the maximum number of inputs pulled up by one resistor should be limited to near 10. Loading of pull-up or pull-down resistors (i.e., the number of inputs pulled up or pulled down by each resistor) should not approach the theoretical limit which is

$$V_{cc\min} - (I_{\text{in total}})(R_{\text{pull-up}}) \geq V_{\text{in high min}}$$

for pull-up resistors and is

$$(I_{\text{in total}})(R_{\text{pull-down}}) \leq V_{\text{in low max}}$$

for pull-down resistors. In most pull-up or pull-down applications, for either advanced CMOS or BiCMOS devices, 1-kΩ resistors are used. Higher values increase the chance for noise pickup (due to higher source impedance) and for problems due to board leakage currents or defective devices with excessive input current.

Current-limited pull-up or pull-down voltages. Other connections that can be used to tie off unused inputs include:

1. Tie unused inputs that require a *high* input to the *high* output of an inverter that has its input tied to ground.
2. When a current-limited *low* level is required, use the output of an inverter or noninverting buffer that has its inputs tied to the appropriate level.
3. Tie unused inputs to used inputs that are functionally the same (on the same device), i.e., parallel inputs.

Paralleling inputs. When unused inputs are tied to used inputs (on the same device) that are functionally the same, the logical operation of the device is not affected and the unused inputs are prevented from floating. However, paralleling inputs can degrade dynamic performance. The capacitance load is increased, which can adversely effect high-frequency performance, and the noise margin may be reduced. Paralleling inputs parallels the capacitance associated with each input and increases the capacitance between the external signal source and the internal structure of the device. The increased capacitance increases the possibility of noise being injected into the internal circuitry of the device and upsetting it. Thus, paralleling inputs is not recommended for high-speed devices.

9.1.3 Derating current drive specifications

In most applications where advanced BiCMOS/CMOS logic families are used in high-performance systems, ac drive capability limits system operation rather than dc drive considerations. However, system designers must be careful to not exceed specified device dc drive limits, and in critical applications, manufacturers' dc output drive specifications should be derated. The input-output voltage levels and current load and drive ratings shown on most logic device data sheets are defined and tested under static conditions. It cannot be assumed that static ratings are adequate for all dynamic situations. If a part is not tested in a configuration exactly as it is used, there is always some uncertainty as to how it may function. Drive derating is one method used to compensate for that uncertainty. It also helps compensate for device degradation with aging and worst-case combinations of electrical and environmental stress.

In benign applications, not exceeding manufacturers' ratings may be adequate to ensure a reliable design. However, in high-performance applications, it is best to not fully load devices so as to leave some margin for uncertainty. In critical applications, such as military or space applications, an extra allowance is needed to ensure that parts will function at the extremes of the operating voltage, frequency, and temperature range and with long-term aging and other degenerative effects. The typical derating criteria used for systems that must operate over the full military temperature range of −55 to +125°C is to limit *fan-out,* or actual load current of devices, to no more than 70 percent of the manufacturer's specified value. For space applications a derating factor of 50 percent is often used.

When using advanced BiCMOS/CMOS devices, care must be taken to not violate dc drive limits when signals are load terminated or when a large number of TTL devices or BiCMOS devices with bipolar inputs must be driven.

9.1.4 Increasing drive capability

When more drive is needed, either ac or dc, the most obvious technique is to split a heavily loaded signal into segments and use multiple devices to drive the separate segments. The next most obvious technique is to use high-current advanced BiCMOS or CMOS bus interface drivers. Finally, if subdividing the load or adding drivers is not practical or does not alleviate the problem, driver outputs can be paralleled.

Enhancement-mode FETs, as used in advanced CMOS logic devices, have a positive ON impedance temperature coefficient, and as a result, paralleled CMOS outputs are self-regulating to some degree. For example, if a large portion of the load current is flowing in one of several paralleled outputs, that output tends to heat up more than the other (paralleled) outputs, and as a result, its resistance goes up and its portion of the current goes down. Thus, paralleled CMOS devices tend to share the load. Still it is best to only parallel CMOS devices in a common package.

Outputs of BiCMOS devices on the same chip or drivers in the same package can be paralleled to increase fan-out capability when there is no other alternative for achieving the needed drive. When paralleling BiCMOS drivers, it is important to do so only with drivers in a common package so that the drivers share the load. Drivers in a common package can be expected to be exposed to nearly the same environmental conditions. They should be exposed to the same temperature and supply voltage. They should also have similar device characteristics because of common processing. Thus, BiCMOS drivers in a common package should tend to share the load, where devices in different packages may not. Devices in different packages may have slightly different characteristics because of different processing and environment, and thus be less likely to share the load. If the devices do not share the load, one device may be overstressed as a result of having to handle a large portion of the load (such a condition is sometimes referred to as "load hogging").

Methods of increasing drive (fan out)

1. Separate the load and use multiple drivers.
2. Use high-current bus interface drivers.
3. Parallel outputs of devices in the same package.

9.2 Board-to-Board Interfaces

At the board-to-board level (and on large boards), signal connections tend to be dominated by bus structures that have large distributed capacitance loads and long runs across backpanels or motherboards (or large boards). Thus, the effects of large loads and the transmission-line effects caused by long runs must be taken into account in the selection of most interface circuits.[8]

9.2.1 Bus interfaces

There are a large number of 8-, 9-, and 10-bit-wide advanced BiCMOS/CMOS drivers, latches, transceivers, and registers that are commonly designated as bus interface devices available. Standard AC and ACT 240 series drivers, 373 latches, 245 transceivers, and 374 registers fall into the bus driver classification, but there exists a wide variance in drive capability depending upon the manufacturer. For example, National FACT AC or ACT 240, 245, 373, and 374 are specified for ±24-mA output drive, but devices with the same basic letter and number designations from other manufacturers are specified for higher current drive (at least for I_{OL}). Most devices described as bus interface devices tend to have I_{OL} ratings of 48 to 64 mA and I_{OH} ratings of 12 to 24 mA depending upon whether output *high* levels are TTL or CMOS; if they are TTL, the current rating tends to be on the low end of the listed range. For example, Texas Instruments' BiCMOS bus interface logic is specified for an I_{OL} of 48 or 64 mA and an I_{OH} of 12 or 15 mA depending upon whether the device is intended for military or commercial operation. However, that is not a universal situation, Advanced Micro Device's 29C800 bus interface family is only specified for an I_{OL} of 24 mA and an I_{OH} of 15 mA.

The main issue when selecting interface circuits is ac drive, not dc drive, and whether first incident switching is required (first incident wave switching is when the initial signal launched into a transmission line is of sufficient amplitude to cross the threshold of the receiving device or devices). Unfortunately, most device manufacturers do not specify dynamic drive. A few list typical values, but typical values are of limited use; system designers need guaranteed dynamic specifications. Some progress is being made in that direction; the FACT Databook states "Fairchild has taken the steps to guarantee incident wave switching on transmission lines with impedances as low as 50 ohms for the commercial temperature range and 75 ohms for the military temperature range."[6] Certain vendors' advertisements for their BiCMOS bus interface logic claim that they have first incident wave switching. However, individual data sheets do not specify dynamic drive.

Drive requirements for buses are complicated functions of the characteristic impedance Z_o of the bus, the location of the loads tied to the bus, and the response needs of the bus. If bus signals have a long time to reach final value, then drive requirements are of little concern, but high-performance systems usually do not have a long time for buses to settle.

It is difficult to maintain 50-Ω effective impedance in multilayer pc boards (see Chapter 6). The typical intrinsic impedance of multilayer pc boards tends to be in the neighborhood of 50 Ω, but buses with distributed loads have an effective impedance Z'_o given by Equation (6.3)

$$Z_o' = \frac{Z_o}{\sqrt{1 + C_{LOAD}/C_{LINE}}}$$

where C_{LOAD} is the total lumped capacitance of each device connected to the line and C_{LINE} is the total line capacitance (see Chapter 6). It is not uncommon for the effective impedance of bused lines or other lines that connect to numerous locations to be reduced to one-half the intrinsic unloaded impedance. If the effective impedance of a bus is low relative to the output impedance of the bus driver, it will take a number of reflections between the ends of the bus and the active source for the bus to charge to a final value (see Figure 9.4). A bus that requires a number of reflections to reach final value has less than optimum performance (i.e., it takes a long time to settle). Ideally, if a bus driver has adequate drive (and neglecting other potential transmission-line effects) at a maximum, a bus should take only two line delays (down

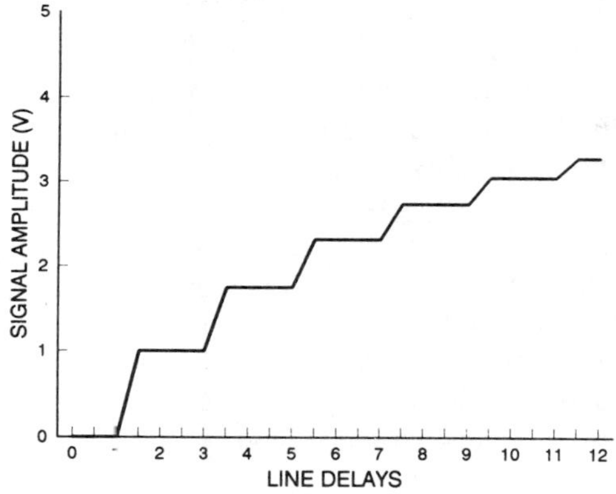

Figure 9.4 Heavily loaded bus signal transitions typically consist of a number of steps instead of smooth fast transitions.

and back) to settle. However, with most present bus interface device specifications the system designer has little information available to predict actual in-circuit worst-case dynamic performance. Even assuming the best case (i.e., down and back), it is difficult to achieve 20-MHz or greater bus operation on large pc boards, but it can be done if the number of loads is limited. It is even more difficult to achieve 20-MHz or greater bus operation on board-to-board interfaces, particularly if it is a requirement that boards must operate on extender boards. In either case, a careful analysis using the bus driver dynamic information that is available or reasonable assumptions for bus driver dynamic characteristics is needed to establish the worst-case operating speed. Optimistic bus operating speed projections based on typical or hoped for characteristics without an understanding of the many issues that affect bus speed are sure to lead to unreliable bus transfers.

9.2.2 General requirements for three-state buses

Three-state signals (buses) should not be allowed to float when unused. Floating lines tend to drift into the threshold region of the receiving circuits, and noise or other disturbances that are coupled into the line tend to cause the input structures of the receiving devices to oscillate. The oscillation tends to increase the overall system noise, particularly on the power and ground distribution systems; often the noise will reach unacceptable levels. Acceptable solutions are to always drive three-state lines or to provide pull-up resistors to ensure that lines do not float near the threshold level of the receiving devices when not driven. The disadvantages of pull-up resistors include added parts and power dissipation. Selecting a pull-up resistor value is often difficult; a tradeoff between power dissipation and rise time of the line when it is not actively driven (i.e., when it is being pulled up) must be made. The use of pull-ups which are too large, to reduce power dissipation, may do more harm than good. Large-value pull-up resistors on lines with large capacitance loads may allow lines to remain in the critical noise-generating area near the threshold region of the receiving devices for long times. Furthermore, large-value pull-ups may pull signals into the threshold region that might have remained near their last active level if not influenced by a pull-up. When power dissipation is of concern and low-value pull-ups cannot be used, it may be best to always drive three-state lines *high* before releasing the lines (letting the drivers go into a high-impedance state). If three-state lines are driven *high,* high-value pull-up resistors can be used to hold three-state lines *high.*

Three-state drivers tied to a common line must be controlled in such a manner that no more than one can be turned ON at the same time, even for short periods during power turn ON or OFF. The safest method of achieving that is to use a single hardware decoder as the source of the enables for all drivers connected to a common line; then only one driver can be ON at one time (unless there is some fault). Where a single decoder for the source of all the line driver enables (for a common line or bus) is not practical, which is often the case in large systems with bus interconnects that interface to numerous units, the system dc reset signal should be used to ensure that the controls to three-state drivers are initialized to the OFF state during power-supply transients and at turn ON. A dc reset should always be used for initialization. Requiring the presence of the system clock, which may not always be present because of faults or other reasons, for correct initialization of three-state buffers is an invitation for disaster. Most of the advanced BiCMOS/CMOS three-state buffers are not rated for continuous short-circuit conditions. When buses go to multiple boards, the bus systems should be designed so that the removal of a board or unit will not result in more than one set of the remaining drivers being turned ON.

9.3 Board and System Interface Guidelines

In high-speed systems, input signals to boards, racks, or systems should be received at one place only to minimize ac loading and transmission-line effects. Signals routed to multiple devices on multiple boards have little chance of settling in a clock period as clock speeds go above 20 MHz. Output signals that are used internally and externally to a board or unit should be buffered before being used externally to isolate the internal signals from transmission-line reflections or other external disturbances, such as shorts, that might occur external to the unit. Signals routed to test points should be buffered to prevent test equipment from interfering with normal system operation. The transmission-line effects of long lines to test equipment are sure to corrupt internal signals given the fast edge transitions of advanced BiCMOS/CMOS devices. In particular, a buffered clock source should be provided for logic analyzers or other test devices so that clocks to test equipment can be appropriately terminated and so that long test lines do not corrupt internal clock signals. The outputs of unbuffered clocked devices (registers, counters, flip-flops, etc.) should be buffered before being sent out of (or off) a unit (boards, boxes, etc.). Line reflections have been known to upset and change the output state of unbuffered clocked elements. Most advanced BiCMOS/CMOS clocked elements have built-in output buffers, but many of the older logic families do not. Devices with three-state outputs have built-in buffers because they are inherent in normal three-state buffer implementations.

Asynchronous input signals should never be run to multiple clocked devices within a functional unit; some of the devices within the unit may sense (sample) the signal at a slightly different time because they may have different line lengths or because of clock phasing at the multiple receiving points, and the unit may react incorrectly. When asynchronous parallel buses must be received and synchronized in a unit, it is best to capture and synchronize the bus control or strobe signals rather than try to "broadside" synchronize the entire bus. There is no assurance that all the signals on a wide bus will be properly captured when the data happen to be changing at the same time that the clock is asserted on the receiving storage elements.

Single-ended signals that traverse noisy areas, such as motherboards, should be received by devices with enhanced noise margin where practical. Some of the bus interface devices, such as the 240 and 244 buffers, are advertised as having input hysteresis and a slight bit more noise margin than standard gates. However, that is not a universal characteristic of all 240, 244, etc., buffers in all BiCMOS/CMOS families. Schmitt trigger buffers such as the 54AC14 or 54ACT14 have higher than normal levels of noise immunity and are capable of operating with slowly changing input levels and thus are useful at noisy interfaces with badly corrupted signal waveforms. CMOS levels provide more noise margin than TTL levels and should be considered for critical signals such as clocks. When BiCMOS/CMOS devices with TTL input levels are used, control signals that traverse noisy areas, such as motherboards, should have their logic polarity configured so that the predominate state (inactive state) is a *high* level. A *high* TTL level has a higher noise margin than a *low* level. If a signal has a state that is tolerant of noise or glitches, the signal should be logically arranged to be *low* for the noise-tolerant state since the *low* state has less noise margin (an asynchronous system reset is a possible example of such a signal).

9.4 Board and System Interface Protection

Isolated board or system interface signals that connect directly to BiCMOS/CMOS devices with CMOS inputs or outputs offer the potential for problems. External connections increase the probability of exposing sensitive CMOS inputs or outputs to out-of-range dc or transient voltages which may induce latch-up. External connections also increase the probability of damage due to static electricity. Advanced BiCMOS/CMOS parts have much higher immunity to latch-up and ESD than earlier CMOS devices, but it is still a prudent design practice to add shunt low-impedance paths to V_{cc} or ground and series current-limiting resistors (see Figure 3.17) to all external board or unit input signals that connect to CMOS inputs.[9] Shunt low-impedance paths help prevent electrostatic buildup when boards or units are isolated (e.g., when a board is out of the chassis). When a unit is in place and powered, shunt resistors prevent open inputs from floating. Series current-limiting resistors provide a means of controlling static and transient currents injected into protection networks. Signals that overshoot or undershoot V_{cc} or ground can induce latch-up if excessive current is allowed to flow in protection networks. Series resistors are also necessary to limit current in protection networks in applications where signal sources may be powered when receiving devices are not. In all CMOS interface applications, input currents under abnormal or worst-case transient conditions must be kept below actual specified device limits. Absolute maximum dc input current limits for most CMOS logic devices are in the 20- to 30-mA range. Transient limits for most advanced BiCMOS/CMOS devices are typically near 100 mA, but caution must be exercised since some CMOS devices may latch up with injected currents as low as 10 mA.

Outputs may need series current-limiting resistors to prevent excessive output currents under abnormal conditions, such as shorts, transients, or large transmission-line reflections. Outputs tend to be more sensitive to latch up than inputs, i.e., they latch up at a lower current. In applications where high-speed signals must cross external board or unit boundaries, series current-limiting resistors may cause excessive *RC* delays. However, if it is not possible to use series current-limiting resistors and the possibility of excessive static or transient output currents exists, other steps must be taken to control the situation.

9.5 Signal Interfaces Between Remote Units

At the unit-to-unit level, significant ac and dc differences in the reference (ground) potential of the various units will generally be present. Furthermore, signals that are routed between units are usually exposed to a more hostile noise environment than signals that remain internal to a given unit or subsystem. Thus, communication between separate units where the actual transmitting and receiving components are *not* directly referenced to a common reference plane, such as the ground plane of a board, motherboard, or backplane, requires a signal transmission means that has a high level of noise tolerance. More noise immunity is required than is provided by standard single-ended BiCMOS/CMOS devices with either TTL or CMOS levels. Single-ended TTL- and CMOS-level devices do not have enough noise margin to function reliably unless both the driving and receiving devices are directly referenced to a continuous common ground plane, which is not possible for drivers and receivers that are in separate units. As shown in Chapter 4, BiCMOS switching currents produce transient voltages across a 1-in length of wire that greatly exceeds static TTL *low*-level noise margin, and CMOS switching currents are even worse. Thus, transient differences that exceed TTL or CMOS levels are to be anticipated between units that require several feet of wire to be interconnected.

Signals transmitted between remote devices (i.e., devices not tied directly to a continuous common ground plane) must be transmitted by some method that will function reliably in the presence of high levels of noise and offset voltages. Single-ended TTL or CMOS levels are not adequate.

9.5.1 Differential unit-to-unit signal transmission

Balanced differential networks are one means of transmitting data between units. A balanced differential interface network consists of a differential line driver connected to a differential receiver by twisted-pair or shielded twisted-pair wires. Differential receivers are insensitive to the absolute input voltage. Data are conveyed by the polarity of the voltage difference between the two inputs, not the absolute input voltage levels.[10] Differential communication links are not upset by source or destination reference offsets which often occur as the result of dc or ac current flow in the ground reference system since reference offset voltages appear common to both lines of a differential interunit connection (see Figure 9.5). Also, most coupled noise appears equally

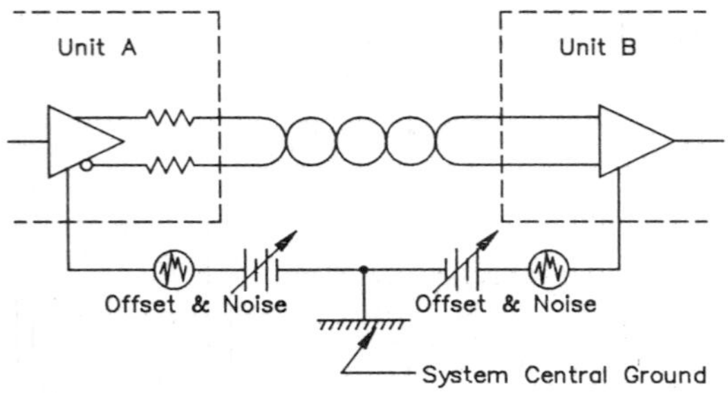

Figure 9.5 Balanced differential unit-to-unit interconnections are insensitive to ground offsets and noise.

on both lines of a differential connection[11] and is rejected since it is common to both lines, but to minimize the coupling and keep the coupling below the common-mode rejection range of the receivers used, shielded twisted-pair lines should be used.[12]

Balanced differential signal transmission has the added advantage of generating little noise in the ground system. The transient line voltages and currents tend to be equal and opposite and hence cancel. Thus, balanced differential communication between units limits the noise pollution of the local environment. Most differential communication links use circuits that meet the requirements of the Electronic Industries Association (EIA) standard RS-422.[13]

Differential unit-to-unit drivers and receivers. Figure 9.6 shows a series-terminated RS-422 unit-to-unit differential interconnection using a CMOS 26C31 driver and a 26C32 receiver. The 26C31 and 26C32 driver-receiver pair (available from National Semiconductor Corp.) meet the requirements of RS-422.[13] Those requirements include: ±7 V of common-mode capability and 0.2-V input sensitivity. In benign ap-

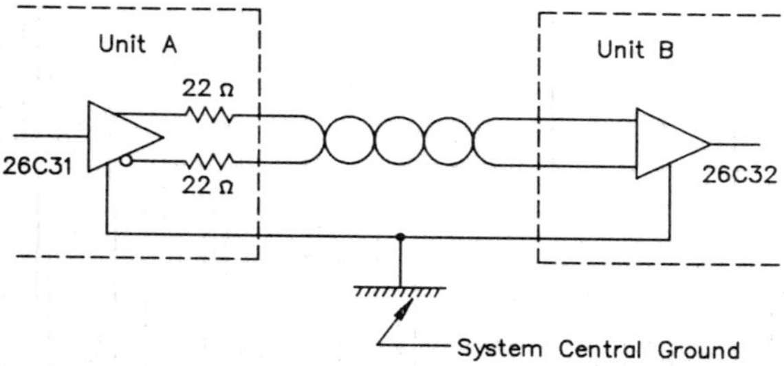

Figure 9.6 Differential drivers and receivers that meet the requirements of RS-422 are often used for differential communication between units.

plications, it is possible to operate the 26C31 and 26C32 pair in the 30- to 40-MHz (actual signal frequency) range. However, operating frequency is dependent on a number of factors, such as interconnecting cable length, cable impedance, and allowable error rate, all of which must be carefully evaluated for each application.

Very high speed differential unit-to-unit drivers and receivers. For communication between separate units at, or near, the upper frequency limits of advanced BiCMOS/CMOS devices (i.e., in the 80- to 100-Mb/s data rate range—40- to 50-MHz clock rates), there are no off-the-shelf BiCMOS/CMOS differential drivers or receivers. If data must be transferred at very high rates, differential ECL line drivers and receivers or custom circuits are required. However, the common-mode noise-rejection capability of ECL receivers (approximately ±1 V) is marginal for most unit-to-unit applications. Where higher common-mode noise-rejection capability is needed, high-speed bipolar analog comparators with input divider networks to increase the common-mode range are a possible solution, but care must be taken in their selection. In general, only those with ECL outputs will operate in the 40- to 50-MHz range.

9.5.2 Low-speed unit-to-unit signal transmission

For low-speed (less than 20 kb/s) data transmission between separate units, single-ended driver-receiver pairs, as shown in Figure 9.7, that have signal levels that conform to the EIA standard RS-232[14] are the most common interface. RS-232 interfaces are intended for short cables (less than approximately 50 ft or 15 m). However, baud rate (i.e., bits per second) and cable length can be traded off; the shorter the cable, the higher the baud rate.

Single-ended RS-232 levels are considerably larger than TTL or

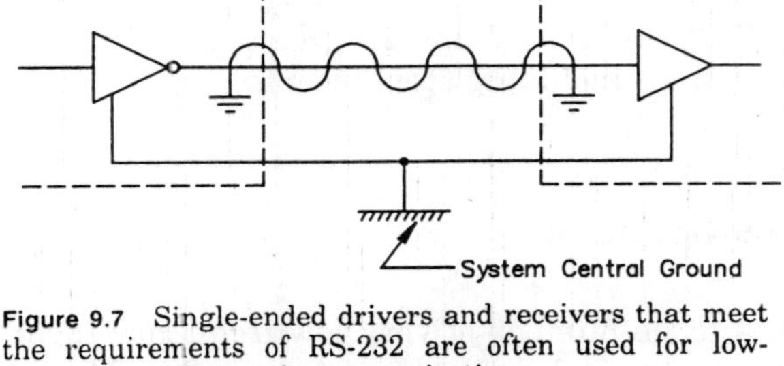

Figure 9.7 Single-ended drivers and receivers that meet the requirements of RS-232 are often used for low-frequency unit-to-unit communication.

CMOS signal levels, and the greater the difference in the signal level and the switching threshold, the greater the immunity to noise. Worst-case minimum RS-232 *high* and *low* levels are +5 V and −5 V, respectively. Typical levels are +9 V and −9 V. The switching threshold of RS-232 receivers is near ground. The large difference between worst-case signal levels and receiver input threshold levels is sufficient to provide an acceptable amount of noise margin for most unit-to-unit applications. Large signal swings provide good noise immunity, but they have disadvantages; slew rate must be kept slow or else large signal swings will create a great deal of noise. The RS-232 specification defines a maximum slew rate for the specific purpose of minimizing switching-edge-induced noise.

The original RS-232 driver, the 1488, required external ±12-V supplies which complicated its use, but there are now a number of RS-232 driver/receivers from a number of manufacturers (Linear Technology, MAXIM, Motorola, and Texas Instruments) with on-board dc-to-dc converts. Parts are available in a number of different combinations of drivers and receivers within a package and with or without on-board dc-to-dc converters. For most applications, some subset of the various device configurations usually provides the exact number of receivers and transmitters required.

RS-232 interfaces require a common ground connection between units. Each unit should have a solid connection to earth ground to ensure that a unit does not float to a potential that might damage the transmitters or receivers. Floating units often take on a static charge or drift toward the ac line potential; RS-232 interfaces cannot withstand 110 V. Ideally, from a dynamic standpoint, each single-ended unit-to-unit signal should have an accompanying ground return line. Ground return lines are required to provide a direct low-impedance path for signal return currents. However, caution must be exercised when direct unit-to-unit ground connections are made.[15] Consideration must be given to the effects of interunit ground connections on overall system reference-level integrity. Large ground-loop currents may flow in the signal return connections, unless care has been taken in the overall system grounding to ensure that large offset voltages do not exist between units.[16] References 17 and 18 give a great deal of guidance on RS-232 applications including standard pin-outs and connector types.

9.5.3 Very low speed unit-to-unit signal transmission

In some digital systems, a large number of very low frequency (near dc) bilevel signals from mechanical switches, relays, solenoids, or other such external sources must be received and translated to the system internal logic levels. Signals from electromechanical devices are generally very noisy and often have large transient excursions. Yet, often little attention is given to the special interface needs of such signals. In many instances, receivers without adequate noise rejection are used with the apparent assumption being that noise is not a problem when signal frequency is low. External signals, even if their frequency is near dc, must be received in such a manner that unambiguous logic decisions are possible and that noise-accompanying signals, or signal references, do not propagate into the receiving system. Cost considerations and the need for simplicity often dictate that such signals be received with single-ended receivers. Single-ended receivers complicate the task of receiving noisy signals. If single-ended communication is used for external interfaces, signal levels must be large enough so that signals can be distinguished from noise, or signals must be heavily filtered to remove noise. Single-ended TTL or CMOS levels are inadequate for noisy external sources not solidly referenced to the same ground as the receiver.

Very low frequency high-level receivers. High-input-impedance CMOS buffers or BiCMOS buffers with CMOS input circuits are ideal for many low-frequency receiver applications. The intrinsic high input impedance of CMOS input stages provide a high level of isolation between source and load and allow the use of a number of filtering techniques that are not possible with bipolar input stages. (*Caution:* Some BiCMOS devices have bipolar input stages.) Buffers with high input impedance allow the use of very high impedance input divider-filter networks (bipolar input stage current requirements severely restrict signal source impedance). For example, a CMOS Schmitt trigger buffer preceded by a divider-filter network, as shown in Figure 9.8, provides a simple, inexpensive means of receiving noisy, low-frequency, high-level, external signals and translating them to TTL or CMOS levels.

The input divider allows signals with level changes greater than the

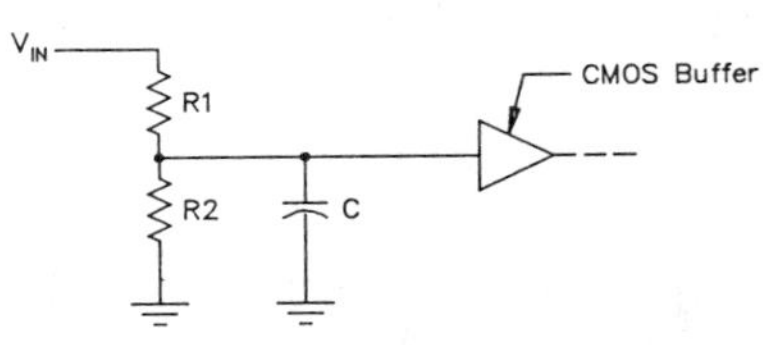

Figure 9.8 Circuit for receiving high-level low-frequency signals.

Schmitt trigger buffer threshold voltages to be reduced to levels compatible with the buffer input requirements, and by adjusting the divider ratio, a large range of input levels can be received. For example, bilevel signals, such as those that might originate from 12- or 28-V relays or solenoids are easily translated to levels compatible with 5-V powered CMOS Schmitt trigger input requirements. The maximum input level is only limited by practical considerations, such as limiting voltages in digital units to safe and reasonable levels. Minimum input signal levels must at least exceed the threshold limits of the receiving CMOS Schmitt trigger buffer. Both minimum and maximum *high* and *low* signal levels must be considered when selecting the divider network ratio to ensure that normal signal level variations at the divider output are sufficient to ensure switching but do not significantly exceed V_{cc} or ground. However, if the divider input resistor $R1$ is large enough, V_{cc} or ground can be exceeded without harm to the buffer. The resistance of $R1$ must be such that under all overvoltage or transient conditions the current injected into the CMOS buffer input protection circuit is less than the specified limits for the device. High-impedance input divider-filter networks also serve a number of other purposes:

1. They reduce the loading of the source.
2. They limit ground currents.
3. They limit fault or transient input currents injected into the receiving CMOS buffers to safe levels.
4. Heavy filtering of signals is possible with small-value capacitors and thus physically small capacitors.

Since filtering will slow the rise time of signals at the buffer input, buffers with minimum rise time requirements cannot be used in such applications. Only devices that can tolerate slow rise time inputs, such as 54ACT14 or 54AC14 Schmitt trigger buffers, are suitable for such applications. Most advanced BiCMOS/CMOS devices cannot be used in such applications since most have minimum rise time requirements. Since speed is of little concern in such applications, advanced BiCMOS/CMOS devices are not required. Devices, such as 4009 or 4010, or 4049 or 4050 level translator buffers, or 54HC14 Schmitt trigger buffers from the older logic families are adequate for most applications.

9.5.4 Unit-to-unit line terminations

Series (source) termination should be used for most unit-to-unit interconnections (see Figure 9.6). Series termination provides short-circuit protection and does not increase static power dissipation as does load termination. For optimum response, the value of the terminating resistors (plus the output impedance of the driver) should be chosen so that the line is slightly underdamped, but not too much. A line that is slightly underdamped will provide a signal at the load that reaches the threshold region of the receiver quicker than a line that is exactly matched or overdamped. The key to successful operation and optimum performance is to not underdampen the line so much that it rings back into the threshold region of the receiving device.

9.5.5 Miscellaneous unit-to-unit considerations

All unit-to-unit signal cables must include ground lines for signal return currents. When single-ended high-level interunit communication is used (low-level single-ended signals should never be used), there should be one ground line per signal line. When balanced differential interunit communication is used, one ground line per four to eight signal pairs is a good rule of thumb. If differential signals are perfectly balanced, no return ground lines are needed, but differential signals are never perfectly balanced. Thus, some grounds are needed to provide a direct return path for the unbalanced portion of the signal currents. However, direct unit-to-unit ground connections offer the potential for large ground-loop currents (see Figure 6.1). Thus, when unit-to-unit signal grounds are required to ensure a low-impedance path for signal return currents, care must be taken to ensure that all units have low-impedance ground connections to prevent large interunit reference voltage offsets and large ground-loop currents. If large offset voltages exist between units, ground current in the signal return lines can be of such a magnitude as to disrupt system operation by introducing noise[15] and in severe cases can overheat and burn out the signal return lines.

Communication between remote units generally must incorporate some form of asynchronous data transfer technique, since clock alignment between remote units is usually difficult to achieve or maintain.

9.6 Summary of Interface Guidelines

1. Input signals to boards, racks, or systems should be buffered at one place so as to minimize ac loading (minimizing ac loading is essential in high-speed systems).
2. Critical single-ended signals that traverse noisy areas, such as motherboards, should be received with 54_240 or 54_244 buffers

that have hysteresis or with Schmitt trigger buffers such as 54_14 or 54_132 that have a higher than normal level of noise immunity.

3. Asynchronous input signals should never be run to multiple clocked devices within a functional unit; some of the devices within the unit may sense (sample) the signal at a slightly different time because of different line lengths or clock phasing at the multiple receiving points, and the unit may react incorrectly.
4. When asynchronous parallel buses must be received and synchronized in a unit, it is best to capture and synchronize the bus control or strobe signals rather than try to "broadside" synchronize the entire bus. There is no assurance that all signals on a wide bus will be properly captured when data happen to be changing at the same time the clock is asserted on the receiving storage elements.
5. Signals that are used internally and externally to a board or unit should be buffered before being used externally to isolate the internal signals from transmission-line reflections or other external disturbances, such as shorts, that might occur external to the unit.
6. When BiCMOS/CMOS devices with TTL input levels are used, control signals that traverse noisy areas, such as motherboards, should have their logic polarity configured so that the predominate state (inactive state) is a *high* level since *high* TTL levels have higher noise margin than *low* levels.
7. If a signal has a state that is tolerant of noise or glitches, the signal should be logically arranged to be in the minimum noise margin state (the *low* state in TTL-level systems) for the noise-tolerant state (an asynchronous system reset is a possible example of such a signal).
8. The output of unbuffered clocked devices (registers, counters, flip-flops, etc.) should be buffered before being sent out of (or off) a unit (boards, boxes, etc.). Line reflections have been known to upset and change the output state of unbuffered clocked elements.
9. Signals routed to test points should be buffered to prevent test equipment from interfering with normal system operation. In particular, a buffered clock source should be provided for test equipment.
10. Three-state signals (buses) should not be allowed to float when they are not being used. Floating lines tend to drift into the threshold region of the receiving circuits causing receiving devices to oscillate and increase system noise. Receivers with CMOS

input stages may go into an excessive current condition that can lead to device destruction.

11. Three-state drivers on a common line should be controlled in such a manner that no more than one be turned ON at the same time, even for short periods during power turn ON or OFF to prevent burnout.
12. Three-state bus systems should be designed so that the removal of a board or unit will not result in more than one set of the remaining drivers being turned ON.
13. Unused inputs on used or spare devices must not be allowed to float to prevent oscillation and excessive system noise or excessive current flow in CMOS input stages.
14. Inputs to BiCMOS/CMOS devices must never be allowed to float to prevent excessive internal feedthrough current flow and burnout.
15. Static pull-up or pull-down voltage sources for unused BiCMOS/CMOS inputs should be current limited to reduce the chance of transients inducing latch-up.
16. Board or system interface connections to BiCMOS/CMOS inputs must have pull-up or pull-down resistors to prevent inputs from floating when normal input sources are not present.
17. Board or system interface lines that originate from sources that might experience large transients connected directly to BiCMOS/CMOS inputs must have series current-limiting resistors to prevent transients from driving inputs into latch-up.
18. Board or system interface lines that originate from sources that have separate or independently controlled power sources must have series current-limiting resistors to limit input current in unpowered inputs to prevent latch-up when power is applied.
19. Static (dc) loading specifications should be derated. A common guideline is: Do not exceed 70 percent of the manufacturer's rated drive capability when system operation is over the military temperature range, or 80 percent when system operation is over the commercial temperature range (unless more conservative drive derating is required by the customer).

9.7 References

1. DiCerto, Joseph: "Poor Packaging Produces Problems," *The Electronic Engineer*, September 1970, pp. 91–93.
2. Southard, Robert K.: "High-Speed Signal Pathways from Board to Board," in 1981 *WESCON Record, Session 18*, Paper No. 2, September 1981.

3. Sokal, Nathan O.: "Designer's Guide to PC-Board Logic Design," Part Two, *EDN,* November 27, 1986, pp. 229–235.
4. Cox, Gerald C.: "Impedance Matching Tweaks Advanced CMOS IC Testing," *Electronic Design,* April 1987, pp. 71–74.
5. *Advanced CMOS Logic Designer's Handbook,* Texas Instruments Inc., Dallas, Tex., 1987.
6. *FACT—Advanced CMOS Logic Databook,* National Semiconductor Corp., Santa Clara, Calif., 1989.
7. *DATABOOK, RCA High-Speed CMOS Logic ICs,* RCA Corp., Somerville, N.J., 1986.
8. Royle, David: "Designer's Guide to Transmission Lines and Interconnections," Part Two, *EDN,* June 23, 1988, pp. 143–150.
9. Walsh, M. J.: *Choosing and Using CMOS,* McGraw-Hill, New York, 1985.
10. Millman, Jacob: *Microelectronics, Digital and Analog Circuits and Systems,* McGraw-Hill, 1979.
11. Harper, Charles A.: *Handbook of Electronic Packaging,* McGraw-Hill, New York, 1969.
12. Ott, Henry W.: *Noise Reduction Techniques in Electronic Systems,* Wiley, New York, 1988.
13. EIA Standard RS-422-A, *Electrical Characteristics of Balanced Voltage Digital Interface Circuits,* Electronic Industries Association, Washington, D.C., August, 1978.
14. EIA Standard RS-232-C, *Interface Between Data Terminal Equipment and Data Communication Equipment Employing Serial Binary Data Interchange,* Electronic Industries Association, Washington, D.C., August 1969.
15. Brown, H. C.: "Get Rid of Ground-Loop Noise," *Electronic Design,* No. 15, July 19, 1969, pp. 84–87.
16. Oates, Edward R.: "Good Grounding and Shielding Practices," *Electronic Design,* No. 1, January 4, 1977, pp. 110–112.
17. Campbell, Joe: *The RS-232 Solution,* 2d ed., SYBEX Inc., Alameda, Calif., 1989.
18. Seyer, Martin D.: *RS-232 Made Easy,* Prentice-Hall, Englewood Cliffs, N.J., 1984.

General reference for bus interfaces

Di Giacomo, Joseph: *Digital Bus Handbook,* McGraw-Hill, New York, 1990.

Chapter

10

Noise-Tolerant Logic Architectures

It is impractical, if not impossible, when using high-speed BiCMOS/CMOS logic devices to keep noise due to reflections, ringing, crosstalk, and ground bounce (or power-supply droop) on all signals to levels that do not exceed input switching thresholds. The only solution is to use logic system implementations, such as synchronous architectures, that tolerate high noise levels.[1]

10.1 Synchronous Design

Synchronous architectures maximize the noise immunity of systems by utilizing the time quantization of their clock to provide a filtering effect. In synchronous systems, clock inputs of storage elements, such as flip-flops, registers, counters, and state machines, are only driven by the system clock.[2,3] Signals generated in combinational logic paths are never used to clock storage elements. Noise-sensitive asynchronous inputs, such as presets or resets, are never used to perform operational system logic functions. Asynchronous inputs are only used for system initialization at power turn ON or for test initialization. In synchronous systems, storage element state changes are controlled by logic conditions on control inputs. Thus, noise on control signals, except at the critical time near a clock transition, will not cause false storage element changes.

Output transitions are the major source of noise in most digital systems. Synchronous systems tolerate a great deal of switching noise without upset or error because switching noise in synchronous systems only occurs following clock transitions.[4] If there is sufficient time for all switching transients to subside before the next clock transition, quiet signals are available to be sampled when the clock transitions

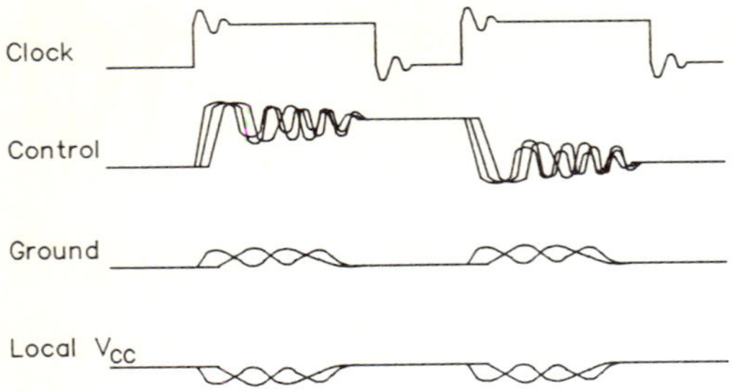

Figure 10.1 Synchronous systems can tolerate noise on signals, power, and ground following the active clock edge if there is time for it to die out before the next active clock edge.

(see Figure 10.1). Ideally, synchronous systems have only one clock frequency and phase to optimize the quiet time.

Synchronous systems are easy to understand, analyze, test, and change if necessary. In synchronous systems, signals flow in an orderly manner from one clocked element to the next. Thus, systems are partitioned into manageable and understandable sections. In the ideal synchronous system, all clocked elements are clocked with the same clock, and hence a fixed and known quantum of time is established for all operations which simplifies timing analysis. Test and design verification are also simplified since clock rates can be increased or decreased to determine timing margins without logic malfunctions occurring because of special timing paths or requirements.

A perceived disadvantage of synchronous design is that decisions can only be made at clock times, and thus timing paths cannot be optimized; i.e., synchronous designs are perceived to be slower than asynchronous designs.[3] In synchronous designs, the clock period must be long enough to accommodate the slowest signal path, and all other signal paths are forced to accommodate that clock period, while theoretically in asynchronous designs each signal path can be optimized.[2] Perhaps, each signal path can be optimized in experimental setups operating in benign laboratory environments, but in production systems that is usually not practical. Production systems, whether synchronous or asynchronous designs, must allow for worst-case timing parameter variations. When proper allowance is made for worst-case timing changes due to aging, process, and environmental conditions, asynchronous designs have little, if any, speed advantage.

10.2 Important Synchronous Design Issues

Three critical issues that must be understood to successfully implement synchronous designs are

1. Clock signal requirements

2. Signal hold-time requirements
3. Synchronization of asynchronous inputs

10.2.1 Clock requirements

The ideal synchronous system uses only one clock frequency and one phase so that all storage elements are clocked at the same time. Such an arrangement establishes when a system will be noisy and when it will be quiet. If the clock period is sufficient to allow for the worst-case signal propagation path, all signals will be quiet at the next clock edge allowing unambiguous logic decisions to always be made (see Figure 10.1). If multiple clock sources or clock phases are used, noise will be present more of the time, and for a given system cycle time, the time for noise to subside is reduced.

When it is necessary to store, process, or transfer data at time intervals that differ from the primary clock frequency, it is usually best to generate enable signals and use devices with clock enable input controls (see Figure 10.2) instead of using different clock frequencies. When enable signals are used, state changes are initiated by the primary clock, not the enable signal, so noise due to state transitions remains locked to the primary clock. Enable signals are easier to use and to generate than multiple-clock frequencies. It is very difficult to maintain the alignment needed for synchronous data transfers between different frequency domains when multiple-clock frequencies are used.

Clock signals require an optimal electrical environment. Clock-signal quality, uniformity of propagation delays (skew), and noise tolerance must be closely controlled throughout clock distribution net-

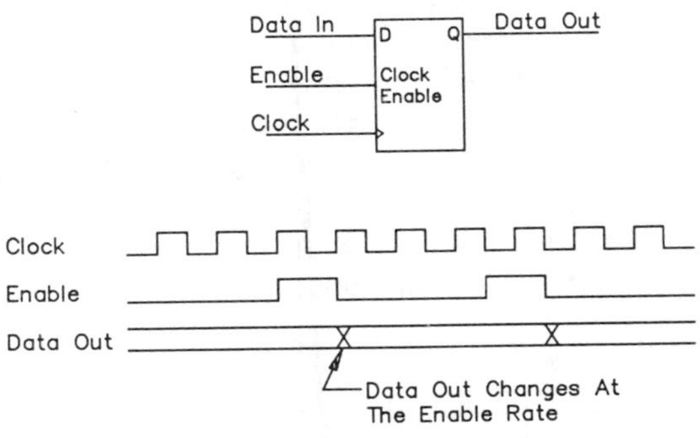

Figure 10.2 When it is necessary to store, process, or transfer data at time intervals that differ from the primary clock frequency, it is usually best to generate enable signals and use devices with clock enable input controls.

works. Clock signals must be as near perfect as possible. Excessive ringing and glitches cannot be tolerated on clock lines. Thus, clock distribution circuitry must incorporate some form of line termination for waveshape control.[2] Clock signals must be routed to their destinations using reasonably well controlled impedance paths so that proper termination can be selected.

Clock distribution networks (trees) must be designed to minimize skew between clock signals. To accomplish the needed alignment, clock distribution circuitry must be standardized. Propagation delays must be matched throughout clock distribution networks. Clock drivers at each level must be identical generic devices of the same logic family. Where possible, all drivers in a given level should be located in a common package. Drivers and loads must be arranged so that loads are equalized. Clock lines must be very short so that wiring propagation delays are insignificant, or lines must be of near equal length in each distribution level to equalize physical propagation delays.

Clock-signal paths must be isolated from other signals to minimize crosstalk. Parallel runs adjacent to other signals or other clocks must be kept as short as possible to minimize the opportunities for cross coupling. When multilayer pc boards are used, clock signals should be isolated and shielded from other signals by locating clock traces between reference planes. On welded-wire or wire-wrap circuit boards or backpanels, each clock line should be twisted with a ground line to provide shielding and control of line impedance, and care must be taken to ensure that clock lines do not get bundled with other signals.

10.2.2 Hold-time requirements

Hold time is the time that an input control or data signal to a clocked element must remain stable after an active clock edge (see Figure 10.3). Ensuring that all signals to clocked devices meet the required hold time is one of the most difficult tasks in a synchronous design. It is usually impossible to show that hold-time violations do not occur under worst-case conditions using worst-case data sheet minimum and maximum timing specifications. However, in practice, when attention is given to the conditions that most often cause hold-time violations, synchronous systems can be built that operate reliably.

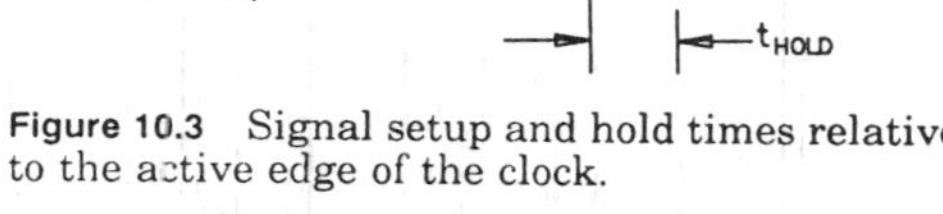

Figure 10.3 Signal setup and hold times relative to the active edge of the clock.

Hold-time violations most often occur where:

1. Clock alignment is poor
2. Signals go directly from one clocked device to another with no combinational logic between them
3. Control signals are gated with clocks to form gated clocks
4. Signals go from a fast device to a much slower device
5. Logic families or technologies are mixed

Good clock alignment is essential for preventing hold-time violations. If clock alignment is poor, control signals from a device with an "early" clock may go away before a receiving device with a "late" clock is clocked, but perfect alignment can never be achieved. Thus, all signal paths must have some allowance for slight misalignment. Adding extra delay elements, such as additional buffers, or requiring that there be some combinational logic between clocked elements can help alleviate hold-time problems. However, care must be taken since minimum propagation times are generally not specified for the exact load conditions of a given situation. Minimum propagation times, when specified, are specified with a standard load (50 pF for most advanced BiCMOS/CMOS devices). In many cases, though, the actual load will be much less than the standard load; particularly when devices are used as delay elements. When the load is less than a standard load, the actual minimum propagation delay will be less than the specified value. Package type can also influence minimum propagation delays. The minimum propagation times shown in the data books for some logic families are specifically for devices in DIPs. Devices in packages with less lead inductance, such as LCCs, will have smaller minimum propagation times than shown in the data book.

Of special concern is the situation where a signal originates from a high-speed advanced BiCMOS/CMOS part and is received by a much slower device from one of the older logic families. When signals flow from fast devices to slow devices, it is generally difficult to meet hold-time requirements. Likewise, when technologies are mixed, for example BiCMOS and CMOS or Schottky TTL and CMOS, hold-time violations may occur at temperature extremes since propagation delays and setup and hold times of different technologies may change at different rates with temperature. Several levels of combinational logic between devices may provide adequate hold time, but care must be exercised in all such cases because of the uncertainty of minimum propagation times.

Off-phase clocking as shown in Figure 10.4, which is undesirable in synchronous systems, is one means of transferring data and preventing hold-time violations where clock alignment is poor or where other

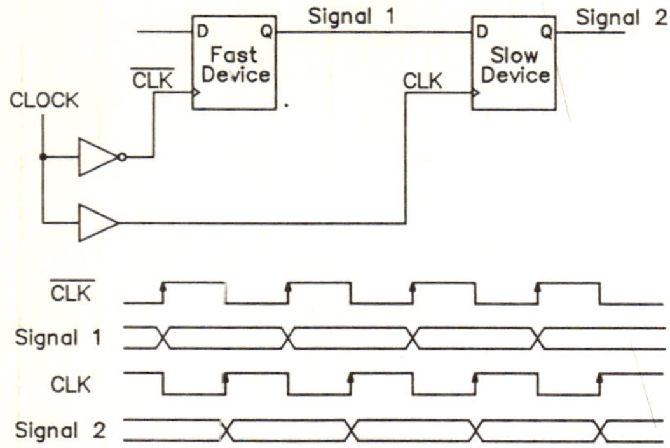

Figure 10.4 Off-phase clocking.

of the above conditions are present. However, off-phase clocking is usually not practical for data transfers where the basic clock rate exceeds 20 MHz. Off-phase clocking more than doubles the effective clock rate since data must be stable within less than one-half of the normal clock period. Ideally off-phase clocking occurs in the middle of the basic clock period, but some timing allowance must be made for setup time and clock misalignment. In most cases where clock frequencies are above 20 MHz, it is difficult to ensure that data will stabilize in less than one-half of a clock period.

10.2.3 Synchronizing asynchronous inputs

Signals coming into synchronous systems are generally asynchronous to the internal system clock and therefore must be synchronized. Most systems must interact with external sources that operate independently with respect to the internal system clock. Signals from independent sources (i.e., with separate clock sources) appear random with respect to the clock in the receiving system. Such signals should be physically isolated from the internal synchronous logic (since the system clock will not act as a filter) and synchronized as soon as possible at one place only. It is important that asynchronous inputs only be synchronized at one point (Figure 10.5). If multiple synchronizers are used, one synchronizer may have significantly different characteristics than another and may detect an input on different clock edges as the internal clock and the external signal shift with respect to one another. Capturing signals at different times in parallel synchronizers can result in portions of a system being out of phase with other portions.

Metastability. At asynchronous interfaces there is always the problem of, and the possibility of, synchronizing devices going into metastable

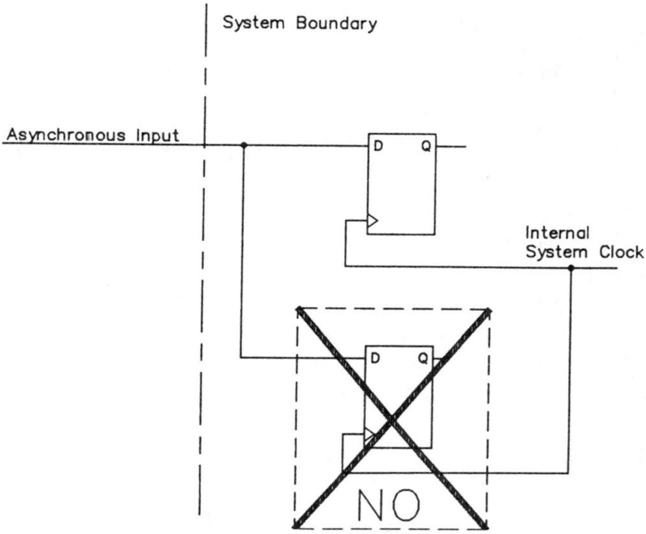

Figure 10.5 Synchronize asynchronous inputs at one place only.

states.[5] Metastable means "in between." When metastable malfunctions occur in digital devices, outputs may linger for some indefinite period in the unknown logic level region (i.e., in between valid logic levels) or they may go to a valid logic level but be unstable and not remain in that state (see Figure 10.6). The possibility of metastable

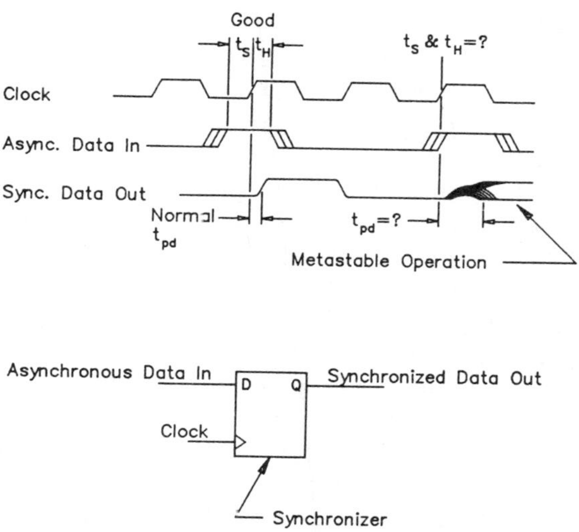

Figure 10.6 Metastable operation of a synchronizing flip-flop.

operation of clocked devices is inherent and impossible to prevent at asynchronous interfaces. All bistable devices, flip-flops, registers, latches, etc., have the possibility of going into a metastable state if their input signals do not meet all the required specifications for input levels or setup and hold times. Since it is impossible to meet such conditions at asynchronous interfaces, care must be taken to ensure that metastable operation of synchronizing devices does not upset system operation. The recovery time for a part that goes into a metastable state can be much longer than the specified propagation delay for the part in a normal operating mode. Thus, at asynchronous interfaces, the system designer needs to know how long to wait before using synchronized data (see Figure 10.6).[6] The higher the system speed, the greater the concern, since the higher the speed, the less time there is for recovery.

In general, faster logic families, such as the advanced BiCMOS/CMOS logic families, have a smaller window of susceptibility to metastable conditions and recover faster from metastable states. Older, slower CMOS logic families, such as HC, have very poor metastable recovery characteristics[6] and should not be used to synchronize data in high-speed systems.[7] Programmable devices such as PALs and PLAs should only be used if their metastable performance is known and meets system requirements (see Reference 8 for metastable data on the 16R4). Programmable devices tend to have a wide range of susceptibility, and there is a tendency to overlook the possibility of asynchronous signals being routed to multiple internal flip-flops and synchronized at multiple places.

The probability of metastable operation occurring depends on the device technology, the clock frequency, and the frequency and phase of the input data. The worst-case data condition is when the data are always changing near the active edge of the clock. That condition occurs when the input data frequency is 0.5 times the clock frequency and the data are aligned so that they are always changing in the critical area where setup and hold times are violated. When a device is triggered into a metastable state, the device technology is one factor that determines how fast the device recovers. Advanced BiCMOS/CMOS devices tend to have a very high probability of recovery after 20 to 25 ns (see Figure 10.7 for typical characteristics). Typically, after 20 to 25 ns the chance of a metastable failure is reduced to 1 in 100 or more years. Note, though, that at 10 to 15 ns the chance of a failure is reduced to approximately one per minute. In most systems, a failure per minute is intolerable. Any failure is undesirable, but metastable failures will occur at asynchronous interfaces. The only recourse is to have the system design allow a long enough recovery time so that the probability of a metastable failure is insignificant. At 20 MHz or

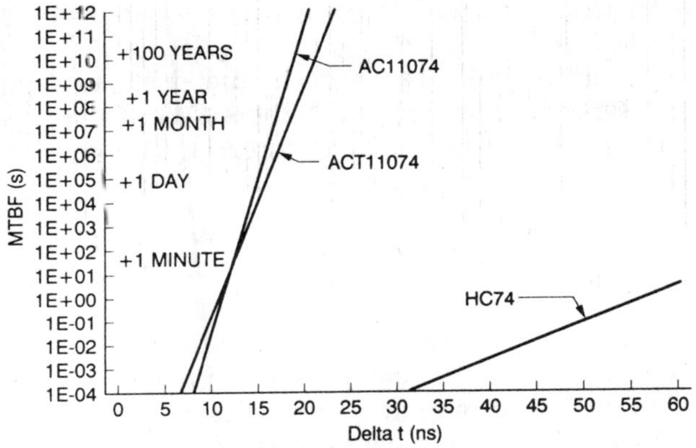

Figure 10.7 Metastability performance of Texas Instruments' advanced CMOS AC11074s and ACT1140s with the metastability performance of HC74s shown for comparison. (*Reprinted by permission of Texas Instruments.*)

higher, however, the recovery time needed for a high probability of not suffering a metastable failure is a significant portion of the clock period. System designers must not forget to add the recovery time to the normal delays in a signal path when determining maximum usable clock frequency of an asynchronous interface. For example, in Figure 10.8 the maximum clock period is the recovery time of *U*1 (not

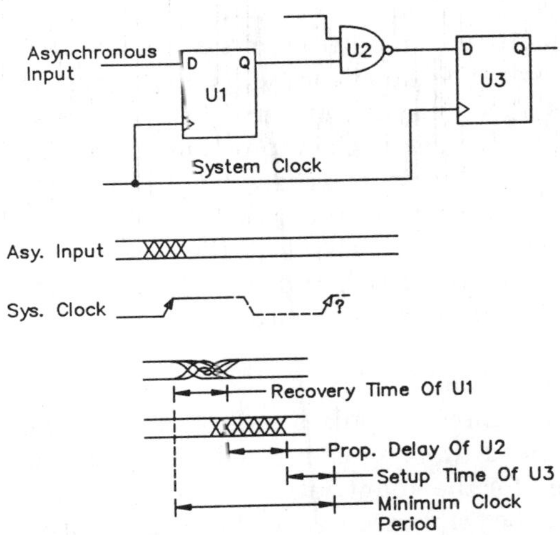

Figure 10.8 The minimum clock period is equal to the recovery time of *U*1 plus the delay of *U*2 plus the setup of *U*3.

the normal maximum propagation time) plus the propagation delay of *U*2 plus the setup time of *U*3.

The recovery time for a given failure rate, data, and clock rate can be calculated if curves of mean time between failures (MTBF) versus recovery time Δt are available for the device of interest. The general equation for the metastable failure rate 1/MTBF is

$$\frac{1}{\mathrm{MTBF}} = K1(f_{\mathrm{data}})(f_{\mathrm{clock}})\epsilon^{-K2\Delta t} \tag{10.1}$$

where *K1* and *K2* describe the metastable characteristics of the synchronizing device and Δt is the time the device is given to come out of the metastable condition. Different manufacturers present metastable data in slightly different forms, so one must be sure when calculating 1/MTBF that the constants, Δt, and the failure rate used in the MTBF equation are compatible. Since metastable characteristics are neither guaranteed nor tested, except in a few isolated cases, it is best to allow enough time to provide a very high probability of recovery. In general, when using advanced BiCMOS/CMOS devices, allowing 25 ns or more for recovery will be sufficient for most applications, but in critical cases or where less recovery time is allocated, the synchronizing devices should be tested for their metastable characteristics (References 7 and 8 show and describe test circuits).

Synchronizing asynchronous buses. Where asynchronous buses are received, it is best to capture and synchronize the strobes or control signals (assuming that the timing of the bus is such that it is possible to do so) rather than try to synchronize an entire bus (see Figure 10.9). Such an arrangement minimizes the chance for metastable operation, since only the control signals must be synchronized. The entire bus is not subjected to the possibility of incorrect setup or hold times. Attempting to synchronize an entire bus widens the window of susceptibility to metastable operation and increases the possibility that one or more bits may go into metastable operation. Not all the receiving devices will be identical; some will be slower, and some will be faster than others, and the clock phasing may be slightly different at each device.

Synchronizing wide pulses. Asynchronous pulses wider than the internal clock period may either need to be synchronized and used with their width unmodified (except for the quantization effect of the internal clock), or they may need to be converted to one clock time pulse at either the leading or trailing edge of the input pulse. If wide pulses only require synchronization, all that is required to synchronize them

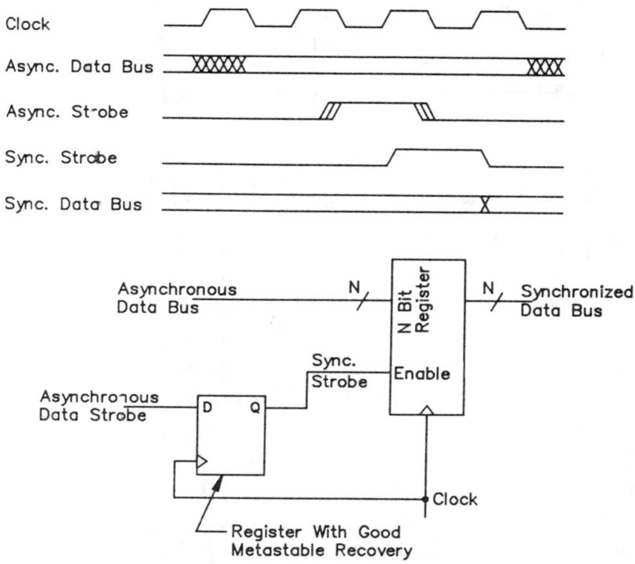

Figure 10.9 It is best to capture and synchronize the data strobe rather than try to synchronize an entire bus.

is to clock them through a resister that has good metastable recovery characteristics. If wide pulses need to be synchronized and converted to a one clock time pulse that follows the leading edge of the input pulse, a leading-edge detector (LED) circuit as shown in Figure 10.10 is typically used. The circuit operates as shown in the timing diagram

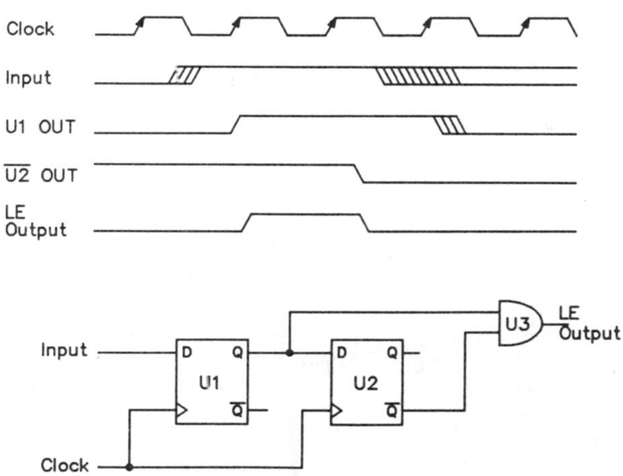

Figure 10.10 Leading-edge detector (LED) and synchronizer circuit.

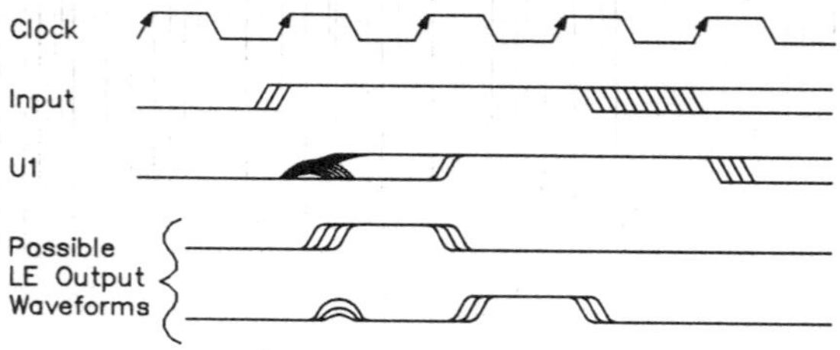

Figure 10.11 Possible LED output waveforms when the asynchronous input changes in the metastable zone of the synchronizing resister.

of Figure 10.10. Register *U*1 in Figure 10.10 must have a metastable MTBF that is sufficient to meet the system reliability requirements at the system clock speed. If the asynchronous input transitions in the metastable zone of *U*1 (near the clock edge), the LED output may glitch because of metastability as shown in Figure 10.11. However, if *U*1 has a sufficiently fast recovery time, the LED output will be stable at the next clock edge. If a clean full clock time wide pulse is required, an additional register must be used between the input signal and the LED circuit to synchronize the signal before it reaches the LED circuit.

A trailing-edge detector (TED) and synchronizer are implemented as shown in Figure 10.12. The TED circuit issues a one clock time pulse during the clock time following the absence of the wide input signal as shown in the timing diagram in Figure 10.12.

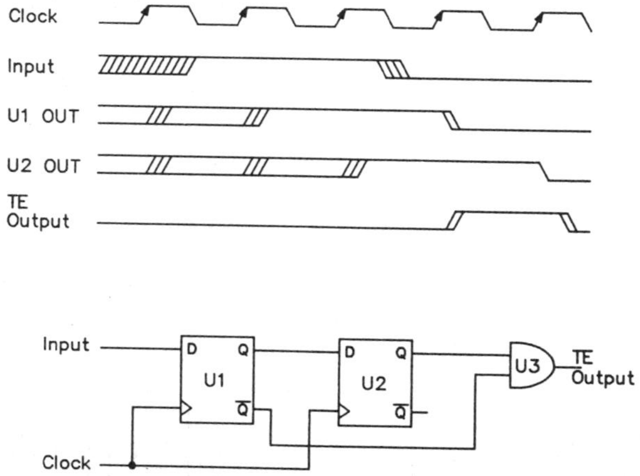

Figure 10.12 Trailing-edge detector (TED) and synchronizer circuit.

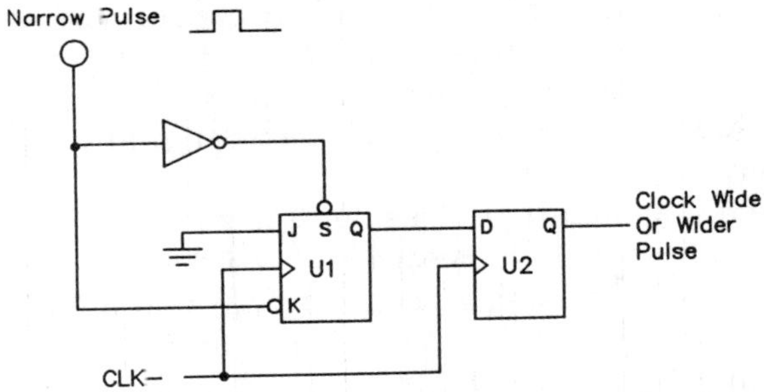

Figure 10.13 Circuit for capturing and synchronizing pulses narrower than the system clock period.

Synchronizing narrow pulses. Asynchronous pulses narrower than the internal clock period cannot be captured and synchronized with a simple clocked register. A narrow pulse may not be present at a clock edge. Thus, some form of asynchronous capture is necessary that can hold the pulse until a clock edge occurs. One circuit that is used to capture and synchronize narrow pulses is shown in Figure 10.13. The circuit shown in Figure 10.13 operates as follows: The presence of a narrow input pulse (but one of sufficient width to be recognized by $U1$) asynchronously forces the output of $U1$ to a "1" where it remains a "1" until the clock edge following the absence of the input pulse. The second register $U2$ captures and synchronizes the changes in the output of $U1$ to the system clock. If the input pulse is wider than the internal clock period, the pulse width is maintained (less quantization variability). If a synchronized one clock time pulse is needed, an LED or a TED circuit can be implemented using $U2$ as the first stage followed by an additional register and gate (see Figures 10.11 and 10.12).

10.3 Devices Incompatible with Synchronous Design

Certain devices are inherently incompatible with synchronous logic design. They include one-shots (multivibrators), transparent latches, and master-slave devices.

10.3.1 One-shots

One-shots should not be used in synchronous systems. One-shot output changes are not clock controlled. Thus, one-shots clearly violate the synchronous design requirement that all signals change relative to the active edge of the system clock. In addition to violating synchronous design requirements, there are several practical reasons for not using one-shots in modern digital systems. One-shot time-out periods are difficult to control or determine accurately, and as a result worst-case time-out limits are generally greatly underestimated. One-shots are also difficult to test. Most automatic testers used for digital hardware cannot test one-shots.

However, there is often a temptation to use one-shots when long time delays or long time-out periods are needed and board space is at a premium. Long time-out periods using standard 4-bit counters may require a number of packages to achieve the same time-out period that can be achieved with a single one-shot (package), a resistor, and a capacitor (most one-shots require an *RC* timing network as shown in Figure 10.14). However, one-shots with long time-out periods tend to have much greater variations in time-out period than is often expected. One-shots with long time-out periods require *RC* timing networks with large-value resistors and capacitors. Yet, large-value resistors and capacitors are generally not available in close tolerances, and they tend to have large tolerance variations with temperature and aging.

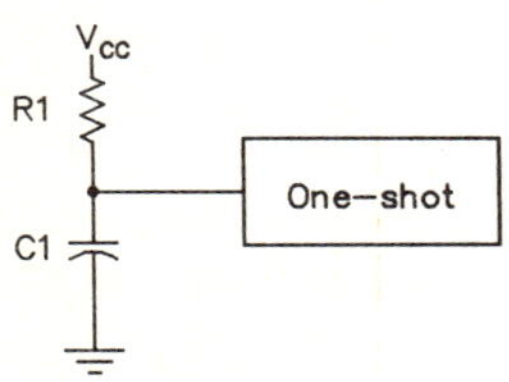

Figure 10.14 Typical one-shot timing network.

Several sources of leakage current exist that contribute to time-out errors. They include capacitor leakage current, board leakage current, and one-shot timing port input current (one-shot input current is technically not a leakage current, but the effect is the same). Worst-case leakage currents for large tantalum capacitors are often in the microampere range. Board leakage depends on environmental conditions and the physical arrangement of the interconnections. It is not uncommon for board impedance to be as low as 100 kΩ in high-humidity conditions and as low as 10 kΩ in some extreme salt-air conditions. One-shot timing port input current is usually not specified, and thus its effect on time-out period is usually unknown. If all the leakage sources are near maximum and a high-value timing resistor is used, time out may never occur. Leakage currents may load the *RC* timing network to the extent that the *RC* network never reaches a level sufficient to trigger the one-shot, and as a result the one-shot never times out (see Figure 10.15).

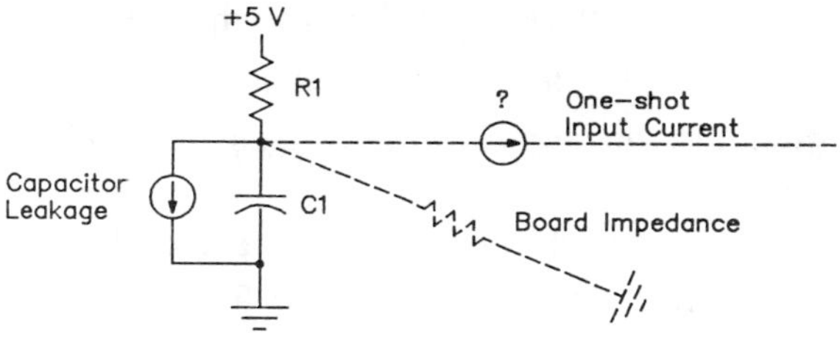

Figure 10.15 Leakage currents and stray board impedance can prevent one-shot time out.

With the possibility that one-shots with long time-out periods may not time out, or if they do time out, the variation in time-out period may be very large, long time out one-shots should be avoided if at all possible. When long time-out periods are needed, there are a number of microprocessor support chips that have long counter chains, and there are a number of long ripple counters (in a single package) in the HC/HCT logic family. Counters are a better solution than one-shots in most applications requiring long time-out periods.

10.3.2 Latches

Latches, depending upon how they are used, may violate the requirement that all signals change relative to the active edge of the clock. Latches also may impose restrictions on latch control signals that are difficult to meet in a synchronous, single-phase clock system. Great care must be taken to ensure that latch control signals are timed so that hold-time violations do not occur at latch inputs or at devices receiving latch outputs.

10.3.3 Master-slave devices

Master-slave devices reduce the amount of quiet time in a system. Master-to-slave transfers are completed on the off-phase edge of the clock which reduces signal settling time by one-half (for a 50 percent duty cycle clock).

10.4 Summary of Synchronous Design Practices

> In synchronous designs, storage elements are only allowed to change states in response to clock transitions on clock inputs.

1. Clocked elements should only be clocked by the system clock.
2. Use only clock signals of a single phase and frequency.
3. All clocked elements should be edge triggered (as opposed to latches or master-slave devices).
4. Clocked devices with TTL input levels should be selected so that all devices clock on the *low*-to-*high* clock transition to maximize the noise margin.
5. Asynchronous presets and resets on clocked elements should never be used for performing operational system logic functions because of their susceptibility to noise.
6. Unclocked feedback paths (such as cross-coupled gates) should not be used because of their susceptibility to being upset by system noise.
7. Never use a counter carry or a decoder output as a clock (such outputs are expected to have spikes).
8. Do not route asynchronous signals to multiple points within a functional unit until they are buffered and synchronized to the internal clock.
9. Monostable multivibrators (one-shots) should not be used.

10.5 References

1. Funk, Richard, and James Nadolski: "Advanced CMOS—Pinouts Are Not the Crucial Factor," *Electronic Engineering Times,* Monday, August 4, 1986, p.33.
2. Mano, Morris M.: *Digital Design,* Prentice-Hall, Englewood Cliffs, N. J., 1984.
3. Fletcher, William I.: *An Engineering Approach to Digital Design,* Prentice-Hall, Englewood Cliffs, N. J., 1980.
4. Hayes, John P.: *Computer Architecture and Organization,* 2d ed., McGraw-Hill, New York, 1988.
5. Kleeman, Lindsay, and Antonio Cantoni: "Metastable Behavior in Digital Systems," *IEEE Design & Test of Computers,* December 1987, pp. 4–19.
6. Chaney, Thomas. J.: "Measured Flip-Flop Responses to Marginal Triggering," *IEEE Trans. on Computers,* Vol C-32, No. 12, December 1983, pp. 1207–1209.
7. *Advanced CMOS Logic Designer's Handbook,* Texas Instruments Inc., Dallas, Tex., 1988, pp. 3-29 to 3-36.
8. *Advanced Micro Devices PAL Device Data Book,* Advanced Micro Devices, Sunnyvale, Calif., 1988, pp. 3-164 to 3-169.

10.6 Bibliography

Fletcher, William I.: *An Engineering Approach to Digital Design,* Prentice-Hall, Englewood Cliffs, N.J., 1980.

Goodrich, Richard: "Pinpointing Metastable Problems Leads to More Reliable Designs," *Communications Systems Equipment Design,* February 1985, pp. 33–35.

Mano, Morris M.: *Digital Design,* Prentice-Hall, Englewood Cliffs, N.J., 1984.

Chapter 11

Worst-Case Timing

At present, most system- and board-level timing analyses must be done by hand. Computer-aided design (CAD) tools that combine logic device timing and pc boards or backpanel interconnection system transmission-line response characteristics exist,[1] but they are not available to most designers. Without sophisticated CAD tools, a thorough timing analysis of complicated systems may be impossible, and without a thorough timing analysis, there is little hope of producing reliable high-speed systems. Thus, high-speed systems must be structured so that a thorough timing analysis is a manageable task by hand. Synchronous design practices help accomplish that objective. In synchronous systems, data flow in an orderly manner from one clocked device to another with signal path propagation requirements clearly defined by the system clock period. In contrast, in asynchronous systems, data tend to flow through long, irregular paths that are dependent on many conditions with the consequence being that a proper timing analysis is usually difficult, or impossible, to achieve.

High-speed BiCMOS/CMOS systems require a careful timing analysis of all signal paths to establish the maximum usable system clock frequency or to establish that critical path delays are compatible with predetermined clock frequencies. Very often unrealizable operating speeds are projected for advanced BiCMOS/CMOS systems. Often no allowance is made for worst-case component delays and little or no allowance for signal wiring propagation delays and transmission-line effects. For a realistic estimate of the operating speed of a system, the worst-case propagation delays of each component in each signal path, plus the physical propagation delays and settling time needed for each interconnection network, must be determined and factored into the projected operating speed. Signal interconnection and settling time delays are a significant part of most signal propagation delay paths when system operating speeds approach 20 MHz or greater. At 20 MHz, interconnection delays average 20 percent of signal delays. At 50 MHz, interconnection delays may average 50 percent or more of the total delay.[2]

11.1 Device Delays

Manufacturers' device specification sheets or data books all contain some timing information, but the information must be interpreted and translated to the actual load and environment conditions of each system application.

11.1.1 Advanced BiCMOS/CMOS timing specifications

Most SSI and MSI advanced BiCMOS/CMOS device data sheets have three columns of timing specifications, typically listing:

1. Minimum, typical, and maximum timing parameters at +25°C
2. Minimum and maximum timing for commercially rated devices
3. Minimum and maximum timing for military rated devices

Traditionally, commercially rated devices are specified for operating conditions of 0 to +70°C and ±5 percent from nominal power-supply limits. However, there is little consistency in the temperature range or power-supply range limits for commercially rated advanced BiCMOS/CMOS parts. Some are specified for the −40 to +85°C which has traditionally been called the industrial range. Some commercially rated advanced CMOS parts are specified with ±10 percent power-supply limits instead of ±5 percent. Specification limits are much more consistent for military rated devices. Most military rated devices are specified for operating conditions of −55 to +125°C and ±10 percent power-supply variations from nominal. The timing information supplied for LSI and VLSI devices is usually less complete than that for SSI and MSI devices; in many cases no typical or minimum data are supplied.

Unfortunately, not all manufacturers use the same load conditions when specifying timing. Most SSI and MSI advanced BiCMOS/CMOS logic devices are specified with a 50-pF load. Many LSI and VLSI devices, such as memories, are specified with a 30- or 35-pF load. The load resistance used to specify timing varies considerably; load resistance does not have a large impact on device timing. However, variations in load resistance make interpreting and comparing timing specifications difficult when differences are small, e.g., when a logic family is claimed to be 1 or 2 ns faster than another family. Load resistors that limit *high* levels when devices are tested to TTL-level test thresholds (1.5 V) tend to improve *high*-to-*low* propagation times. It takes less time to transition a shorter distance (across a smaller voltage difference). If the actual load varies significantly from the load used to specify the device timing, the timing must be adjusted to reflect the actual load conditions. Loading of control signals and local data lines seldom exceeds 50 pF. Thus, in most cases, manufacturers' maximum and minimum specified timing can be used without adjustments since most signals fall into the control signal or local data categories. Buses usually exceed 50 pF, but most buses are long line and thus are not lumped loads and should not be treated as such (they should be treated as transmission lines—see Chapter 7).

11.1.2 Timing parameter adjustments for worst-case conditions

If worst-case timing is not specified, some derating criteria must be applied for worst-case operating conditions and possible process variations. The manufacturer is the best source of information on parameter limits when they are missing and should be contacted for the missing information. However, for those cases where the manufacturer does not have the specific parameter limits needed or is unwilling to supply them, the following sections describe some of the major device variables and lists typical variability rates that can be used to estimate timing parameter limits.

Timing changes due to process variations. Advanced BiCMOS/CMOS device timing changes due to process variations are typically in the range of ±30 to ±50 percent.[3,4] Process variations increase the spread in device timing parameters more than any other single factor, and there is little that the system designer can do to influence process limits. For ICs to be cost-effective and producible, a great deal of processing tolerance is required. System designers are often misled into believing that devices with tighter process tolerances can be purchased, but it is usually neither practical nor cost-effective to do so.

Timing adjustments for temperature. Propagation delays of advanced CMOS devices typically change by a factor of 0.3 percent per degree Celsius[5,6] which corresponds with the rate of change in transconductance g_m and output current of MOSFETs with temperature.[7]

performance systems, it is often best to set the supply voltage at 5.2 V to optimize device speed and to help compensate for ac and dc losses in the power distribution system.

Propagation delays of BiCMOS devices are less sensitive to power-supply variations than are those of CMOS devices. The majority of the change in propagation delay due to power-supply level changes is a result of the change in output drive capability of the output stage. Output drive characteristics of BiCMOS bipolar outputs do change with power-supply voltage, but not as much, nor in as consistent a manner as does the drive of CMOS outputs. Output *high* drive tends to go up with higher power-supply levels, but the turn OFF time of *low* outputs tends to increase. Because of the conflicting trends, propagation delay changes cannot be relied on to track with power-supply voltage changes, but propagation changes due to power-supply changes are not a major factor when BiCMOS devices are used.

Rules of thumb for quick estimates of device timing parameter limits. For a quick estimate of worst-case timing parameters, the following rules of thumb can be used to convert +25°C, +5 V, timing parameters, i.e., propagation delays or setup and hold times, to worst case over temperature and supply limits:[9]

Rules of Thumb for Timing Parameter Conversion Factors
To convert typical +25°C timing parameters to worst case over the commercial temperature range, multiply typical +25°C timing parameters by a factor of 1.5.
To convert typical +25°C timing parameters to worst case over the military temperature range, multiply typical +25°C timing parameters by a factor of 2.
To convert maximum (or minimum) +25°C timing parameters to worst case over the commercial temperature range, multiply maximum (or minimum) +25°C timing parameters by a factor of 1.25.
To convert maximum (or minimum) +25°C timing parameters to worst case over the military temperature range, multiply maximum (or minimum) +25°C timing parameters by a factor of 1.5.

Timing adjustments for actual load conditions. When signal line length is less than the critical line length, line loading appears to be lumped.[10] When lines are less than the critical length, transmission-line effects are not significant, but signal transition and propagation times may be slowed as a result of capacitance. If the capacitance loading is different (which it will usually be) from the value at which the timing is specified, in critical applications the timing must be adjusted to take into consideration the actual loading. Propagation delays for most SSI and MSI advanced BiCMOS/CMOS devices are specified with 50-pF loads. In contrast, propagation delays for most memory devices and many other LSI devices are specified with 30- or 35-pF loads. However, in many situations actual loads will exceed 50 pF, particularly in many bus driving applications. In those cases where loads exceed 50 pF (or 35 pF for memories), some adjustment must be made to the specified output response time of the driving device. In some cases, device manufacturers provide graphs that show the typical change in output response versus load capacitance. For most BiCMOS and CMOS devices, the typical change in output propagation time versus capacitance load for advanced BiCMOS/CMOS devices is in the range of 15 to 25 ps/pF.[3–5]

> The typical change in propagation delay versus load capacitance is 15 to 25 ps/pF for most advanced BiCMOS and CMOS devices.

Worst-case response versus load capacitance is seldom provided. Multiplying typical values by a factor of 1.5 for commercial operation and a factor of 2 for military operation provides a reasonable estimate of worst-case change in response versus load capacitance. The factors of 1.5 and 2 are based on possible device characteristic changes due to process variations and operation at the limits of the temperature and supply voltage range. When minimum propagation delay is of concern, for example to ensure hold time, minimum specified propagation times should be adjusted using the propagation delay versus load capacitance factors.

When delay versus load capacitance is not provided (and a verification of the typical number above is needed), an estimation of the increased delay due to larger-than-specified load capacitance can be determined by Equation (4.14).

$$\frac{dv}{dt} = \frac{I_{OS}}{C}$$

where I_{OS} is the output short-circuit current of the output driving device and C is the increase in load capacitance beyond the specified load value. For BiCMOS devices, or CMOS devices with TTL output levels (those with source follower pull-up output stages), the value of I_{OS} is different for *high* and *low* outputs. In most cases, *low*-to-*high* transition times are slower than those of *high*-to-*low* transitions because *high*-level I_{SC} is less than *low*-level I_{OS} because of the construction of TTL-level output stages. BiCMOS totem-pole outputs have current-limiting pull-up resistors, and most CMOS devices that have TTL output levels use *n*-channel source followers for output pull-ups. Source follower circuits tend to have less *high* output drive than conventional switched CMOS outputs.

Load capacitance. Total signal load capacitance must be established to determine signal response. Total signal capacitance consists of the input or output capacitance of each device connected to the line plus the line capacitance.

Device input or output capacitance is usually included on data sheets. If it is not, capacitance loading due to standard SSI or MSI device connections can be estimated using the following values:

- 5 pF for inputs
- 7 pF for outputs
- 15 pF for bidirectional ports

However, the input-output capacitance of some parts will vary considerably for the above, so data sheets should be consulted to ensure that the proper values are used. Input and output capacitance of LSI and VLSI devices and of devices in large packages may be much larger than the values listed above.

Typical interconnection (wiring) capacitances are as follows:

- 2 to 4 pF/in of pc board track
- 1 to 2 pF/in for welded-wire or wire-wrap wire

Actual interconnection capacitance should be used in timing calculations when available. However, in many applications, the actual capacitance of a given signal interconnection is difficult to determine because there are great variations in conductor and ground reference spacing. Wire spacing relative to reference planes or other conductors on welded-wire or wire-wrap boards varies a great deal. Even where conductor topology is known, it is difficult to determine exact line capacitance because of the interaction of nearby conductors. Thus, most line capacitance calculations are made assuming only an isolated conductor and a nearby reference plane.

For most applications, lumped loading due to interconnection (wiring) capacitance can be estimated using the following values:

- 3 pF/in of pc board track
- 1.5 pF/in for welded-wire or wire-wrap wire

Guidance for calculating more exact interconnection capacitance values can be found in a number of the references covering transmission lines listed at the end of Chapter 7.

Timing adjustments for simultaneous switching. Some degradation of propagation delay normally occurs in multiple-output devices when several outputs switch simultaneously. Most data sheet propagation times are for only one output switching. Propagation delay is typically increased by 0.2 to 0.3 ns per simultaneously switched output.[11]

> Propagation delay increases approximately 0.2 ns per simultaneously switched output for multiple-output advanced BiCMOS and CMOS devices.

However, many factors influence the actual change. Load conditions, device technology and inherent speed, internal and external noise levels, the efficiency of the package and external power and ground distribution, and decoupling all impact speed degradation. Thus, 0.2- to 0.3-ns degradation per switched output should only be used as a rough guideline. Actual system degradation may be much worse. Octal and wider parts may have serious degradation if all devices switch at once. In time critical applications, system designers need specific data for the actual devices they are using.

Timing effects of package style. Package style and lead inductance influence the response time of devices. Surface-mount packages with short leads tend to respond faster than devices in large DIPs. Some data books specify (in very fine print) that the listed times are for devices in DIPs and that the minimum times shown may be shorter for devices in other package styles. Quantitative data for package style effects on device timing are not readily available. Nonetheless, the system designer should be aware that package style does influence timing parameter limits.

11.1.3 Caution—beware of F_{MAX}

The parameter F_{MAX}, which is listed as the maximum toggle rate or maximum clock rate on data sheets for clocked devices such as counters, flip-flops, and shift registers, should never be used as an indication of the useful speed of a device.[12] It is a measure of what might be achieved with an individual part under ideal conditions with no restrictions on input pulse widths or load conditions.[12] Since most digital systems must operate under conditions that vary greatly from ideal, and since devices must communicate with other devices to serve a useful purpose, F_{MAX} is of little use for actual system timing. Actual signal path propagation times must be used to determine the maximum operating speed of systems, not F_{MAX}.

11.2 Circuit Board Interconnection Delays

Interconnection delays consist of conductor propagation delay and signal settling time delay. Signal propagation delay is the time required for a signal to traverse a physical conductor. Signal settling time is the time required for a signal to transition to a proper logic level or to settle to a proper logic level if it overshoots and rings. Signal settling time is most often of concern when signal lines are long, i.e., where lines behave as transmission lines.

11.2.1 Interconnecting line propagation delay

Calculating interconnection delays due to load and transmission-line effects is not straightforward, yet both effects may add significant time to signal delays. For pc boards the intrinsic propagation delay t_{pd} is approximately 2 ns/ft, and for welded-wire or wire-wrap boards t_{pd} is approximately 1.5 ns/ft. For extremely time critical circuits or very long lines (several feet) more exact intrinsic propagation delay times should be calculated (see Chapter 7 and the various transmission-line references listed at the end of Chapter 7). However, for most applications the above per-unit-length intrinsic signal propagation values are sufficient. Distributed loads along a line modify the propagation time of signals. The actual delay is called the effective delay t'_{pd}. Actual conductor delay is equal to the per-unit-length effective propagation delay t'_{pd} of electrical energy in the interconnection media multiplied by the length of the interconnection.

11.2.2 Line propagation delay with distributed loads

Lines with distributed loads slow the propagation of signals by a factor equal to

$$t_{pd}(\text{slow down factor}) = \sqrt{1 + \frac{C_{\text{LOAD}}}{C_{\text{LINE}}}} \tag{11.1}$$

Thus, where loads are distributed, the actual propagation time t'_{pd} is given by Equation (6.4)

$$t'_{pd} = t_{pd}\sqrt{1 + \frac{C_{\text{LOAD}}}{C_{\text{LINE}}}}$$

When lines have distributed loads, t'_{pd} should be used in Equation (7.1) to determine the critical line length.

11.2.3 Line delay due to transmission-line effects

Transmission-line effects become of concern in high-speed BiCMOS/CMOS systems when line lengths exceed the critical line length (see

Chapter 7). The common definition of the critical line length is given by Equation (7.1)

$$\text{Critical line length} = \frac{1}{2}\frac{t_r}{t'_{\text{pd}}}$$

where t_r is the rise time of the driving source (20 to 80 percent) and t'_{pd} is the actual loaded propagation delay of the line.

If a line does not fall into the critical line category, i.e., it is shorter than the critical length, the line delay used in the timing budget is the one-way propagation delay of the line (t'_{pd} times the line length). If a line exceeds the critical line length, some allowance for transmission-line effects must be incorporated into the timing analysis unless the line is terminated in such a way that the line is stable at all loads after a one-way delay. For the general category of control and data signals, it is impractical to terminate all lines. Thus, additional time beyond the one-way line delay must be allowed for most signals to settle to acceptable logic levels. For most lines in excess of the critical line length, allowing five line delays will provide adequate time for transmission-line reflections to subside (see Chapter 7).

> A rule of thumb that will expedite most timing analysis is: Allow five line delays for unterminated lines to settle.

Assuming five line delays for long unterminated lines, which is a safe assumption in most cases, simplifies and expedites the timing analysis task. However, it is difficult to generalize very complex phenomena. Critical situations must be carefully analyzed.

CAUTION: An allowance for any additional signal length that might be added to signal paths during system test or troubleshooting should be included in the worst-case timing analysis where appropriate. One example is the added circuit board-to-motherboard signal length that occurs when extender boards are used to extend circuit boards for troubleshooting.

11.3 Backpanel Interconnection Delays

The same device and line delay considerations must be addressed for backpanel or motherboard signal connections as for circuit board interconnections (see Section 11.2). However, most backpanel connections must be treated as transmission lines. Most backpanel interconnections will exceed the critical line length.[13] In addition to normal backpanel signal path interconnection distance, backpanel signal interconnection length and delay calculations must include an allowance for extender cards (extender cards allow circuit boards to be extended out of a chassis so that test equipment can be connected for troubleshooting). All systems will have faults, so it is essential that board-to-board signals have adequate timing margins so that extender cards can be used for troubleshooting. The extra length of an extender board is certain to increase backpanel interconnections beyond the critical line length when advanced BiCMOS/CMOS signals are routed between boards. Thus, additional time beyond the one-way line delay must be allowed for unterminated backpanel signals to settle to adequate logic levels, or signals must be terminated to reduce the uncertainty of the settling time. For most backpanel lines, allowing five line delays will provide adequate time for transmission-line reflections to subside (see Chapter 7). It will generally expedite the timing analysis to make the assumption that all backpanel interconnections will require five line delays to settle. If system timing requirements are met with five line delays, then no further analysis is required. If five line delays are unacceptable, then further analysis is required. It may be determined that the signal or signals of interest will settle faster, or steps, such as terminating lines, may be required to improve settling time.

11.4 Unit-to-Unit Interconnection Timing

Timing analysis of signal interconnections between remote units (i.e., units that do not have a common backpanel or motherboard) must allow for transmission-line effects. In most cases, unit-to-unit signal paths will exceed the criteria for critical line length, regardless of the logic family being applied. Thus, signals traversing unit boundaries must be properly terminated, or time must be allowed for signals to settle to valid logic levels (see Chapter 7).

Most unit-to-unit interconnections consist of some form of cabling using twisted-pair or shielded twisted-pair lines, coaxial cable, or flat cable. Cable manufacturers' specifications should be consulted for cable propagation times.[14] *Caution:* Cables with polyvinyl chloride (PVC) dielectrics may have very long per-unit-length propagation delays—on the order of 3.5 ns/ft.

Synchronous transfer of data between remote units is generally not practical because of the difficulty of maintaining clock alignment which means extra time must be allotted for synchronization. If synchronous transfers are planned, some allowance for clock alignment uncertainty must be included in the timing analysis.

11.5 Worst-Case Timing Examples

The signal path shown in Figure 11.1 is representative of signal paths commonly encountered in large synchronous systems where data and control signals flow between boards mounted on a common motherboard or backpanel. A signal originates at the output of a clocked device (flip-flop *U*1), proceeds through buffer *U*2 and off circuit board 1 on to the motherboard. It then goes to circuit board 2, where it goes through buffer *U*3 and then goes to a control input on a clocked device (flip-flop *U*4). Large systems typically require a number of similar data paths for board-to-board data transfers. In some cases, the output buffer *U*2 and receiving buffer *U*3 may not be needed, but they are required in many cases. Even if they are not required from a functional standpoint, they may be needed to ensure adequate hold time at the receiving flip-flop *U*4. In high-speed systems, it is difficult to keep clock signals on separate boards sufficiently aligned to make reliable direct register-to-register transfers. Buffers or other delay elements may be needed to ensure hold time.

In the example circuit (Figure 11.1), the worst-case signal propagation time is determined by the time required for a signal to travel from flip-flop *U*1 on board 1 to flip-flop *U*4 on board 2. The incremental signal path delays are as follows:

1. The maximum propagation delay from the clock to the output of flip-flop *U*1
2. The time required for the signal to propagate from the output of flip-flop *U*1 to the input of buffer *U*2 and the time required for the signal to stabilize at the desired logic level
3. The maximum propagation time of the output buffer *U*2
4. The time required for the signal to traverse the backpanel (mother-

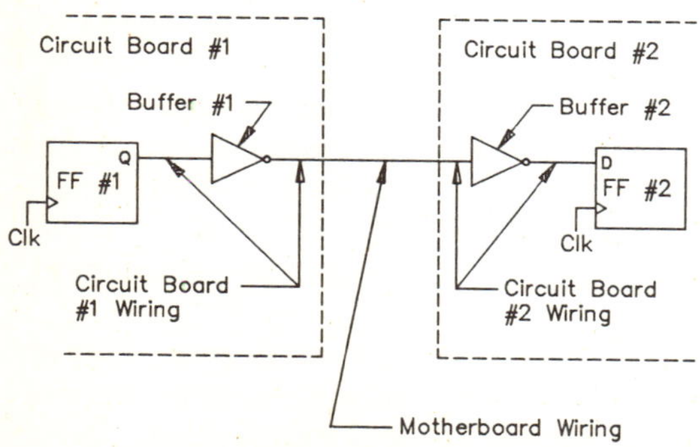

Figure 11.1 Typical board-to-board signal path.

board) and to stabilize at the desired logic level at the input of the receiving buffer $U3$ on board 2 (transmission-line effects, such as ringing, must be considered)

5. The maximum propagation time of buffer $U3$ on board 2
6. The time required for the signal to propagate from the output buffer $U3$ to the control input of the receiving flip-flop $U4$ and to stabilize at the transmitted logic level
7. The setup time required by flip-flop $U4$

Tables 11.1 through 11.6 show the worst-case signal path delay for several advanced BiCMOS and CMOS logic families when used in the example configuration. The tables are useful for comparing logic family speeds and as a guide for maximum single clock time signal transfer rates. The flip-flop and buffer propagation times listed in these tables are for the 374 register and the 240 buffer. The propagation delay and setup times listed are worst-case manufacturers' data sheet specifications for operation over the military temperature range (−55 to +125°C) and power-supply variation limits of ±10 percent from nominal 5 V (except Table 11.1 which is based on the −40 to +85°C temperature range specifications). The use of commercial device specifica-

TABLE 11.1 Example Circuit Worst-Case Delay with Motorola 74BC240s and 74BC374s[15]

Signal path delays	Time, ns
1. Flip-flop $U1$ propagation delay Max value of t_{PHL} or t_{PLH} over T_A = − 40 to +85°C	10.0
2. Circuit board 1 track delay Output of flip-flop $U1$ to input of buffer $U2$ track delay	1.0
3. Output buffer $U2$ propagation delay Max value of t_{PHL} or t_{PLH} over T_A = − 40 to +85°C	7.5
4. Backpanel propagation delay Assume a 1.5-ft interconnection: 5 × 2 ns/ft × 1.5 ft =	15.0
5. Input buffer $U3$ propagation delay Max value of t_{PHL} or t_{PLH} over T_A = − 40 to +85°C	7.5
6. Circuit board 2 track delay Output of receiving buffer $U3$ to input of flip-flop $U4$	1.0
7. Flip-flop $U4$ setup time Largest value of minimum t_{SH} or t_{SL} over T_A = − 55 to +125°C	2.0
Total propagation delay	44.4

TABLE 11.2 Example Circuit Worst-Case Delay with TI BiCMOS 54BCT240s and 54BCT374s[16]

Signal path delays	Time, ns
1. Flip-flop $U1$ propagation delay Max value of t_{PHL} or t_{PLH} over $T_A = -55$ to $+125°C$	11.6
2. Circuit board 1 track delay Output of flip-flop $U1$ to input of buffer $U2$ track delay	1.0
3. Output buffer $U1$ propagation delay Max value of t_{PHL} or t_{PLH} over $T_A = -55$ to $+125°C$	6.4
4. Backpanel propagation delay Assume a 1.5-ft interconnection: 5 × 2 ns/ft × 1.5 ft =	15.0
5. Input buffer $U3$ propagation delay Max value of t_{PHL} or t_{PLH} over $T_A = -55$ to $+125°C$	6.4
6. Circuit board 2 track delay Output of receiving buffer $U3$ to input of flip-flop $U4$	1.0
7. Flip-flop $U4$ setup time Largest value of minimum t_{SH} or t_{SL} over $T_A = -55$ to $+125°C$	6.5
Total propagation delay	47.9

TABLE 11.3 Example Circuit Worst-Case Delay with IDT FCT-A 54FCT240As and 54FCT374As[17]

Signal path delays	Time, ns
1. Flip-flop $U1$ propagation delay Max value of t_{PHL} or t_{PLH} over $T_A = -55$ to $+125°C$	7.2
2. Circuit board 1 track delay Output of flip-flop $U1$ to input of buffer $U2$ track delay	1.0
3. Output buffer $U2$ propagation delay Max value of t_{PHL} or t_{PLH} over $T_A = -55$ to $+125°C$	5.1
4. Backpanel propagation delay Assume a 1.5-ft interconnection: 5 × 2 ns/ft × 1.5 ft =	15.0
5. Input buffer $U3$ propagation delay Max value of t_{PHL} or t_{PLH} over $T_A = -55$ to $+125°C$	5.1
6. Circuit board 2 track delay Output of receiving buffer $U3$ to input of flip-flop $U4$	1.0
7. Flip-flop $U4$ setup time Largest value of minimum t_{SH} or t_{SL} over $T_A = -55$ to $+125°C$	2.0
Total propagation delay	36.4

TABLE 11.4 Example Circuit Worst-Case Delay with GENERIC AC 54AC240s and 54AC374s

Signal path delays	Time, ns
1. Flip-flop $U1$ propagation delay Max value of t_{PHL} or t_{PLH} over T_A = − 55 to +125°C	11.0
2. Circuit board 1 track delay Output of flip-flop $U1$ to input of buffer $U2$ track delay	1.0
3. Output buffer $U2$ propagation delay Max value of t_{PHL} or t_{PLH} over T_A = − 55 to +125°C	8.0
4. Backpanel propagation delay Assume a 1.5-ft interconnection: 5 × 2 ns/ft × 1.5 ft =	15.0
5. Input buffer $U3$ propagation delay Max value of t_{PHL} or t_{PLH} over T_A = − 55 to +125°C	8.0
6. Circuit board 2 track delay Output of receiving buffer $U3$ to input of flip-flop $U4$	1.0
7. Flip-flop $U4$ setup time Largest value of minimum t_{SH} or t_{SL} over T_A = − 55 to +125°C	2.0
Total propagation delay	46.0

TABLE 11.5 Example Circuit Worst-Case Delay with TI ACL 54AC11240s and 54AC11374s[18]

Signal path delays	Time, ns
1. Flip-flop $U1$ propagation delay Max value of t_{PHL} or t_{PLH} over T_A = − 55 to +125°C	10.9
2. Circuit board 1 track delay Output of flip-flop $U1$ to input of buffer $U2$ track delay	1.0
3. Output buffer $U2$ propagation delay Max value of t_{PHL} or t_{PLH} over T_A = − 55 to +125°C	9.0
4. Backpanel propagation delay Assume a 1.5-ft interconnection: 5 × 2 ns/ft × 1.5 ft =	15.0
5. Input buffer $U3$ propagation delay Max value of t_{PHL} or t_{PLH} over T_A = − 55 to +125°C	9.0
6. Circuit board 2 track delay Output of receiving buffer $U3$ to input of flip-flop $U4$	1.0
7. Flip-flop $U4$ setup time Largest value of minimum t_{SH} or t_{SL} over T_A = − 55 to +125°C	2.5
Total propagation delay	48.4

TABLE 11.6 Example Circuit Worst-Case Delay with FACT 54AC240s and 54AC374s[4]

Signal path delays	Time, ns
1. Flip-flop $U1$ propagation delay Max value of t_{PHL} or t_{PLH} over $T_A = -55$ to $+125°C$	12.0
2. Circuit board 1 track delay Output of flip-flop $U1$ to input of buffer $U2$ track delay	1.0
3. Output buffer $U2$ propagation delay Max value of t_{PHL} or t_{PLH} over $T_A = -55$ to $+125°C$	8.5
4. Backpanel propagation delay Assume a 1.5-ft interconnection: 5 × 2 ns/ft × 1.5 ft =	15.0
5. Input buffer $U3$ propagation delay Max value of t_{PHL} or t_{PLH} over $T_A = -55$ to $+125°C$	8.5
6. Circuit board 2 track delay Output of receiving buffer $U3$ to input of flip-flop $U4$	1.0
7. Flip-flop $U4$ setup time Largest value of minimum t_{SH} or t_{SL} over $T_A = -55$ to $+125°C$	5.0
Total propagation delay	51.0

tions would give slightly more optimistic results, but the objective is to show worst-case limits.

The interconnection delay calculations are based on pc board 1 and 2 track lengths of 6 in and a backpanel plus extender card interconnection length of 1.5 ft. Two nanoseconds per foot is used for board and backpanel signal track propagation delays. It is assumed that the two signal interconnections on the boards do not ring and are stable after a one-way line propagation delay. It is assumed that the interconnection between boards does ring and that five line delays are required for the backpanel interconnection to settle. All on-board signal capacitance loads are assumed to be 50 pF.

Summary of times listed in Tables 11.1 through 11.6. The path delays for the various advanced BiCMOS and CMOS logic families listed in Tables 11.1 through 11.6 are not a great deal different. Most are in the 40- to 50-ns range. Only the FCT-A family is below 40 ns. Since the circuitry and interconnections used represent the minimal circuitry needed for board-to-board data transfers, the resulting total signal path propagation times shown in the tables are indicative of the maximum clock frequency that could be safely used for single clock time data transfers in systems with similar signal paths (see Table 11.7).

The maximum clock frequencies listed in Table 11.7 represent the

TABLE 11.7 Maximum System Clock Frequency for Listed Logic Families

Logic family	Maximum signal path delay, ns	Maximum system clock frequency, MHz
BC	44.4	22.5
BCT	47.9	20.9
FCT-A	36.4	27.5
AC	46.0	21.7
ACL	48.4	20.7
FACT	51.0	19.6

best-case system clock speed for the listed logic families assuming all data transfers must occur in one clock time. To build systems with the logic families listed that will operate reliably with a higher clock frequency, signal path lengths must be reduced, or all board-to-board signals must be terminated to reduce settling times. Some may argue that the assumptions made are too pessimistic and that all elements in a delay chain will not exhibit worst-case propagation delays simultaneously, and that is normally true. However, if worst-case timing numbers are used for all elements, a system will have some margin to compensate for those items that are overlooked, and there are always some of those.

The maximum clock frequencies listed in Table 11.7 are much below the frequencies being projected for advanced BiCMOS and CMOS systems. Why the discrepancy? What is the upper clock frequency limit for BiCMOS and CMOS systems? Possible clock frequency depends on system size and architecture. Small compact systems where all lines are short can achieve higher operating speeds than large systems with long interconnections. It is possible to operate small compact groups of advanced BiCMOS/CMOS devices in the 40- to 50-MHz range, but as Tables 11.1 through 11.6 illustrate, 20 to 25 MHz, depending upon the logic family, is near the limit for point-to-point single clock time data transfers in large systems. Actual systems tend to have much more complex data paths, such as buses, which will require longer transfer times. Interconnection delays limit the operating speed of large systems.

11.6 Signal Timing Summary

A great deal of care and attention to details must be given to system timing when advanced BiCMOS/CMOS devices are used.[1] The optimistic typical timing specifications prominently displayed on the first page of most data sheets are of little value and should be disregarded. The parameter F_{MAX}, which is listed as the maximum toggle rate or maximum clock rate on data sheets for clocked devices such as counters, flip-flops, and shift registers, should never be used as an indication of the useful speed of a device.[12] In all high-speed applications, actual in-circuit worst-case device timing parameters must be used to establish system timing limits. System timing must allow for signal interconnection delays since interconnection delays are generally a major portion of most signal delays in high-speed systems.

Carefully follow synchronous design practices so that system timing requirements can be understood and analyzed. Structure the design so that a measure of actual operating speed can be determined. Provide a method for inserting a variable-frequency test clock so that actual system performance can be determined during development and so that individual unit performance can be verified during production. Under ideal conditions (i.e., +25°C and +5 V), a well-designed system with the margin to allow for worst-case military operating conditions and degradation with time should operate at near twice the design speed. Systems with less than 20 percent speed margin under ideal conditions, even if intended for benign conditions, present a high level of risk because of the variability of parts, actual operating conditions, and degradation of parts with age.

11.7 Signal Timing Checklist

1. Use worst-case device and interconnection timing parameters based on the required system operating conditions (i.e., temperature, load, and supply voltage levels). Never base a design on typical device timing information.
2. Adjust manufacturer's maximum timing values for load capacitance in excess of that at which the timing is specified. Adjust minimum timing values when the load is less than the specified load. Use manufacturer's timing adjustment guidelines or estimate change in delay for advanced BiCMOS and CMOS devices using 20 ps/pF × ΔC where ΔC is the actual load minus the specified load.
3. Signal line capacitance needed for timing calculations is the sum of the input and output capacitance of all devices connected to the line plus the line capacitance. Actual device input or output capacitance should be used when available. If actual input-output capacitance is not available, for standard SSI and MSI parts use
   a. 5 pF per input
   b. 7 pF per output
   c. 15 pF per bidirectional port
   For VLSI parts such as PLDs and gate arrays, manufacturer's data must be consulted. Estimate line capacitance using

   a. 1.5 pF/in for welded-wire and wire-wrap interconnections
   b. 3 pF/in for pc board traces
4. To estimate unloaded interconnection propagation delays use
   a. 1.5 ns/ft for welded-wire and wire-wrap interconnections
   b. 2 ns/ft for pc board traces

   For cables or twisted-pair lines, consult the manufacturer's data.

   Caution: Some PVC insulated cables have very slow propagation times (greater than 3 ns/ft).
5. Take into account the slowing effect of distributed loads on line propagation delay. As a very rough rule of thumb, allow an additional 0.15 ns per load.
6. Allow for the additional length of extender boards or test cables in the timing analysis when appropriate (e.g., board-to-board signals).
7. Allow five line delays for unterminated long lines to settle.
8. Terminate long signal lines (greater than the critical length) that have critical timing or waveshape requirements. In the timing analysis
   a. Allow one line delay for load-terminated lines.
   b. Allow two line delays for source-terminated lines.

11.8 References

1. Nass, Richard: "PC-Board Speeds Skyrocket," *Electronic Design,* September 28, 1989, pp. 31–38.
2. Meredith, Mike: "Analyzing Interconnection Timing," *Electronic Engineering Times,* September 11, 1989, p. T6.
3. *Motorola FACT Data,* Motorola Inc., Phoenix, Ariz., 1988.
4. *FACT—Advanced CMOS Logic Databook,* National Semiconductor Corp., Santa Clara, Calif., 1989.
5. *Advanced CMOS Logic Data Book,* VTC Inc., Bloomington, Minn., 1988.
6. *1.5-Micron Compacted Array Technology Databook,* LSI Logic Corp., Milpitas, Calif., 1987.
7. Frederiksen, Thomas M.: *Intuitive CMOS Electronics,* McGraw-Hill, New York, 1989.
8. *FAST—Advanced Schottky TTL Logic Databook,* National Semiconductor Corp., South Portland, Maine, 1988.
9. Buchanan, James E.: *CMOS/TTL Digital Systems Design,* McGraw-Hill, New York, 1990.
10. Royle, David: "Transmission Lines and Interconnections," Part One, *EDN,* June 23, 1988, pp. 131–136.
11. *Bus Interface Products 1988 Data Book,* Advanced Micro Devices Inc., Sunnyvale, Calif., 1987.
12. *Bipolar Microprocessor Logic and Interface Data Book,* Advanced Micro Devices Inc., Sunnyvale, Calif., 1981.
13. Royle, David: "Transmission Lines and Interconnections," Part Two, *EDN,* June 23, 1988, pp. 143–150.
14. Crouch, Ronald A.: "Choose Cable with Care to Optimize System Design," *EDN,* November 5, 1978, pp. 113–116.

15. *Motorola Bi-CMOS Logic Data,* Motorola Inc., Phoenix, Ariz., 1989.
16. *BiCMOS Bus Interface Logic,* Texas Instruments Inc., Dallas, Tex., 1988.
17. *High Performance CMOS Data Book Supplement 1989,* Integrated Device Technology Inc., Santa Clara, Calif., 1989.
18. *Advanced CMOS Logic Data Book,* Texas Instruments Inc., Dallas, Tex., 1988.

Chapter

12

System Initialization and Low-Voltage Sensing

Most digital systems need some means of initialization to a known starting condition when power is turned ON, following low-voltage transients, and for test. Many systems also need some means of preventing unsafe or possibly destructive conditions, such as bus contention or random outputs, and preventing false writes to nonvolatile memories as power turns ON or OFF and during power transients.[1] Overvoltage needs to be sensed to protect logic circuits from damage, and low or undervoltage conditions need to be sensed to prevent erroneous operation.

In most systems, overvoltage and undervoltage sensing circuitry is located in the power supply, and the power supply provides a "volts good" signal to the system. The volts good signal is used to initialize the system at power turn ON or to shut down the system under abnormal voltage conditions. If the power supply does not have circuitry to generate a volts good signal, an equivalent signal must be generated elsewhere. In most applications, the volts good signal is ORed with software-generated reset signals, hardware time-out signals, and in some cases manual reset signals to form a "master reset" signal. The master reset signal is buffered and distributed to each unit, board, etc., in the system that requires a reset, i.e., initialization,* signal.

A central master reset signal is needed to ensure that all sections of a system are initialized at the same time and under the same condi-

*Reset and initialization are often used in an interchangeable manner. In general, initialization is the action required which is accomplished with a reset signal.

tions. Distributed reset signal generators, particularly those implemented with *RC* networks, should not be used because of the uncertainty of the operation of *RC*-generated resets. Distributed reset signal generators introduce the possibility that sections of a system may not be initialized or may not be initialized at the same time.

The master reset signal, controlled by the volts good signal and any required time-out counters, should be asserted (i.e., applied) when system power is turned ON or when the supply voltage V_{cc} falls below the actual level at which the logic devices used in the system no longer operate. When the master reset signal is applied, it should remain ON until the system V_{cc} stabilizes at the proper operating level. During power turn OFF or during low-voltage transients the master reset signal should be applied immediately to prevent false outputs or random conditions from occurring. In most system applications, it is not practical to set the trip point (threshold) of the voltage-level sensing circuit (i.e., the volts good circuit) at exactly the minimum rated digital logic device supply voltage level; some allowance for system noise and an allowance for inherent accuracy limitations of practical level sensing circuits is needed. In most cases, it is acceptable to set the trip point 0.25 to 0.5 V below the rated low power-supply operating level of the system logic components with the highest minimum rating. Some allowance is needed to avoid excessive tripping due to noise and to provide an allowance for error in the level sensing circuitry. Most BiCMOS/CMOS systems designed with reasonable safety margins will continue to operate with supply levels slightly below rated minimum levels. However, when a system must either operate properly or be shut down to prevent hazardous outputs or damaging internal conditions when low voltage occurs, the reset circuit trip point must be set above the point at which the system stops operating.

The recommended low V_{cc} limit is 4.5 V for most BiCMOS and CMOS devices with TTL input levels for both commercially rated devices and for military rated devices (the low limit is 4.75 for most commercially rated TTL devices). However, most BiCMOS or CMOS devices with TTL inputs will operate at much lower voltages, but device input-output levels may not meet TTL specifications and speed is reduced. Most advanced CMOS devices with CMOS input levels (BiCMOS SSI or MSI devices are typically not available with CMOS input levels) have a rated low supply operating limit of either 2 or 3 V, but device speed decreases significantly at reduced supply levels. Thus, it is best to sense a minimum supply level well above the allowable low supply limit. Most systems will consist of a mixture of TTL, CMOS, and BiCMOS. Thus, in most cases the system reset signal generator threshold should be set near the low supply operating limit for the TTL devices or the BiCMOS/CMOS devices with TTL levels used in the system.

12.1 Initialization Signal Generation

A volts good signal used to initialize or reset a digital system should be generated using an analog voltage-level sensing circuit—not an *RC* circuit. Simple *RC* networks should not be used to initialize or reset systems because of the uncertainty of their delay and the unpredictability of their response to power-supply transients or power-supply turn ON or OFF rates. Simple *RC* circuits may operate as expected in the lab, but when exposed to actual system power turn ON or OFF rates, they may not function as expected; actual system power-supply turn ON rates tend to be much slower than those of breadboard setups or test station power turn ON rates. If the power supply turns ON very slowly, simple *RC* delay circuits may not reset the connected registers, etc. Furthermore, most automatic test equipment cannot test *RC*-generated resets that time out as a function of the application of V_{cc} to the unit under test. In contrast to crude *RC* networks, analog comparator circuits can be designed to accurately sense a predefined V_{cc} level.

Figure 12.1 shows an analog comparator network configured to generate a volts good signal when the supply voltage V_{cc} is above a predefined level.[2,3] The comparator (Figure 12.1) operates as follows: The LM193 IC analog comparator *U*1 compares the 1.22-V reference input from the LM113 voltage reference to the output of the divider network *R*1 and *R*2.[4] When the voltage level out of the divider network is

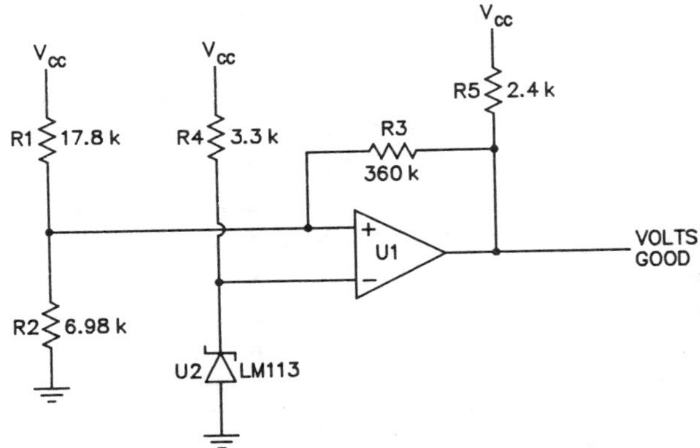

Figure 12.1 Volts good signal generator implemented with an analog comparator.

greater than the reference, the output of the comparator is a *high* level (pulled up to V_{cc} by $R5$). When the output of the divider network is lower than the reference, the output of the comparator is a *low* level (near ground). The selection of the values of $R1$ and $R2$ controls the level of the sensed voltage, V_{cc} in this case, at which the comparator output switches. With the resistor values shown in Figure 12.1, the comparator switches states at about 4.3 V. Resistor $R3$ provides a small amount of hysteresis to prevent oscillation of the comparator when the sensed voltage is near the threshold level. Unless hysteresis is used, oscillation tends to occur as a result of amplification of noise riding on the sensed voltage or the reference voltage. Spurious oscillations need to be avoided where possible to keep system noise as low as possible to reduce the chance of system malfunctions.

In noisy applications (which most applications are), filter capacitors should be used at the two inputs of the comparator to reduce the chance of noise triggering the comparator to the "volts bad" state. When filter capacitors are used, a diode may need to be added in parallel with $R1$ (cathode to V_{cc}) to quickly discharge the filter capacitor on the divider input to speed up the reaction of the comparator when power is lost. A capacitor in parallel with $R2$ can also serve to delay the volts good state and provide a small amount of extra time for V_{cc} to stabilize. If volts good or reset must be delayed a significant time, a digital time-out circuit (i.e., a counter) started with volts good should be used; RC delays are not reliable.

Two areas of the comparator circuit are critical with respect to threshold accuracy: the divider network $R1$ and $R2$ and the reference $U1$. To achieve an accurate threshold level, resistors with a tolerance of 1 percent or better should be used for $R1$ and $R2$. The tolerance of the other resistors in the circuit is not critical. Reference uncertainty tends to be the largest source of error in a level-sensing circuit. Low-voltage precision IC references, such as the LM113, are required to meet reasonable threshold tolerance limits. Low-voltage zener diodes do not have sufficient accuracy for most applications. Divider networks connected to system V_{cc}, or other supply voltages, should not be used as a reference; normal supply voltage levels usually are not precise enough and are often noisy. In all applications, the worst-case threshold limits should be determined using the worst-case tolerances of the components selected. If the results are unsatisfactory, higher-quality components should be selected. A worst-case threshold tolerance of ± 0.25 V is achievable with low-cost standard components.

In most applications, the system reset signal generator must be powered by the same supply V_{cc} that is being sensed.[5] There is often no other (higher) supply voltages available. If there are higher voltages, for example ± 12 or ± 15 V, available and they are guaranteed to be stable when V_{cc} is transitioning, there are a number of analog comparators that can be used to sense V_{cc}.[4] If higher voltages are not available, it is important to select a comparator and reference that will operate predictably when their supply voltage V_{cc} is less than 4.5 V. It would serve no useful purpose to employ a low-voltage sensing circuit that may not work correctly when the supply level that it must sense (and be powered by) is lower than the nominal power-supply operating range (4.5 to 5.5 V or 4.75 to 5.25 V). The sensing circuit should be designed to operate from a low supply level of 2 to 2.5 V to a high level of at least 7 V (which is the absolute maximum upper limit for most TTL, CMOS, and BiCMOS devices—the sensing circuit should have at least as much high-voltage tolerance as the digital components). The LM193 comparator and the LM113 voltage reference used in the low-voltage sensing circuit shown in Figure 12.1 have a suitable range for such applications. The LM193 is specified for operation with a supply voltage as low as 2 V and as high as 36 V. The LM113 provides a 1.2-V reference level for a wide range of input current. Resistor $R4$ must be selected so that the current supplied to the LM113 is adequate for the LM113 to be in regulation at a V_{cc} level much below the desired low V_{cc} threshold level.

If a volts good signal must be combined with other reset signals, and the combined reset signal must be asserted under marginal power conditions to prevent unsafe operation (e.g., false outputs, bus contention, or write signals to nonvolatile memories), CMOS gates that operate with low supply levels should be used to combine the signals (for an example see Figure 12.2).[6,7] Most bipolar devices are not suitable since they are only specified for operation above 4.5 or 4.75 V, but CMOS OR gates are available that are specified for operation with supply voltages as low as 2 V.[8]

If the only objective is to initialize a system to a known starting point, and it is certain that it is not necessary to guard against unsafe

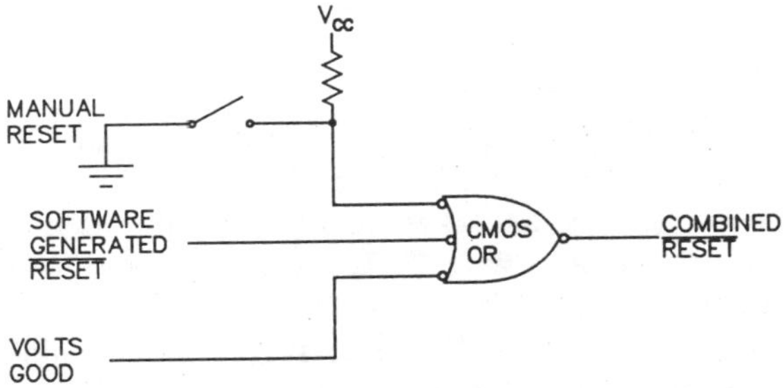

Figure 12.2 CMOS gates that operate with low power-supply voltage should be used to combine reset and volts good signals.

conditions during power excursions, then there is no need to use special components to combine multiple-reset signals. In those cases where it is sufficient for the reset signal to be issued a short time after V_{cc} stabilizes at the operating level, devices from any of the logic families can be used to combine reset signals. For commonality of parts, the device or devices used should be selected from the logic family being used to implement the system.

12.2 Reset Signal Distribution

Great care must be taken in the design of the electrical and physical network used to distribute reset signals. Reset signals must not be degraded by excessive loading or noise. Reset signals should be buffered at a single central point in each system unit (board, chassis, rack, etc.) and redriven to each internal subunit (component, board, etc.) so as to minimize the length and load of reset signal lines. Where a reset signal enters a unit or board, it should be buffered so that units or boards present only a single load to the central source.

Reset signal buffers and signal buffers driving heavy loads (greater than 15 pF) should not be mixed in a common device (package). Large transient load currents may cause local power-supply droop, which will appear as negative-going spikes on the reset lines. If a *low* inactive reset signal polarity is used (in TTL-level systems), it is even more important that reset signal buffers do not share packages with other buffers driving heavy loads. Ground-bounce spikes on inactive *low* reset lines are more serious than on *high* inactive lines since *low* TTL-level signals have less noise margin than *high* signals. However, in spite of the desirability of isolating reset signals, board space limitations often make it impractical to isolate reset signal buffers. If reset signal buffers must share a package with other active buffers, the key to avoiding system operational problems is to ensure that the other buffers are very lightly loaded (the ac load is of most concern).

12.3 Reset Signal Phasing

The inactive state of reset signals should be the logic state with the maximum noise immunity since most reset (set, clear, or preset) inputs on BiCMOS/CMOS logic devices are asynchronous and thus are sensitive to noise spikes. When devices with TTL levels are used, the inactive state should be the *high* state since devices with TTL levels have more noise margin in the *high* state. Since most systems consist of a mixture of logic families and technologies, reset signals should be phased to be in a *high* inactive state throughout the interconnection system to take advantage of the higher, *high*-level noise margin of those devices with TTL input levels. The *inactive* level of all TTL-level reset signals that must pass through noisy areas, such as motherboards, should be the *high* level. Most reset (clear) or preset (set) inputs on most devices take advantage of the higher noise margin of the *high* state at the device level by requiring an active *low* input level for resetting (or presetting). Either polarity is acceptable when devices have CMOS input levels since most CMOS devices have symmetrical noise margins.

To further enhance the noise margin of *high inactive* reset signals, 1-kΩ pull-up resistors should be used in each electrically isolated reset signal line segment to provide additional noise margin. A pull-up resistor will ensure that an inactive reset line is near 5 V, which will provide an additional 1.5- to 2-V noise margin under typical conditions and near 2.5 V under worst-case conditions (without a pull-up resistor the worst-case minimum output *high* level for most devices with TTL output levels is 2.5 V). Pull-up resistors also serve to keep open reset lines (a line not connected to a source) in the *inactive* state. For example, pull-up resistors are often necessary at major interfaces to ensure that reset signals are not asserted during checkout or test when a unit may not have all signal sources present.

12.4 Reset Signal Loading

It is important to limit the number of loads on reset lines. There is often a temptation to connect an excessive number of loads to reset lines since reset signals are perceived as "slow" dc signals. The rationale is that advanced BiCMOS/CMOS devices have the capability to drive many loads and that speed is of no concern. The number of loads should be limited to near 10 for practical purposes. Troubleshooting a shorted signal line that runs to a great number of loads can be very time-consuming and difficult. Also, driving a large number of loads results in very long signal runs which increases the possibility of coupled noise. Since reset inputs are asynchronous, reset lines must be noise-free.

12.5 Reset Signal Timing

Reset signals should be asynchronously asserted and synchronously removed. Reset signals need to be asynchronously asserted so that under fault conditions, such as the absence of the system clock, the systems can be reset to a benign state with no bus contention, false nonvolatile memory writes, or false outputs that could initiate unsafe conditions. Thus, reset signals must be asserted through combinatorial logic paths that have no clocked elements in the path.

The presence of the system clock should not be a condition for asserting reset signals.

In contrast to being asynchronously asserted, reset signals should be removed synchronously to ensure that all clocked elements are properly initialized on the same clock edge. If reset signals are removed from clocked devices, such as multiple-stage counters, in an asynchronous manner, there is no assurance that all the clocked elements will react uniformly.[9] A metastable reaction may cause some stages to not remain reset (see Chapter 10). If some stages remain reset and some do not, the system may not react as expected. Two logic configurations that provide asynchronously asserted and synchronously removed reset signals are shown in Figure 12.3.

The circuits shown in Figure 12.3 operate as follows:

Circuit A. When RESET- goes *low,* the LOCAL RESET- signal is asynchronously activated through AND gate *G*1. When RESET- goes *high,* the transition of LOCAL RESET- to a *high* level is delayed until *F*1 is clocked *high.*

Circuit B. When RESET- goes *low,* LOCAL RESET- is asynchronously activated through the asynchronous reset input of *F*2. When RESET- goes *high,* the transition of the output of *F*2, LOCAL RESET-, to a *high* level is delayed until the next clock edge.

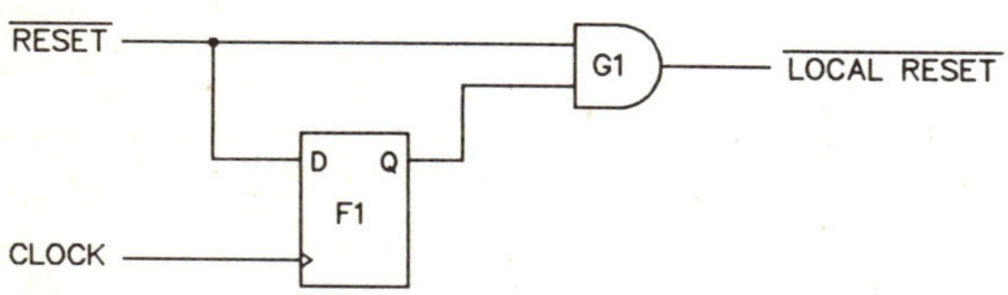

CIRCUIT A

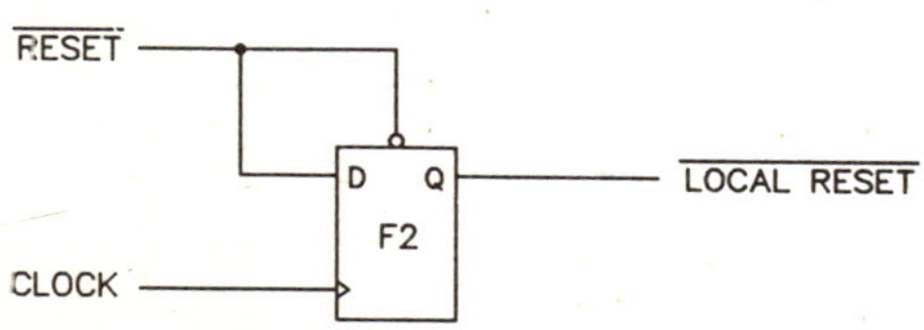

CIRCUIT B

Figure 12.3 Two circuits for asynchronously applying and synchronously removing reset signals.

Note that either circuit (in Figure 12.3) may glitch because of metastable behavior if RESET- goes away at the same time that the active clock edge occurs, but high-speed BiCMOS/CMOS devices should recover by the next clock. A glitch following a clock edge that recovers by the next clock edge should not cause a problem in most applications since the LOCAL RESET- signal was stable (at the other clocked elements on the board) until after the clock edge. If a glitch could cause a problem, RESET- can be double registered which will remove the possibility of LOCAL RESET- glitching (see Figure 12.4).

In low-frequency applications with good clock alignment, the reset synchronizing circuitry can be located at the central source of the reset signal or in a central location in each given unit. However, when a system is operating near the upper end of the speed capability of the logic family being used, reset signals may need to be synchronized at the local circuit board level because of clock phasing uncertainty. If a synchronizing circuit must operate, i.e., supply an active reset signal, during undervoltage conditions, the circuit must be implemented with advanced CMOS devices that will operate with a low supply voltage.

The synchronizing circuit also provides a convenient location to add a time-out circuit. Some reset signals require a fixed amount of time to have elapsed after V_{cc} stabilizes at the operating level. A timer synchronously started when LOCAL RESET- goes inactive provides one

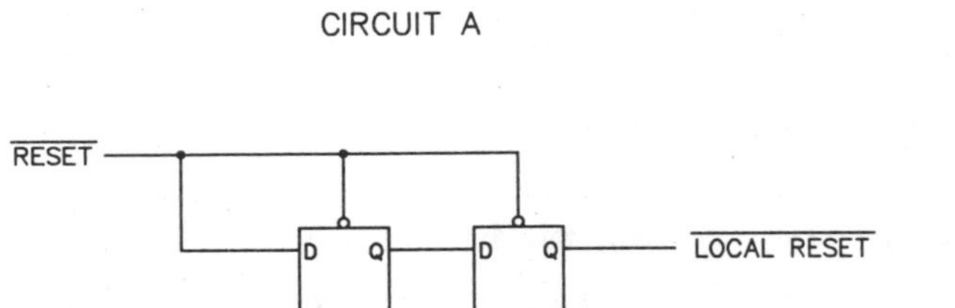

CIRCUIT A

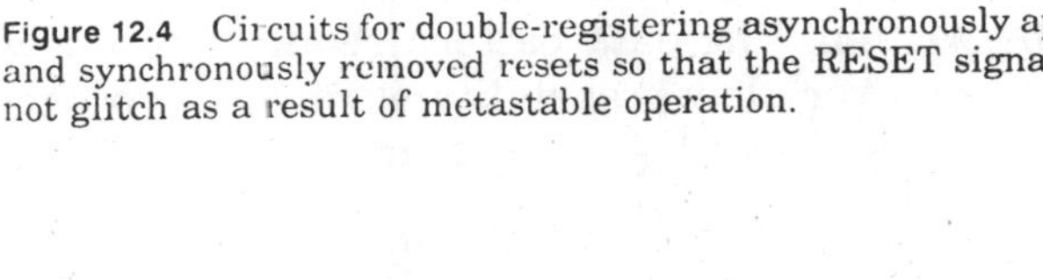

CIRCUIT B

Figure 12.4 Circuits for double-registering asynchronously applied and synchronously removed resets so that the RESET signal does not glitch as a result of metastable operation.

means of adding a fixed time-out to the removal of the system reset after V_{cc} has reached the normal operating level.

12.6 Write Protection of Nonvolatile Memories

In most system applications, nonvolatile electrically alterable memory devices that are alterable with normal system 5-V supply voltage levels require some form of write protection during power turn ON or OFF and during low-voltage transients. Most EEPROMs and NOVRAMs fall into that category. Most UV EPROMs and flash EPROMs do not; they typically require a programming voltage higher than 5 V, but 5-V-only flash EPROMs are available. In most cases where UV and flash EPROMs are used, the higher voltage required for programming (assuming the higher voltage parts are used) is not present in the actual system application, so false writes cannot occur. However, if the voltage needed for programming is present in the system and the memory devices are connected to control signals that allow in-circuit alteration, some form of write protection is needed during power sequencing. When V_{cc} is below the normal operating range, memory interface signals may be out of control and randomly glitch to a state that indicates a memory write should occur. It is also possible for some microprocessors to be out of control and inadvertently issue write commands during the first few clock cycles after power-up. Random glitches on write control lines may start write cycles even when they do not meet minimum pulse width requirements. Although data sheets define minimum pulse widths, under most conditions much shorter pulses will cause writes.

Most of today's 5-V-only alterable memories have some form of built-in protection. Some have both internal hardware and software protection. Internal hardware protection typically consists of one or more of the following:

1. A power-supply level sensor that locks out writes when V_{cc} is low
2. A noise filter on the write (WE-) input that prevents a write cycle unless WE- is of at least a certain width (typically 15 to 20 ns)
3. A power-up delay that prevents writes until a fixed time after V_{cc} stabilizes at the operating level

Internal software write protection typically requires that certain data patterns be written to several internal write protection control registers before data can be written. The data pattern required to allow writing is such that it has a very low probability of randomly occurring. Devices with software and low V_{cc} lockout circuitry usually do not require further write protection, but devices that only have low V_{cc} and other of the hardware lockout features generally require some form of external protection. Internal V_{cc} sensors cannot be built economically with the accuracy needed to provide complete protection. To ensure that internal V_{cc} level sensors do not interfere with normal operation, their thresholds must be set much below the normal V_{cc} operating range. Most internal V_{cc} sensors lock out writes below 3 to 3.5 V. A danger zone exists for V_{cc} levels above the internal protection circuit's threshold and the minimum operating V_{cc} level of the logic devices controlling the nonvolatile memory (usually 4.5 or 4.75 V). When V_{cc} is in the danger zone, the condition of system and memory control signals are undefined; write and chip select signals may become active and initiate unintended writes.[6]

Inadvertent data writes to electrically alterable memories without adequate internal protection must be prevented by forcing the signals that can initiate writes to nonwrite states during the critical portion of power transitions. Writes to most EEPROMs can be prevented by keeping either chip enable (CE-), which is called chip select (CS-) on some data sheets, write (W- or WE-), or output enable (OE-) in the nonwrite state. The nonwrite state for CE- and W- is *high,* and for OE- *low* for most EEPROMs. If CE-, W-, or OE- is forced into a nonwrite state when V_{cc} is between 3.0 and the V_{cc} level at which the control logic is stable and under control, writes cannot occur. The forcing circuitry has the same basic requirements as a volts good signal generator. It must:

1. Operate at low voltage
2. Be insensitive to power-supply turn ON or OFF rates
3. Be insensitive to noise
4. Have a precise threshold

Simple *RC* networks are inadequate; they do not meet requirements 2 and 4. A circuit similar to the reset signal generator shown in Figure 12.1 combined with a CMOS gate is needed that will operate at low voltage.[6,7] For example, Figure 12.5 shows how CMOS gates are used to combine system-generated CE-, W-, or OE- signals with a RESET-signal from a low-voltage sensing circuit to force the CE-, W-, or OE-signals applied to nonvolatile memories, such as EEPROMs or flash EPROMs, to a nonwrite state during critical power-up or power-down times. The CMOS gate or gates used and the low-voltage sensing comparator must maintain control until V_{cc} is below the trip point of the internal voltage level sensor, i.e., below 3 V in most cases.

Holding CE- *high* is generally the safest method of preventing mem-

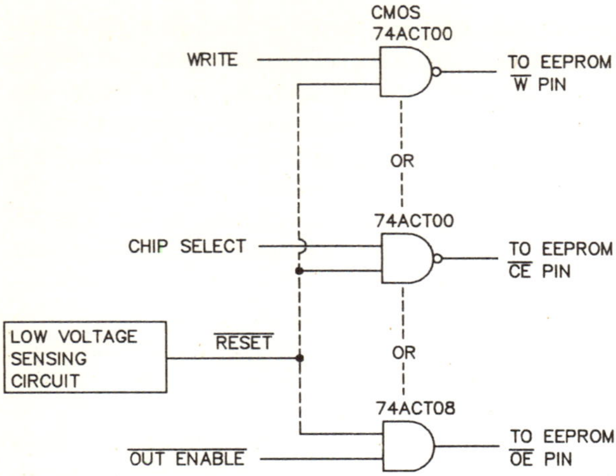

Figure 12.5 CMOS gates that operate with low V_{cc} should be used with the system RESET- signal to force system-generated CE-, W-, or OE- signals to nonvolatile memories, such as EEPROMs, to a nonwrite state during critical power-up or power-down times.

ory writes. Holding CE- *high* prevents writes and ensures that EEPROM data outputs remain in a high-impedance state during power ON or OFF, thereby eliminating any possibility of output data drivers being in contention. Holding OE- *low* causes outputs to be in the driven state during power excursions which could cause contention between data outputs and other devices that might share the data bus.

12.7 Logic Initialization to Prevent Hazardous or Damaging Conditions

Many systems need some means of initialization to prevent unsafe or possibly destructive conditions, such as bus contention or random outputs, during power excursions.[10]

If a system has three-state buses or three-state control signals, the system reset signal should be used to ensure that the controls that enable the three-state drivers are initialized to the OFF state during power-supply transients and at turn ON. Requiring the presence of the system clock, which may not always be present because of faults or other reasons, for correct initialization of three-state buffers is an invitation for disaster. Most advanced BiCMOS/CMOS three-state buffers are not rated for continuous short-circuit conditions. Ideally, three-state drivers connected to a common line are controlled in such a manner that no more than one can be turned ON at the same time, even for short periods during power turn ON or OFF. The safest method of achieving that is to use a single hardware decoder as the source of the enables for all drivers connected to a common line; then only one driver can be ON at one time (unless there is some fault). Where a single decoder for the source of all the line driver enables (for a common line or bus) is not practical, which is often the case in large systems with bus interconnects that interface to numerous boards, the system reset signal must be used to ensure that three-state drivers are in the high-impedance state during power-supply transients and at turn ON.

Some systems have critical outputs that must be controlled at all times to prevent harm to personal or physical damage. One means of controlling critical outputs when V_{cc} is low and the logic devices that are the source of the critical signals may be out of control is to use the reset signal to force the signals to harmless states. The reset signal and all circuits beyond the point where the reset signal is introduced to a critical output data path must be capable of operating at low voltage. The reset signal used to control critical outputs must have a direct asynchronous path. The presence of the system clock, which may not be present because of faults or other reasons, must not be required to ensure control of potentially hazardous outputs.

12.8 Summary of Initialization and Low-Voltage Sensing Techniques

1. Use a single volts good circuit as the central source for system master reset signals.
2. Generate volts good signals using analog comparators that can accurately sense when V_{cc} is out of tolerance.
3. Do not use *RC* networks to generate reset signals. Their response is unpredictable.
4. Do not overload reset signal drivers or allow reset signal buffers to share packages with buffers driving large loads to prevent ground bounce or power-supply droop interference.
5. Buffer reset signals where they enter units, boards, etc., so that each destination presents only a single load to the central source.
6. Pull up reset signals where they enter units, boards, etc., so that reset will not be asserted when the normal source is not present.
7. The phasing of TTL-level reset signals should be such that reset lines are *high* in the inactive state to maximize the noise margin.

8. Pull up *high* inactive TTL-level reset signals for extra noise margin.
9. Reset signals should be asynchronously asserted to prevent damage and synchronously removed to prevent metastable conditions.
10. Where it is important for reset signals to be asserted to prevent unsafe conditions or false writes to nonvolatile memories during transient power-supply conditions, CMOS logic components that will operate with low supply voltage should be used throughout the reset signal path.
11. The system reset should be used to ensure that three-state drivers are in the OFF state during power transients and at power turn ON.

12.9 References

1. Buchanan, J. E.: "A Normally ON FET Control Circuit for TTL Logic Systems," *Computer Design,* June 1972, pp. 83–85.
2. Frederiksen, Thomas M.: *Intuitive Operational Amplifiers,* McGraw-Hill, New York, 1988.
3. *Linear Applications Databook,* National Semiconductor Corp., Santa Clara, Calif., 1986.
4. *Linear Databook 1,* National Semiconductor Corp., Santa Clara, Calif., 1988.
5. Austin, W. H., J. E. Buchanan, and C. W. Nelson: "Precise Voltage Level Detector," *Digital Design,* August 1973, p. 30.
6. *1988/89 SEEQ Data Book,* SEEQ Technology Inc., San Jose, Calif., 1988.
7. *Memory Products Data Book,* Advanced Micro Devices Inc., Sunnyvale, Calif., 1989.
8. *FACT—Advanced CMOS Logic Databook,* National Semiconductor Corp., Santa Clara, Calif., 1988.
9. Chaney, Thomas J.: "Measured Flip-Flop Responses to Marginal Triggering," *IEEE Trans. on Computers,* Vol. C-32, No. 12, December 1983, pp. 1207–1209.
10. Buchanon [*sic*], James E.: "How to Use the On/Off Relay Action of Junction Field-Effect Transistors," *Digital Design,* July 1977, pp. 26–34.

Chapter

13

Memory Subsystem Design

The speed and performance of most high-performance digital systems are limited by the speed and storage capability of their memory subsystems. In turn, memory subsystems are limited by available memory device speed and density. Even though memory device technology continues to make huge strides, system requirements stay slightly ahead of memory device technology. In most cases today, available memory devices are more likely to meet density requirements than speed requirements. That is fortunate since system designers have little choice with regard to memory device density; they are limited to what is commercially available. That is also true of device speed, but device speed does not give a complete picture of memory subsystem speed. Memory subsystem speed is determined by both memory device speed and memory subsystem design. A common mistake is to assume that a memory subsystem can operate at a cycle time near the access time of the memory devices used. Memory data paths or loops have many other sources of delay. Interconnection delays and interface circuitry such as address drivers and data buffers or transceivers add significant delays in high-speed systems.[1] System designers must guard against accepting or perpetuating unrealistic memory cycle time expectations. Memory devices keep getting faster, but memory interface circuitry and interconnection delays are near their limits, and both become a larger percentage of total delay as memory devices and memory subsystems get faster. Poor interface and interconnection designs often severely degrade memory subsystem speed and waste inherent memory device speed.

Not long ago most memory subsystems consisted of several boards that contained mostly memory devices and buffers for address and data lines. That is no longer the case. Today with denser memory devices and higher speed requirements most memory devices are distributed among the circuitry they support. In most cases, the needed density can be achieved with a small number of memory chips located near the processor or other memory user. Speed requirements also force memory devices to be colocated with the devices that use them. Long signal paths, such as are required to go from one board to another, cannot be tolerated when 50-ns or less memory cycle times are needed.

13.1 Semiconductor Memory Devices

Semiconductor memory devices fall into two broad categories: volatile and nonvolatile. Volatile devices lose their contents (data) when power is removed; nonvolatile devices do not. Dynamic random-access memories (DRAMs), static random-access memories (SRAMs), latches, registers, and register files are examples of volatile memory devices; they can store data as long as power is present, but once power is removed they lose their content and must be reloaded when power is restored. Examples of nonvolatile memory devices (i.e., memory devices that retain their data contents when power is absent and do not have to be reloaded when power is restored) are read-only memories (ROMs) and programmable read-only memories (PROMs) such as electrically erasable (EE) and (flash), and ultraviolet (UV) erasable PROMs (EPROMs). Power is required to read a nonvolatile device but not to retain data.

Volatile memory devices usually have relatively fast *read* and *write* cycle times (i.e., the time it takes to read or write a location and prepare for the next operation), and the control signals required to initiate a read or write are relatively simple to generate. Read and write cycle times are approximately the same. Nonvolatile devices, on the other hand, have relatively fast read cycles but have long write cycles and require special procedures to write or load data. ROMs are the least flexible nonvolatile memory devices from a system design standpoint; data can only be entered, or programmed as it is called, once. Some are mask programmable at the factory. Others can be programmed by the purchaser but require special programming tools of which there are a number of commercial ones available. In either case, once data are entered or programmed in ROMs, they cannot be changed. PROMs on the other hand can be reprogrammed, and in most cases in-circuit. EEPROMs and flash EPROMs are designed to be in-circuit programmable. EEPROMs are the most flexible nonvolatile memory devices; individual location or words can be modified. Write cycles take much longer than read cycles, but today's

EEPROMs require no special voltages for writing or programming. Flash EPROMs are next in flexibility. They can be modified in-circuit but not on an individual word or location basis. Most, but not all, flash devices must be completely erased before they can be written, which means all locations must be rewritten. However, some flash devices can be erased on a sector basis. Most flash EPROMs require a voltage higher than 5 V, such as 12 V for writing, but some are available that only require 5 V. UV PROMs are the least flexible of the reprogrammable memory devices; they must be erased with a UV light source before being rewritten which means the devices or the board they are mounted on must be removed from the equipment. Removing devices or boards from equipment is an undesirable and risky procedure. The chance for problems is greatly increased when equipment is opened or devices are handled.

How often data must be changed and initial device cost are two of the factors that determine which type of nonvolatile memory should be used in a given application. Mask programmable ROMs are the least expensive if they are used in large volume and the contents do not have to be changed. In the reprogrammable category, UV PROMs are the least expensive and EEPROMs the most expensive, but device cost is not the only consideration. The time required and the simplicity of reprogramming are also important cost factors. As a rule of thumb, UV EPROMs are most cost-effective if the data only have to be changed once every 2 or 3 years; flash EPROMs are most cost-effective if the data only have to be changed once or twice a year, and EEPROMs are most cost-effective when the data have to be changed on a daily, weekly, or monthly basis.

13.1.1 Typical Memory Device

From the system designer's viewpoint, semiconductor memory devices are places to store and retrieve large amounts of data. To apply them, system designers must understand how they function logically and what their interface requirements are. Most LSI and VLSI memory devices, regardless of whether they are RAMs or ROMs, are structured as shown in the block diagrams of Figures 13.1 and 13.2. They consist of an array of memory storage elements surrounded by address buffers, address decoders, data buffers, and controls. The address buffers and decoders are used for selecting particular locations in the memory element array. Data in buffers (if an alterable device) minimize the load for the source and isolate the memory array from external disturbances. Data out buffers isolate the memory array and provide drive for external loads. Alterable memory devices come in two basic data in-out configurations: Some have separate input and output

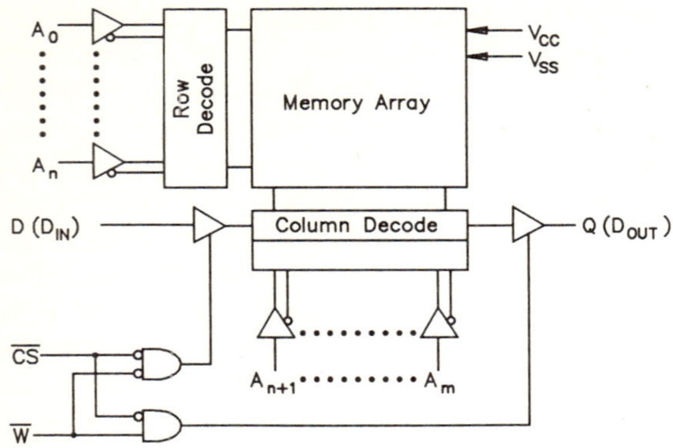

Figure 13.1 Block diagram of a typical alterable memory device with separate input and output data paths.

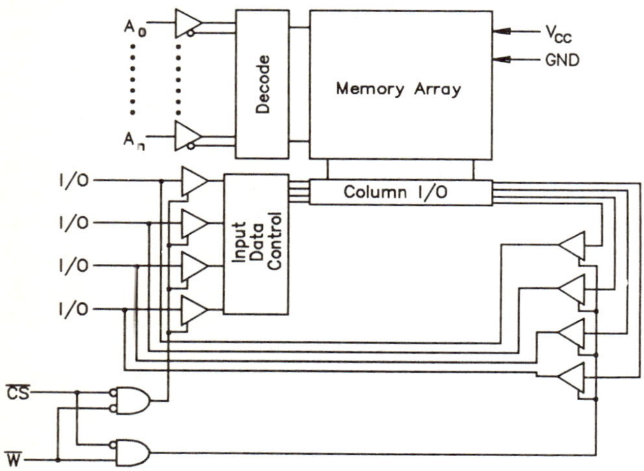

Figure 13.2 Block diagram of a typical alterable memory device with common input and output data paths.

pins as shown in the block diagram in Figure 13.1, and others have common input-output (I/O) data connections as shown in the block diagram in Figure 13.2. Memory devices with separate I/O data pins are generally easier to use if the system architecture is such that advantage can be taken of separate data paths; separate I/O memories require more package pins. Common I/O memories save package pins but complicate data bus control. They compound the risk of data bus contention between memory device outputs and data input sources.

Control signals common to most memory devices. Memory control signals fall into three basic categories: chip selects, out enables, and write controls. Not all memory devices have all three; most have chip selects; some have out enables, and only those that are in-circuit alterable have write controls. In general, these controls function as follows (check data sheets for specific devices).

Chip select signals (sometimes called chip enable signals) typically serve to activate memory devices for either read or write cycles. Most memory device chip select inputs are level sensitive, not edge sensitive. Most SRAMs, PROMs, UV EPROMs, and flash EPROMs require chip select signals to be active throughout a read or write cycle (if alterable). Most EEPROMs require chip select signals to be active throughout a read but may not require the chip select to be active throughout a write cycle (only at the start). Synchronous SRAMs typically use a clock or latch enable signal to capture the chip select signal (as well as address, data, and write signals). Chip select signals to synchronous RAMs only have to be valid within the setup and hold-time window around the clock edge or latch enable signal. Synchronous RAMs greatly simplify chip select signal generation in high-speed applications. Dynamic RAMs typically require multiple chip select or chip enable control signals (e.g., they typically require a row chip select signal called RAS- and a column chip select signal called CAS-). Specific timing of DRAM chip select controls tends to be device-specific, and it is difficult to generalize the requirements. It is suggested that readers consult data sheets for specific chip select timing requirements.

Memory devices are typically implemented so that low chip select signals activate I/O circuitry and start memory cycles, i.e., when chip select signals are *low,* most memory devices are active, and when chip selects are *high,* most memory devices are inactive. Memory access time is usually measured from the time chip select is applied or when all address lines are stable (whichever is last). In most applications, chip select signal timing controls memory access time because chip select signals are generally decoded for upper address signals, and as a result chip select signals usually stabilize some time later than address signals. For most memory devices, chip select must be active for outputs to be active; when chip selects are not active, outputs are typically in a high-impedance (tristate) condition. When chip select inputs are level-sensitive (in contrast to edge activated), chip select signals cannot be removed before data is read.[2] Memory deselect time is usually measured from the time chip select is removed (i.e., goes inactive). When chip selects are inactive, most peripheral circuits on most memory devices power down to conserve power and to minimize chip temperature.

Out enable signals, on those memory devices that have them, control whether outputs are in an active or high-impedance state. Out enable circuits in memory devices are usually implemented so that a *low* out enable signal causes outputs to be active and a *high* out enable signal causes outputs to be in a high-impedance or tristate mode. Out enable controls greatly simplify data bus control when memory devices have common I/O.

Write control signals (on those devices that have them) control when data is written into in-circuit alterable memory devices such as DRAMs, SRAMs, and EEPROMs. Flash EPROMs and UV EPROMs, although alterable in-circuit, typically do not have a write control input as such, but instead writes are controlled by a certain combination of other control signals and the application of a special programming voltage to a V_{pp} pin. If V_{pp} is not present in normal operation, then data cannot be modified by erroneous signals or operation. In applications where devices can be modified in normal operation, write signals are extremely critical signals. Noise or glitches cannot be tolerated in those circumstances. Any disturbance on a write signal may cause an inadvertent write to memory which corrupts the memory data. Noise, glitches, or other disturbances on write signals to high-speed SRAMs are a particular problem since most SRAMs have asynchronous write inputs and are sensitive to very narrow pulses at any time. Data sheets may call out a minimum write pulse width, but under typical conditions much narrower pulses may cause writes. Write signal timing with respect to address and data is extremely critical and must be given a great deal of attention to ensure that address and data setup and hold times relative to write signals are not violated. Write signal timing violations account for a large percentage of memory problems.

13.1.2 High-speed advanced BiCMOS/CMOS memory devices

A brief status of size, speed, and a description of the basic characteristics of several of the major memory device categories utilizing advanced BiCMOS/CMOS technology is provided for those readers with little exposure to semiconductor memory devices and their applications. Memory technology moves so fast that an attempt to report current memory device status in a book such as this is certain to be out of date before the book is published. The best that can be accomplished is to identify the present state of art in each major memory device category and indicate near-future expectations. A comprehensive coverage of all memory device types and the many subgroups and their characteristics is beyond the scope of this book as well as an impossible task in a single chapter. A chapter or more would be needed for each major category and perhaps for several of the subgroups, which means the book would have to more than double in size because of the great variety of memory devices. Again, the best that can be accomplished is to provide a broad overview of the basic characteristics of several of the major categories.

Static RAMs. Static RAMs are used for high-speed random-access memory (RAM) applications such as cache, scratchpad, and control memories. Density and speed requirements drive SRAM technology, and advanced CMOS and BiCMOS process technologies have allowed density to be pushed up and access time down.[3] Static RAMs are now available in production quantities with 1M densities, and 4M devices will soon be available. Smaller RAMs are available with 10-ns access times, and 1M parts will soon be available with 25-ns access time. Static RAMs are the fastest memory devices available. From the system designer's standpoint, SRAMs are one of the essential building blocks for high-speed digital computers and processors. No other memory device can match SRAM speed or simplicity of operation for either read or write cycles. Static RAMs are required to keep up with processors running in the 20-MHz or higher range unless extra clock times[4] (wait states) or other special techniques such as interleaving are used for memory cycles.

Static RAMs come in a number of configurations:[5] by 1, 4, 8, and even 16 data bits wide and with depths (i.e., number of words or bits deep) of up to the limit of technology which is 1 M to 4M bits at this time. Smaller devices tend to be faster than large devices. One-bit-wide devices tend to be faster than multiple-bit-wide devices. Multiple-bit-wide SRAMs generally have common I/O pins (some 4-bit-wide ones are available with separate I/O). Most SRAMs are asynchronous devices. Address, data, and controls are not latched and must be present throughout a memory cycle. Because they are asynchronous devices, they are sensitive to noise or erroneous signals at any time which means extra care is required to prevent crosstalk or glitches on signals. Extra care is also required to ensure proper write pulse widths and timing of write pulses to avoid setup and hold-time violations. In the past when system clock periods were much longer, write pulse requirements could usually be met with a minimum of effort, but as memory speed increases and clock periods decrease it becomes extremely difficult to meet write pulse requirements. In very high speed applications it becomes almost impossible to shape write pulses.[6] Self-timed or synchronous RAMs[7] of which there are now some available that are less sensitive to noise and require no external write pulse shaping. The write pulse shaping is done internal to the device. Synchronous RAMs latch or register address, data, and control signals depending upon their exact implementation so that address, data, and control signals including write signals only have to meet setup and hold times relative to a clock edge which greatly reduces the window where signals must be valid. Because of the reduced timing constraints on inputs, self-timed or synchronous RAMs offer a great advantage in high-speed applications. The disadvantage of synchronous RAMs is that they usually must be used in a "pipeline" fashion (i.e., address and controls must have a clock time to setup before being clocked in, or data require an extra clock time to be clocked out depending upon whether the device has input or output data registers—see synchronous RAM data sheets).

Most, but not all, advanced BiCMOS/CMOS SRAMs are designed to power down the peripheral or I/O circuitry when not chip selected to conserve power and keep chip temperature as low as possible. Some SRAMs designed specifically for very high speed applications keep the I/O circuitry powered up and active to gain a few nanoseconds in speed. Where memory devices are in use most of the time, there is little advantage to powering down, but in most applications that is not the case. In most applications where a large number of RAMs are used only certain banks or subgroups are active at any one time, and in those cases RAMs that power down should be used. Static BiCMOS/CMOS RAMs that power down when not chip selected (as well as those that are always active) typically dissipate 300 to 600 mW when active. When not chip selected and in the powered-down standby mode, they typically dissipate under 100 mW with TTL-level inputs, and some dissipate under 10 mW with CMOS-level inputs. Note that standby power is higher with TTL input levels than CMOS because TTL input levels do not completely turn OFF the complementary input pairs. When input pairs are not completely cut off, dc current flows between V_{cc} and ground and power goes up. Wherever possible, RAMs that do power down when not chip selected should be used. They offer a significant power savings where some of the memory chips are not active at all times. Memory subsystem power dissipation and memory device reliability are often major factors in determining overall system reliability, so it is important to keep memory power dissipation as low as possible. Another possible technique that sometimes can be used to conserve system power and maintain memory data integrity when SRAMs are not in use is to reduce the RAM power supply to a holding level which is typically in the 2- to 3-V range. Most advanced BiCMOS/CMOS SRAMs will maintain data at reduced power-supply levels, but data sheets must be consulted for specific limits and reduced voltage operating requirements.

Dynamic RAMs. Dynamic RAMs are used where very large quantities of alterable memory are required. Dynamic RAMs are in general the most dense commodity semiconductor devices available. They push the state of the art in semiconductor process technology development.[8] Today 4M-bit DRAMs are commercially available; 16M-bit parts will be shortly, and 64-M bit parts are in development. Because DRAM cells only require one transistor where SRAM cells require at least four transistors, DRAM density will lead SRAM density by a factor of 4 until the limits of technology are reached. Dynamic RAMs have typically been available in 1-bit- and 4-bit-wide versions. In most applications, several are used in parallel to store a word or byte of data, but as density goes up there is need for wider versions in many applications. Often 16M words or more of memory is not needed, but denser (and wider) devices are needed to reduce component count. Most 16M and larger DRAMs are expected to be 8 bits wide or wider.

Dynamic RAMs are not as fast as SRAMs, but are faster than most other memory devices. Dynamic RAMs are available with access times under 50 ns, but access time is not the complete story. Dynamic RAM data sheets specify both access times and cycle times, and cycle times are longer than access times. Even access time specifications are confusing and can be misleading since the typical DRAM data sheet lists two or three access times measured from different control signals. Cycle time is usually a better measure of memory subsystem performance when DRAMs are used. Cycle time is the time required to make a true random read or write. Cycle time usually takes approximately twice the longer of the listed (on data sheets) access times to complete. Cycle time is longer than access time because certain DRAMs internal signals must be precharged before each random access. Today some 1M devices are available with random-access cycle times as low as 100 ns, but most 1M and 4M devices have cycle times in the 125- to 180-ns range which translates to 150- to 200-ns system cycle times when some allowance is made for practical limitations of generating the required control signals. In general, it is very difficult to meet all the address, data, and control signal timing requirements needed for minimum cycle times. Most systems do not have clock signals of the frequencies or shapes needed to meet all the interrelated control, address, and data timing specifications when running at the minimum cycle time.

Dynamic RAMs, even without a precharge time penalty, are slower and more complicated to use than SRAMs because address information is loaded in two steps or words. For example, 22 address bits are required to decode 1 of 4,194,304 storage locations. Parallel entry of that many address bits would mean a very large package, so to conserve pins and package size DRAMs have traditionally loaded address information in two steps (or words). To overcome cycle time limitations imposed by address input limitations, most DRAMs have some form of what is often called "page-mode operation" where they can be accessed at a higher rate than the cycle time for a limited number of address locations (for example, a page). Page-mode operation can be used to advantage in pipeline sequential operations where only a limited amount of the address information (bits) have to be changed to access a new location. However, at page boundaries and for true random accesses, time must be allotted to allow loading of multiple address words and for precharging. Interleaving is another technique used to compensate for the slower speed of DRAMs. Interleaving banks of DRAMs allows overall memory system speed to match that of high-speed processors.[9]

Aside from the speed issue, the main disadvantage of DRAMs from the system users' standpoint is their requirement for refresh cycles. Dynamic RAMs, as their name implies, must be constantly accessed or the data go away. Data is stored on very small capacitors, and unless the charge on each capacitor is restored periodically, the charge leaks off and the data is lost. Most modern high-density (1M or greater) DRAMs require the charge to be refreshed at a 4- to 16-ms rate. Refresh circuitry adds system complexity and in certain cases slows memory response, but in most cases refresh cycles can be made to happen in the background and be transparent to the memory user. A number of semiconductor manufacturers have developed and now market LSI/VLSI DRAM control circuits that provide refresh timing and control and in most cases are much better solutions than discrete implementations.

UV EPROMs. UV EPROMs have served as the mainstay of reprogrammable nonvolatile memories for a number of years. They have lead the way in density; 2M and 4M units are available now,[10] and 8M and 16M are in development. They are also the least expensive of the reprogrammable memories on a per-bit basis, and they are available from a large number of sources which simplifies procurement and keeps cost low. Until recently most UV EPROMs were relatively slow (most larger EPROMs had access times above 200 ns), but that is no longer the case. Now there are 35-ns 256K and 45-ns 1M units available. In the past, in most high-speed applications special techniques had to be used to compensate for the slow access times of the available EPROMs. They were often interleaved to decrease the average access time, or the data stored in them were down-loaded to high-speed RAM at startup (or at other times) to allow the operational memory to keep up with the processor or other circuits it might be serving. Interleaving memories decreases flexibility and increases complexity. Downloading to RAM increases hardware; both RAMs and EPROMs are required. Now that under 50-ns UV EPROMs are available and faster ones are in development and will be available soon, the opportunity exists to run directly from EPROMs and eliminate the RAM used for down-loading.

Most UV EPROMs come in 8-bit-wide configurations, but as density goes up, 16-bit-wide units are also available. For example, 131,072 × 8 bits, 65,536 × 16 bits, and 262,144 × 8 bits, 131,072 × 16 bits are standard configurations for 1M and 2M EPROMs. UV EPROM packages tend to be large. Address inputs are not multiplexed as they are for DRAMs, which means as density goes up, a large number of pins are required for address inputs; 256K EPROMs fit in 28-pin packages, but 1M units need more than 28 pins and are typically packaged in 32- or 40-pin DIPs or 32- or 44-pin LCCs.

One of the advantages of UV EPROMs compared to fuse-link-type PROMs is that data in UV EPROMs can be changed a limited number of times—normally the limit is about 100 times. The small number of erase and write cycles allowed is usually not a problem since UV EPROMs are intended for applications that do not require frequent data changes, but care must be exercised during board or system test to not inadvertently exceed or approach the limit. In general when data are changed, the complete chip must be erased or cleared (except for one special case) with a UV light before being rewritten or programmed as it is called. An erase cycle is not required when data only have to be changed from the erased or neutral state to the programmed state, but if data bits have to be returned to the neutral state, the complete chip must be erased and all bits or words reprogrammed. Most UV EPROMs require a voltage higher than +5 V for writing data in or programming. Older devices often required +20 V or a level in that vicinity to program, but the newer, higher-density devices typically require a voltage in the neighborhood of +12 V. Programming voltage requirements also vary from vendor to vendor; even for similar parts. Because of the variations and special voltage levels required for programming, UV EPROMs are generally programmed before installation using one of the commercially available programming tools. They can be programmed in-circuit if provisions are made for applying the proper programming voltage, data, and controls. If the programming voltage is normally present in the system, care must be taken to ensure that it and the control signals needed for programming do not get applied during transient conditions such as might occur during power turn ON and OFF times (see Chapter 12). Another complication with in-circuit programming is that parts from different vendors often have different programming sequences and timing requirements as well as programming voltage level requirements which means parts from different vendors cannot be mixed. Also, some means of identifying the type of part that is in the circuit must be provided. When in-circuit programming is used, data sheets must be studied thoroughly to ensure that all programming requirements are met.

Flash EPROMs. Flash EPROMs are expected to displace UV EPROMs and be used in most nonvolatile memory applications in the future. The technology used to build flash EPROMs is similar to that used for UV EPROMs and their operation and speed are similar, but flash EPROMs have the advantage that they are electrically erasable and do not have to be removed from the circuit or equipment to be erased as do UV EPROMs. In-circuit erase capability is viewed as a significant advantage in most quarters. Anytime equipment is opened or components are handled the chance for problems increases significantly. Mechanical damage, ESD, and a host of other potential sources of damage are always possible when equipment is opened.

Flash EPROMs are programmed similarly to UV EPROMs. They must be erased before being programmed. Like UV EPROMs, flash EPROMs are specified for a very limited number of erase-write cycles—first-generation flash EPROMs are specified for as low as 100 cycles, but second-generation parts are expected to have a duration of 10,000 or more erase-write cycles. Most first-generation flash EPROMs required approximately +12 V for programming, but +5-V-only devices are available now.[10] Five-volt-only devices are a mixed blessing. They simplify the programming task, but they open the possibility for inadvertent write during power transients or normal power-supply ON-OFF transitions unless special precautions are taken. Regardless of what the programming voltage level is, if it is normally present in the system, care must be taken to ensure that it and the control signals needed for programming do not get applied during transient conditions such as might occur during power turn ON and OFF times (see Chapter 12). Another complication with in-circuit programming is that flash devices from different vendors have different programming sequences and timing requirements as well as programming voltage level requirements which means parts from different vendors cannot be mixed. The situation with respect to standardization is even worse for flash devices than for UV devices. To date, flash EPROM users have had to deal with a complete lack of standardization. Thus, when flash devices from different vendors are used, some means of identifying the type of part that is in the circuit must be provided as well as some means of changing the programming algorithm. When in-circuit programming is used, data sheets must be studied thoroughly to ensure that all programming voltage levels and timing requirements are met.

CAUTION: When in-circuit programming of flash EPROMs is possible, care must be taken to ensure that the specified erase-write cycle limit is not inadvertently exceeded or approached by unnecessary mode changes or test routines. Particular care must be exercised during board or system test.

At present, flash technology and device density is lagging UV technology slightly; 1M devices are available in production quantities, and sample quantities of 2M devices are available, but 4M units are still in development. Flash EPROMs are available with pin-outs compatible with both UV EPROMs and EEPROMs (see the UV EPROM section above for typical package size). As with UV EPROMs, most low-density flash EPROMs come in 8-bit-wide configurations, but as density goes up, 16-bit-wide units will also be available.

EEPROMs. EEPROMs are the most versatile of the alterable nonvolatile memories. They can be written on a word-by-word basis. Write cycles are much longer than read cycles, and some devices require that a word be erased before it is written, but the flexibility exists to modify individual words in a true random fashion. EEPROM read times have typically been close to 200 ns, but now devices are available with access times under 100 ns. Write times are typically 1 to 10 ms. Read cycle operation is similar to that of SRAMs and UV or flash EPROMs. Address and chip select inputs are not latched or registered and must be present for the duration of the cycle. In contrast, during write cycles most EEPROMs latch or register address, data, and some or all control inputs. Some EEPROMs are self-timed and latch address and all control signals which allows the controlling circuitry to do other tasks while write cycles are in progress. EEPROMs have limitations on the number of writes allowed just as do other alterable memory devices. Some are rated for 10,000 cycles, but some are rated for 100,000 cycles, and a few are rated for 1,000,000 cycles.[10]

CAUTION: When EEPROMs are used, care must be taken to ensure that specified erase-write cycle limits are not exceeded or approached during board or system test.

EEPROM internal circuits and storage cells are more complicated than those of UV or flash EPROMs, and as a result EEPROMs lag UV and flash EPROMs in density. The progression to each new generation of EEPROMs has been slow; 1M EEPROMs are being sampled, but 256K EEPROMs, which have been available for some time, remain the mainstay in the EEPROM arena. Most EEPROMs come in byte-wide configurations, but there are some serial EEPROMs and other special-purpose devices.

13.2 High-Speed Memory Subsystem Design

It would seem that memory subsystem design should get easier as memory devices get larger and faster, but device characteristics never seem to catch up with system requirements. Memory subsystem or circuit design is extremely critical to the success of most systems. System speed and performance are generally a direct function of memory circuit speed. System reliability and owner satisfaction are generally a direct function of memory reliability. An old and true adage is, "A digital system is only as reliable as its memory subsystem or circuits."

Memory subsystem or circuit design is not as easy as it may first appear. Most memory devices are asynchronous devices and thus are sensitive to noise at any time. At the same time, memory circuits generate a great deal of noise. Many address and data lines tend to switch at once, which in turn causes large transient currents, which causes lots of noise. Timing constraints are often difficult to meet and are often misunderstood and violated, particularly when system requirements push memory subsystems and devices to their limits.[11] Noise and timing violations typically do not cause outright failure but do cause intermittent memory problems. Intermittent troubles are the worst kind. They are typically data-pattern sensitive and depending upon the degree of noise or timing violations may occur very infrequently. Memory errors that occur very infrequently are extremely difficult to find and correct, but until the trouble is found, the equipment is unreliable and in most cases useless. A single bit error in a memory storing millions of bits of instruction code usually means the machine gets lost and ceases to perform its function.

13.2.1 Typical memory subsystem

A typical memory subsystem consists of several memory devices interconnected to a source of address and control and a source for data (if an in-circuit alterable device) and a destination for data. For example, a block diagram of a typical memory subsystem is shown in Figure 13.3.

When high-speed memory devices are used, regardless of whether they are RAMs or ROMs, and regardless of what type of memory function they are used to implement, certain fundamental issues must be addressed for their successful application. Those fundamental issues include

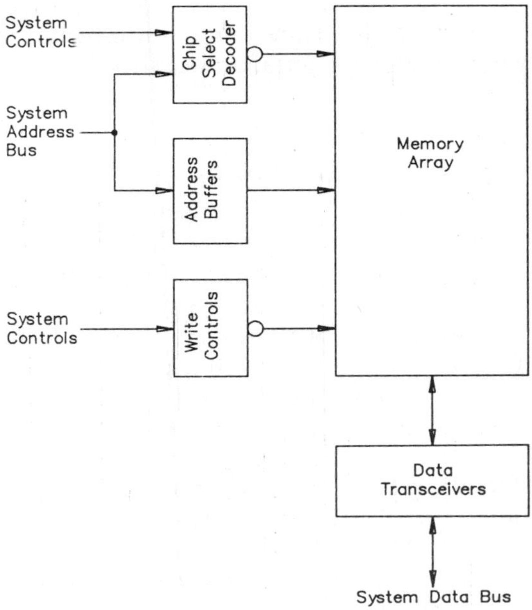

Figure 13.3 Block diagram of a typical memory subsystem.

1. Board layout for minimum signal routing
2. Control signal generation and timing (including write signal generation if RAM)
3. Worst-case address, data, and control path timing
4. Crosstalk control
5. Data bus contention considerations
6. Power and ground distribution and decoupling

Physical arrangement, timing, and crosstalk control are interrelated. It is impossible to do the timing analysis until the approximate signal line lengths and physical arrangements are known. Interconnection delays are a direct function of line length, and crosstalk is a direct function of the physical arrangement of memory interconnections. Not only must normal line delays, including delays caused by transmission-line effects, be considered in the timing analysis, but crosstalk may also impact timing. Extra time may be required for crosstalk to subside. Thus, the first task in a memory subsystem design is to make a rough estimate of the layout or physical packaging of the memory devices. The layout must minimize all signal interconnections to minimize line delays while at the same time minimizing the chance for cross coupling between address, data, and control lines.

13.2.2 Memory circuit layout

The prime objective when laying out memory circuits is to minimize interconnecting line lengths. The shorter the interconnections, the quicker signals settle and the less the chance for crosstalk. When only one or two memory devices are used, the strategy for minimizing line length is straightforward. Locate the memory devices as close to the source and destination of interface signals as possible.[12] When a large number of memory devices function together (sometimes referred to as an array of devices), the typical strategy is to locate the source and destination circuitry in the center of the array of memory devices. For example, Figure 13.4 shows a typical memory-processor layout used when maximum possible speed is needed.[1] By locating the processor in the center between two equal groups of memory devices, address, control, and data lines are minimized. When address and data lines have separate sources and destinations, an arrangement such as shown in Figure 13.4, where address buffers are located in the center of the memory array and data buffers or transceivers are located above or below the memory array, tends to be the best strategy for minimizing line lengths. Once a basic layout is determined, the next step is to interconnect the circuitry using the techniques for minimizing signal settling time and controlling crosstalk described below. Often several iterations are necessary before the required performance is achieved.

13.2.3 Control signal generation and timing

The first step in any memory device application is to thoroughly read and understand the control signal timing requirements specified on

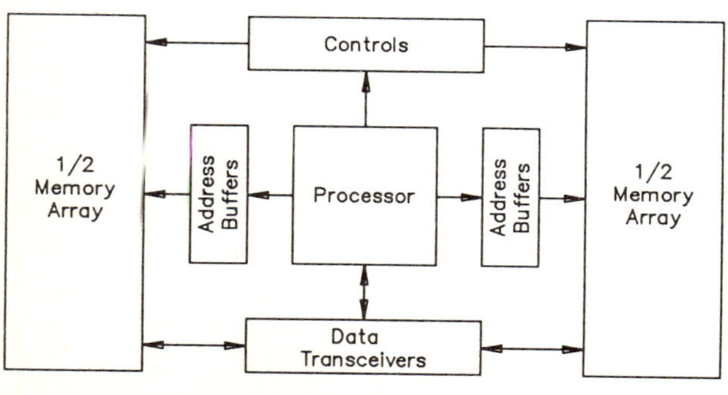

Figure 13.4 Memory and processor layout that minimizes signal length.

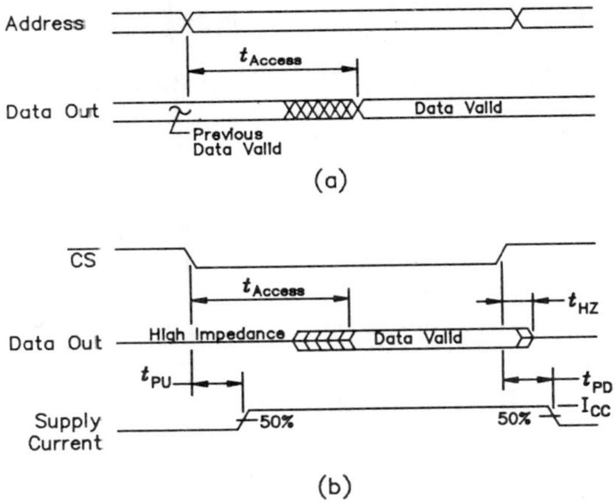

Figure 13.5 Functional timing for a typical memory read cycle. (*a*) Address controlled; (*b*) chip select controlled.

the memory device data sheet. That is often not an easy task. Many data sheets have several pages of timing specifications. Dynamic RAMs in particular have many critical timing requirements. Failure to read and understand data sheet requirements remains one of the leading causes of memory system problems. It is easy to be lazy when faced with several pages of timing diagrams, but unless all timing requirements are met, there is no hope for a reliable memory system. *Read the timing requirements over and over and over.*

The functional timing for a typical memory device read cycle is shown in Figure 13.5. Good data appear at the destination some time after both the address and chip select inputs to the memory array stabilize and the memories and the output data paths have time to react. Delays through the subsystem include

1. The propagation delays from clock to output of the address and chip select sources
2. The interconnection delays between the source of address and chip select signals and the address and chip select buffers
3. The address and chip select buffer delays
4. The interconnection delay between the address and chip select buffers and the memories
5. The access time of the memories
6. The interconnection delay between the memories and the data buffers

7. The data buffer delays
8. The interconnection delays between the data buffers and the final destination
9. The setup time of the receiving devices

The functional timing for a typical write-signal-controlled write cycle (for those devices that are in-circuit alterable) is shown in Figure 13.6. Address and chip select functionality and timing are the same as for a read cycle. The difference is that (in a write cycle) data must be supplied to the memories as well as a write signal. Write signal timing is always critical. Input data and all address lines must be stable and meet setup and hold times relative to write signals. In high-speed memory subsystems, that is usually not easy to accomplish.[2] Also, data contention between memory outputs and data input buffers is difficult to avoid during the start of write cycles when common I/O memories are used, and the high currents caused by data contention can upset memory devices and change data (see Section 13.2.6). In a write-signal-controlled write, memory devices first react as though a read cycle is starting; outputs become active and drive data out. Active memory outputs may conflict with data being driven in unless care has been taken to ensure that the data in drivers are not enabled until the write signal is active. To avoid problems with data contention, chip select controlled write cycles as shown in Figure 13.7, which are possible with most memory devices, are preferred and safer in most cases. In a chip select controlled write, memory outputs remain in a high-impedance state throughout the memory write cycle eliminating any possibility of contention.

In high-speed SRAM subsystems such as cache memories, write cycles usually must be completed in one clock cycle of the highest-frequency clock available which means there are no other clock frequencies or phases to shape write signals to satisfy address and data setup and hold times. In those cases where there are no other clock

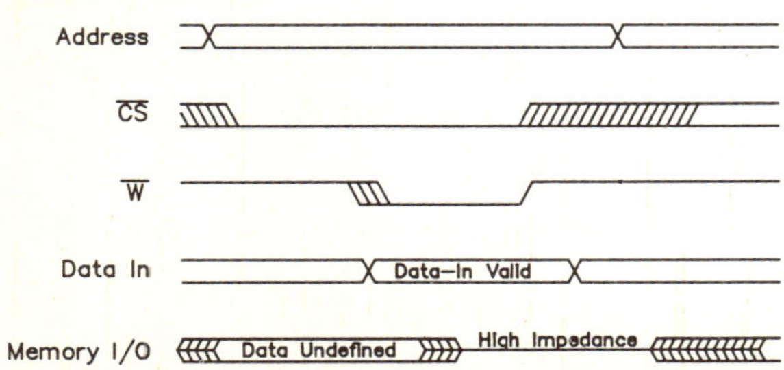

Figure 13.6 Functional timing for a typical write-controlled memory write cycle.

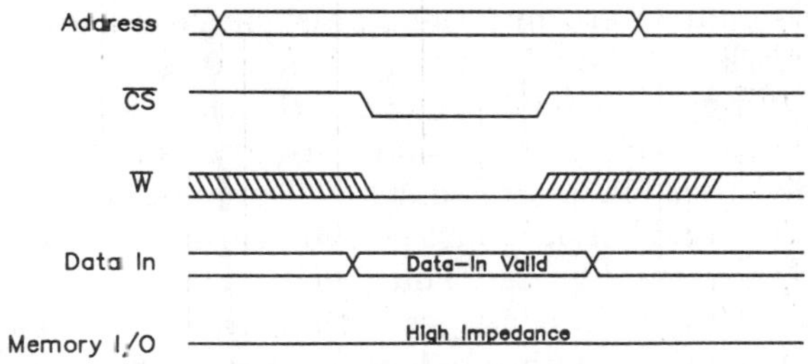

Figure 13.7 Functional timing for a typical chip select controlled write cycle.

frequencies or phases, write signals are typically generated by gating the clock with the write control signal as shown in Figure 13.8. Such an arrangement works fine at the start of a write cycle. Address and data have one-half of a clock period to stabilize before the write pulse becomes active. The danger with such an arrangement is that the address or data may go away before the write pulse goes away. If the address goes away first, data may be written to incorrect locations. If the data goes away before the write pulse, incorrect data may be written to the selected address. Usually only one level of gating is possible without skewing the write signal into the danger zone. The gating

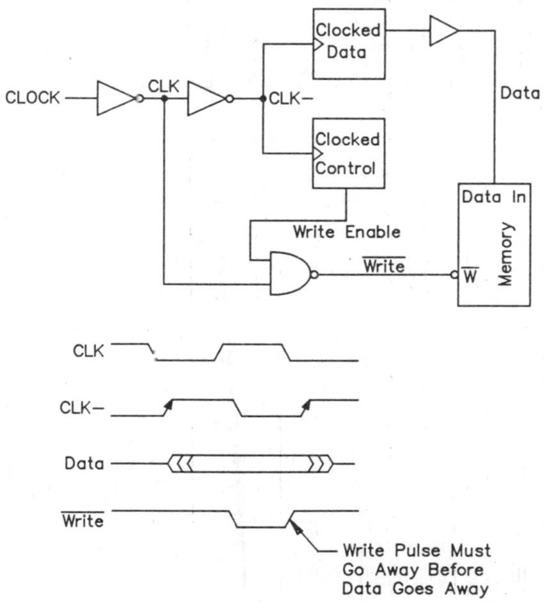

Figure 13.8 Typical circuit for generating write signals to RAMs.

structure and the write signal interconnections must be faster than the address and data buffers and their interconnections. It is usually not practical to use PALs to gate write signals with the clock for two reasons: PALs are usually much slower than discrete advanced BiCMOS/CMOS gates, and there is greater danger of ground bounce due to signal interaction. Ground bounce must be avoided since spikes on write lines can cause inadvertent writes. Thus, write control signal gates or buffers must not be mixed in packages with address, chip select, or other heavily loaded buffers. Other potential sources of spikes are decoders and multiplexers. Decoders must not be used for generating write signals unless a decoder disable signal is available that is guaranteed to disable the decoder during all possible times when decoder data inputs are changing. Multiplexers must never be used for write signals. Most multiplexers will spike when different inputs are selected even when the input signals are in states that do not call for an output change.

The selection of memory address and data buffers or transceivers requires a great deal of care. Address and data buffer characteristics must be matched to the application. Typically, they must have the capability to drive large capacitances in a short time. Most advanced BiCMOS/CMOS buffers are suitable for such tasks, particularly the bus interface devices. Board space limitations usually dictate octal or wider parts. Octal 240 or 244 drivers with series damping resistors (see Section 13.2.4) are often used for address drivers and octal 245 transceivers for data interfaces. Today a number of similar devices with built-in series resistors are available. Termination of address lines is discussed in Section 13.2.4.

When large banks of memories must be driven, tradeoffs must be made between the number of loads per driver and the number of drivers. As a rule of thumb in very high speed applications (50-ns cycle times or less), about 16 memory devices is the maximum number of loads (memories) that should be driven by one driver. When speed is important, it is usually best to split signals with larger than 16 loads into two groups and use two buffers. The tradeoff that must constantly be made in high-speed memory applications is speed versus hardware. The above rule of thumb is often not followed in DRAM applications. Often 30 or more DRAMs are driven by one address buffer. A number of special-purpose address generators and multiplexers are available for DRAMs that are designed to drive large numbers of DRAMs, but there is so much variation in the operation and functionality of these special-purpose parts that it is impractical to cover them in a book such as this. The reader is referred to manufacturer data books.

Chip select timing often limits memory subsystem speed. Chip select signals are typically decoded from upper address bits and as a re-

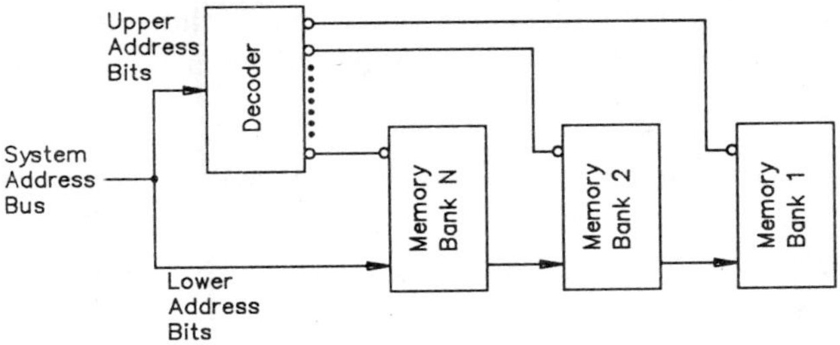

Figure 13.9 Chip select signals are typically decoded from upper address bits which causes the start of memory cycles to be delayed by the propagation delay of the decoder.

sult are delayed beyond the address signals as a minimum by the propagation time of the decoder used (see Figure 13.9). Special high-speed decoders are available for address decoding, but they still have some delay. Some advanced BiCMOS/CMOS decoders are down into the 5- to 7.5-ns range, so if the address signals can be picked up at the same level as the inputs to the address buffers and if possible decoder glitches do not have to be masked, it is possible to approximately match address buffer and chip select decoder delays.

A note of caution. In some applications, it is important that chip select signals do not glitch. For example, it is important that row access strobe (RAS) and column access strobe (CAS) signals to dynamic RAMs do not glitch and that chip selects to synchronous SRAMs do not glitch. In general, it does not upset asynchronous memories if their chip select glitches, but it is best to check with the manufacturer. Even if a glitch does not directly upset a memory device, glitches increase system noise directly and glitches by temporally selecting banks of memory cause transient power-supply disturbances. Power supply current goes up significantly when chip select inputs are active in most cases. To minimize equipment temperature and improve long-term reliability, chip select control circuitry should be designed so that none of the chip selects are active when none of the memory subsystem is being used (if memory devices that power down when chip selected are being used). That means if a decoder is used to generate chip selects, it must be controlled by an enable signal that forces all decoder outputs to the inactive state when none of the memory subsystem is in use.

13.2.4 Memory worst-case timing analysis

A careful analysis of memory address, data, and control path timing is essential for high-speed memory applications.[1] Usually system operating speed depends on memory loop speed, and usually there is little room for error. The analysis is typically approached from one of two possible standpoints:

1. To establish the system operating speed given a certain memory device speed
2. To establish that the memory devices available are adequate for the required system speed

In either case, the designer's task is to carefully determine the worst-case delay for each device in each signal path and the worst-case delay for each segment of the interconnection network and be able to show that the total delays are less than the required memory cycle time. To ensure that address, data, and control signal timing is met, a complete timing diagram that shows the worst-case *maximum* and *minimum* timing relationship of each control signal relative to worst-case address and data timing must be generated. It usually must be generated by hand since most engineers do not have access to CAD tools that account for both device and interconnection delays, and interconnection delays are generally significant in high-speed memory applications.

Worst-case device delays. Data sheet device propagation times and memory access times must be adjusted for worst-case environmental conditions and actual load conditions. Propagation times for most advanced BiCMOS/CMOS bus interface devices such as might be used for address and data buffers are specified for both worst-case commercial and military conditions and require no further derating for temperature or power-supply level limits. Propagation times for most BiCMOS/CMOS devices (excluding memories and some special address drivers) are specified with 50-pF loads. In many memory system applications, actual load capacitance will exceed 50 pF. In those cases, device timing should be adjusted using the techniques described in Chapter 11. BiCMOS/CMOS memory device access times are usually given as worst-case values over the specified operating conditions and do not require any adjustment for temperature or power-supply voltage variations. However, access time is usually specified with a 30- or 35-pF load which is typically less than actual load conditions. Memory access times are adjusted for actual load conditions using the same techniques used for discrete devices.

Interconnection delays. Interconnection delays often account for a large part of memory loop delays. Address and control lines, and data lines to a lesser extent, typically go to a large number of devices which means they are inherently long lines with long propagation times. Line delays are further increased because of large distributed loads (where there are a large number of devices connected to the line) which cause the intrinsic propagation time t_{pd} to be increased by the factor (see Chapter 11)

$$t_{pd}\,(\text{slow down factor}) = \sqrt{1 + \frac{C_{LOAD}}{C_{LINE}}}$$

where C_{LOAD} is the sum of the input and output capacitance of all devices connected to the line and C_{LINE} is the line capacitance. Determining C_{LOAD} is straightforward; C_{LOAD} is the sum of the input, or output capacitance, as the case may be, of all the devices connected to the line. Determining C_{LINE} requires that per-unit-length line capacitance C be known ($C_{LINE} = C \times$ line length). For pc boards, C is typically 2 to 4 pF/in, and for welded-wire and wire-wrap boards C is typically 1 to 2 pF/in. When the correction factor is applied, the actual interconnection propagation times t'_{pd} are given by Equation (6.4):

$$t'_{pd} = t_{pd}\sqrt{1 + \frac{C_{LOAD}}{C_{LINE}}}$$

In most cases where a large number of memory devices are interconnected, it will be found that the interconnection delay is substantially greater than the intrinsic line delay. In many cases, the actual delay may be 4 or 5 ns/ft instead of the intrinsic 2 ns/ft of pc boards.

Line delay due to transmission-line effects. Signal delays caused by transmission-line effects must also be considered when calculating memory interconnection delays. Transmission-line effects become of concern when line lengths exceed the critical line length (see Chapter 7). The common definition of *critical line length* is given by Equation (7.1)

$$\text{Critical line length} = \frac{1}{2}\frac{t_r}{t'_{pd}}$$

where t_r is the rise time of the driving source (20 to 80 percent) and t'_{pd} is the actual loaded propagation delay of the line.

If a memory interconnection exceeds the critical line length, and many will, some allowance for transmission-line effects must be incorporated into the timing analysis unless the line is terminated in such a way that the line is stable at all loads after a one-way delay. In most memory applications, it is impractical to terminate all lines. Thus, additional time beyond the one-way line delay must be allowed for most signals to settle or reach acceptable logic levels. Two conditions must be considered. They are

1. Underdamped lines
2. Overdamped lines

When only a small number of memory devices (perhaps one to five) are interconnected, most lines will be underdamped and will ring unless terminated. The rule of thumb for settling time for underdamped unterminated long lines is to allow five line delays for signals to settle (see Chapter 7). For pc boards that means approximately 1 ns/in [(2 ns/ft/12 in ft) × 5] must be allowed for interconnection delays. Assuming five line delays for long unterminated lines simplifies and expedites the timing analysis task. If the memory loop timing meets the system requirements with that assumption, it is usually safe to proceed. However, it is difficult to generalize a very complex phenomenon. Critical situations must be carefully analyzed. If the timing requirements are not met with the five-line-delay assumption, the next step is to determine the actual line response. It may not require five line delays for the lines of interest to stabilize. If after the actual settling time is determined the system timing requirements are still not met, there is little choice but to use faster memories and interface devices or find some way of matching impedance levels so that the lines responsible for the largest delays can be effectively terminated.

When a number of memory devices (perhaps 6 to 15) are interconnected, most interconnecting lines are overdamped. When a large number of memory devices (16 or more) are interconnected, signals are normally severely overdamped. Overdamped signals normally require at least one trip down the line and back to reach threshold levels. Severely overdamped signals typically require several reflection cycles, as shown in Figure 13.10, to reach the threshold level of the receiving devices.

Whether a signal is underdamped or overdamped depends on the ratio of the signal dynamic source impedance and the effective line impedance. If the source impedance is less than the line impedance, signal waveforms will be underdamped. If the source impedance is more than the line impedance, signal waveforms will be overdamped. The ideal case is when source and line impedance are equal, but that case seldom occurs in the real world. Dynamic source impedance is seldom given on buffer or memory data sheets. However, output short-circuit current I_{OS} or dynamic output current I_{OD} is usually given and can be used to calculate the voltage step size V_{STEP} if the effective interconnection impedance is known. That is,

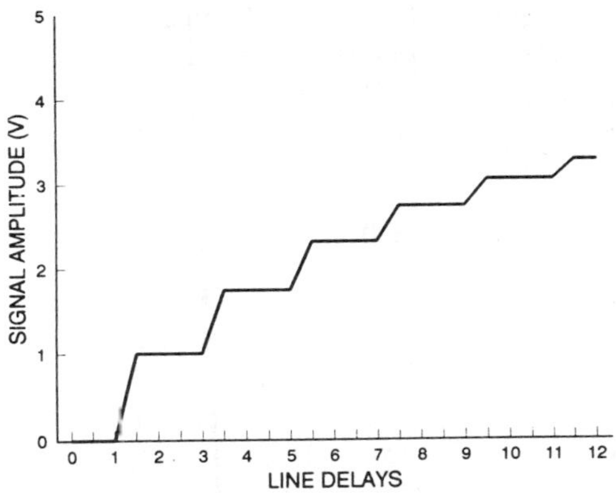

Figure 13.10 Address lines connected to a large number of memory devices are typically severely overdamped and require several reflection cycles to reach threshold levels.

$$V_{\text{STEP}} = I_{\text{OS}} \times Z'_o \tag{13.1}$$

Effective interconnection impedance Z'_o is found with Equation (6.3)

$$Z'_o = \frac{Z_o}{\sqrt{1 + C_{\text{LOAD}} / C_{\text{LINE}}}}$$

where C_{LOAD} is the total lumped capacitance (inputs and outputs) of each device connected to the line and C_{LINE} is the total line capacitance (see above and Chapter 6 for further definition of C_{LOAD} and C_{LINE}). Where a large number of memory devices are connected to a line, the effective line impedance may be as low as 20 Ω. For example, consider the case of a memory address line that originates at a buffer and goes to 16 memory address inputs. Assuming 7 pF for buffer output capacitance and 5 pF for memory address input capacitance, C_{LOAD} is 87 pF (7 pF + 16 × 5 pF) and assuming the line is 9 in long and is a pc board with an intrinsic impedance of 40 Ω and a per-unit-length line capacitance of 3 pF/in, C_{LINE} is 27 pF (3 pF × 9 in). The effective impedance Z'_o of the line is

$$Z'_o = \frac{40\ \Omega}{\sqrt{1 + 87\text{ pF} / 27\text{ pF}}} = \frac{40\ \Omega}{\sqrt{4.22}} = 19.5\ \Omega$$

The example is not unrealistic. It is difficult to maintain high intrinsic line impedance in high-density multilayer pc boards, and it is common to have address drivers driving 16 or more memory devices.

Once Z'_o is known, the worst-case step size, i.e., the minimum step size, is found using Equation (13.1) and the minimum short circuit current I_{OS} or dynamic output current I_{OD}. Unfortunately, maximum not minimum short-circuit current I_{OS} is usually given on data sheets. Dynamic output current I_{OD} is specified as a minimum value and is typically more appropriate (for minimum step size calculations). For example, FACT 54AC240s or 54ACT240s octal buffers have a minimum dynamic output current rating of 50 mA (75 mA for 74AC or ACT 240s). The initial voltage step for a 50-mA buffer driving a 20-Ω line is

$$V_{\text{STEP}} = 50\text{ mA} \times 20\ \Omega = 1.0\text{ V}$$

which means the signal must propagate down the line and back several times depending upon the starting level of the signal and the thresholds of the receiving memory devices before a new level is recognized. If the signal starts near V_{cc} and the receiving devices have TTL levels, it must transition 4.5 V which means 10 line delays (five round-trips) are needed for the signal to safely cross the *low* TTL threshold. Ten line delays at approximately 3 ns each (4 ns/ft × 9 in) is a significant and intolerable delay in most applications. The situation may be even worse where signals originate from memories or special-purpose devices, processors, ASICs, etc. They often have very low drive and their dynamic output drive may not be specified. Where short-circuit output current I_{OS} or dynamic output current I_{OD} is not given on data sheets, it can be estimated using the following rule of thumb: Typically the dynamic output current is approximately 10 times the static (dc) output current rating, that is

When I_o dc is 4 mA, I_{OD} is typically 40 mA

When I_o dc is 8 mA, I_{OD} is typically 80 mA

When I_o dc is 16 mA, I_{OD} is typically 160 mA

Etc.

Low effective line impedance is not always a problem in memory systems. Sometimes line impedance is too high relative to address or data line drivers. High line impedance means signals are underdamped and ring which causes extra delay. When line impedance is high, one solution is to terminate lines. Either source or load termination can be used. Source termination has the advantage of requiring less parts (than load termination) and of dissipating less dc power. The disadvantage of source-terminated lines is that at least two line delays are required for all points to transition to valid new logic levels (see Chapter 7). Signals start at half amplitude (assuming an exact match in source and line impedance) and must propagate to the load and reflect back to the source to reach final value. Load-terminated lines have the advantage of requiring only one line delay to reach final value (assuming the source impedance is much less than the line impedance). When the signal reaches the end of the line or the last load on the line, all loads have new valid logic levels. The disadvantages of load-terminated lines are high dc power dissipation and more parts. Load termination typically must be implemented with split or Thevenin's terminations which means at least two resistors are required per load termination. Split termination is required because, few advanced BiCMOS/CMOS buffers, even those designed for memory address driving applications, have the capability to drive 60-Ω or less terminations connected directly to ground or V_{cc}. The advantage of a Thevenin's termination is that it provides a means of matching lower line impedances while minimizing driver current source and sink requirements. However, Thevenin's or split terminations cannot be used with true CMOS levels that must switch between V_{cc} and ground. A split termination by definition is a divider between V_{cc} and ground and divides or limits signals to some value less than V_{cc}. In general, the use of split terminations is not a problem in memory applications since most BiCMOS or CMOS memory devices have TTL levels.

Another solution when signals are underdamped is to lower the impedance of the interconnections until they more nearly match the signal source output impedance. The simplest technique and the one most frequently used is to locate the signal source, for example an address buffer, in the middle of the line. The source (buffer) sees two lines in parallel which means the effective impedance seen by the source is one-half the normal line impedance. The source must be near the middle of the line or reflections may not cancel out. Another technique, which is similar and has the same effect as center driving lines, is looping lines back to the source. The advantage of looping lines to lower impedance is that it removes any question as to whether the driver is in the middle of the line. The disadvantage is that most pc board autorouters will not loop lines. If center driving or looping does not lower line impedance sufficiently, lines to memory arrays (i.e., groups of memory devices) can be connected in a grid to lower the effective impedance. Connecting lines in a grid, in effect, is the same as paralleling a number of lines which means the effective impedance of a grid interconnection is lower by a factor equal to the number of lines paralleled. One disadvantage of lowering line impedance to control ringing is that it increases the switching currents in the signal sources and in the power and ground system. Center driving, looping, and gridding memory interconnections tend to reduce worst-case signal propagation paths (as well as reducing ringing) which helps improve memory system response.

In high-speed memory applications each interconnecting path must be carefully analyzed and optimized. When it is found that performance does not meet requirements, definitive steps must be taken to improve the situation; lines must be shortened or broken into segments so that multiple drivers can be used; higher current drivers and memories may be required along with some or all of the above described special termination and interconnection techniques. High-speed memory operation does not come easily.

13.2.5 Crosstalk control

The faster edge speeds of advanced BiCMOS/CMOS address and data buffers and transceivers and of BiCMOS/CMOS memory devices greatly increase the possibility of memory subsystem being degraded by crosstalk. Crosstalk is of special concern in memory subsystems because most memory devices are asynchronous devices and are subject to disruption at any time. Crosstalk may not only cause erroneous memory operation; it may also slow memory access times. Extra time may have to be allotted to allow crosstalk to subside if it is not controlled. Extra time is usually not available when BiCMOS/CMOS memory devices are applied; the goal is usually to achieve the highest possible operating speed.

Cross coupling between any combination of memory input or output signals as well as crosstalk nearby unrelated signals can be disruptive and must be guarded against. Crosstalk to RAM write control signals is of particular concern; it may cause inadvertent writes. Crosstalk between address and data buses is a common problem; bused signals tend to change at the same time which means there is more coupling energy. Crosstalk between buses tends to be data pattern sensitive and is worse when all address or data lines change in the same direction at the same time. Excessive cross coupling from memory data lines to address lines during read cycles can result in positive feedback that degrades the response time of the memory device and in extreme cases can cause unstable oscillatory operation. During write cycles, the danger is that data to address line cross coupling may upset address lines sufficiently to cause writes to incorrect memory locations.

Crosstalk is a function of the separation between signal lines, the linear distance that signal lines run parallel with each other, and the height above a ground, or other, reference plane. Standard techniques for controlling cross coupling between signals include

1. Run signals at right angles to each other
2. Keep lines as close as possible to ground or reference planes
3. Isolate signals from one another by ground traces and reference planes

Multilayer pc boards simplify the task of isolating and controlling crosstalk in memory interconnection systems; as a minimum, address, data and control signals can be routed at right angles on different layers, and in more critical applications, voltage and ground planes can provide isolation for critical memory signals. Where high-speed RAMs are used, multilayer pc boards are essential; address, data, and control signals (particularly write signals) must be isolated by reference planes. All write control signals must be isolated from other signals by reference planes and extra-wide line-to-line spacing. Critical write signals can be further protected by running a ground trace on each side of signal traces. If each signal layer in the pc board is not separated by a reference plane, care must be taken to ensure that there is not a noisy signal running parallel and directly above or below a critical signal.

On welded-wire or wire-wrap boards or backpanels, memory interconnection wiring must be as direct as possible between points so as to randomize the routing, and the wiring must be kept as close as possible to the board or backpanel ground (or voltage) plane. Care must be taken to ensure that wiring is not channelized. Particular care must be taken to ensure that critical write signals do not get channelized with address or data buses or any noisy signals for that matter.

13.2.6 Data bus contention prevention

Bus contention occurs when two or more devices with opposing output states are enabled on to the same line at the same time. Contention, even for short times, causes high currents in the device outputs in contention. High currents cause noise that can cause system and device upsets and destroy devices.[13] Even if immediate destruction does not occur, extended periods of bus contention can impact long-term device reliability. Bus contention problems in memory subsystems are most apt to occur when common I/O memory devices are used, but contention can be a problem anywhere two or more tristate devices are connected to a line.

Write enable controlled write cycles to common I/O memories are rife with opportunities for data bus contention[14] and are the most common cause of contention problems. In a write enable controlled write cycle (see Section 13.2.3), the chip select and write timing is such that the chip select is applied before and removed after or coincident with the write signal. Problems occur because until the write enable is applied, memory outputs are active following the application of chip select. If the input data buffer is enabled before the write signal forces the memory outputs to a high-impedance state (see Figure

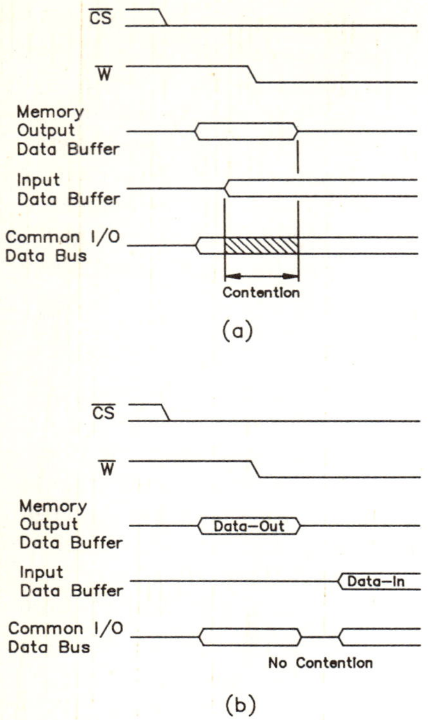

Figure 13.11 (*a*) Data contention at the beginning of a write cycle; data-in enabled before memory outputs are disabled. (*b*) No data contention; data-in enabled after memory outputs are disabled.

13.11), contention exists for the period until the write signal is activated and the memory outputs go to a high-impedance state. Contention can also exist at the end of the cycle if the memory address changes, if the input data changes, or if there is faulty operation (see Figure 13.12). As long as the memory address and input data remain the same and the device is operating correctly the output data are the same as the input data (since it should be reading out what was just written in) and no contention should occur. Applying chip select at the same time as write, or slightly after write and removing it slightly before removing write, or using a chip select controlled write cycle as described in Section 13.2.3 eliminates most contention problems and should be used where possible. Write-controlled write cycles should be used unless the timing is such that it can be guaranteed that no contention will occur; often write-controlled write cycles lead to devices being in contention for half or more of a clock cycle because the problem is not understood.

In high-speed applications it is usually very difficult to eliminate all possibilities for contention at cycle boundaries. In the past, when memory devices were much slower, a few nanoseconds of contention

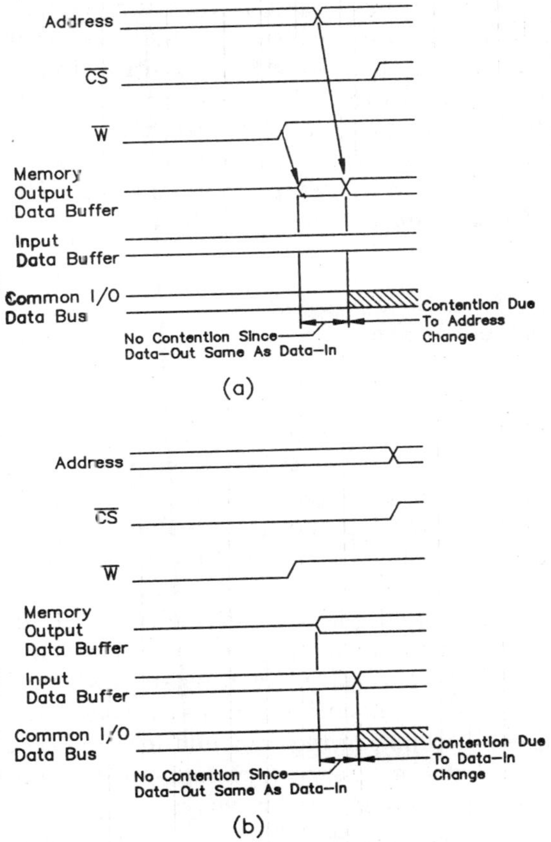

Figure 13.12 Data contention at the end of a write cycle. (*a*) Contention due to address changes; (*b*) contention due to data-in changes.

seldom caused a problem, but that is no longer the case. Many high-speed RAMs, particularly those with access times under 25 ns, are extra sensitive to data contention; very short periods of contention will cause data errors. When high-speed memories are used, steps must be taken to control memory chip selects or out enables and bus driver enables so that contention does not occur. In some cases contention problems can be cured by limiting the contention current. Contention currents can be limited by series resistors in the 22- to 100-Ω range, but series resistors slow response and may cause voltage offsets that in turn may cause logic level errors.

It is important to prevent bus contention at power ON to prevent damage to bus drivers and memory chips. Some memory devices have built-in circuits to limit output currents at start-up, but not all do, and most bus interface drivers do not. Thus, it is essential that tristate buses be controlled in such a way that all drivers are initialized OFF or only one is ON when power first comes ON. All bus control circuits should be asynchronously initialized by the system reset signal so that all bus control signals start up in the inactive state (so that all buses start up in a benign state with no contention). The presence of the system clock should not be required for a safe start-up at power-up. Under fault conditions, and during initial system check-out, the clock may not be present.

13.2.7 Power and ground distribution and decoupling

High-speed memory subsystems require low-impedance power and ground planes to distribute power. No other means of power distribution should be considered. Most memory devices have very large transient-current demands when first selected, often in the 100- to 200-mA range. Only very low impedance planes are capable of supplying such large transient currents without severe voltage drops. If universal boards are used, package power and ground pins must be connected directly to power and ground planes with solder clips or washers (see Chapter 5). The inductance of any length of wire sufficient to make a connection will be excessive. Even when power and ground planes are used, decoupling capacitors are also needed. The memory device manufacturer's decoupling recommendations must be followed exactly to ensure a reliable memory subsystem. Most manufacturers recommend one decoupling capacitor for each memory device of at least 0.1 μF for memory devices below 256K and of at least 0.22 μF for memory devices of 256K and above. All address and data buffers must also be decoupled since they tend to have very high transient-current demands.

Ceramic capacitors with good high-frequency response are appropriate for memory decoupling (see Chapter 5). Leads need to be as short as possible to minimize inductance. Surface-mount ceramic capacitors are ideal since they have the shortest possible leads. Decoupling capacitors must be connected directly to power and ground planes; no wiring or long pc board tracks should be used. They must be located as close as possible between the power and ground pins of the package being decoupled. A row of capacitors on the edge or center of a board does little good when high-frequency devices are used.

In addition to local decoupling, when a large number of memory devices are colocated and a large number of them are accessed at the same time, bulk storage capacitors that are as large as physically possible should be located near where power enters the board and around the perimeter of the memory devices (see Chapter 5 for guidelines on selecting bulk decoupling capacitors). Bulk decoupling capacitors help replenish the charge on local decoupling capacitors so that they are back to full charge at the start of the next memory cycle.

13.2.8 Failure rates and error detection and correction

Errors are always of concern in memory systems, and the larger the memory subsystem, the greater the chance for errors. An early decision that must be made in any memory subsystem design is whether some means of flagging and reporting errors is needed, or whether some form of error detection and correction circuitry is needed to meet error rate requirements. Memory subsystem error rates are mainly a function of memory device error rates since most of the devices in a memory subsystem are memories. Memory devices are usually pushing the state of the art, and devices that push technology are often more prone to failures or errors. The shear size and number of transistors in large memory devices greatly increase the chance for manufacturing defects including latent defects that do not show up until after devices are installed in systems. Some of the large memory devices are built with redundant sections that the manufacturer activates as needed during initial test to overcome fatal manufacturing faults and improve yield. Others have built-in error detection and correction circuits that are transparent to the user that become active when needed. Regardless of the manufacturing approach used to assure the manufacturer a reasonable yield, most of today's LSI/VLSI memory devices have a reasonably low failure rate considering the complexity of the devices. Failure rates, or more correctly projected failure rates, since state-of-the-art devices usually have little actual failure rate history, are specified in failures per billion device hours (FITs). With the assumption that the memory devices are the main contributors to memory subsystem failure, memory subsystem failure rate is the memory device failure rate times the number of memory devices. If it is possible to find and use memory devices that have a sufficiently low failure rate to meet system requirements, that is a much better solution than adding error detection and correction circuits. Error detection and correction do not come free; extra circuitry is required and the cost and the benefits of the extra circuitry must be weighed carefully. In some cases, the size and complexity of the extra circuitry may negate the perceived benefits. The extra circuitry may also slow through-put. Extra time is required for error detection and correction circuitry to make decisions. The simplest and most used technique for detecting memory subsystem errors is to use byte or word parity checking. Byte or word parity checking requires one additional bit of storage per byte or word and likewise in the data paths to and from the memory subsystem which is not a severe overhead. Actual error correction significantly increases the overhead. Logic implementations for detecting and correcting errors are out of the scope of this book. They are covered in most books on logic design.

13.3 Memory Subsystem Testing

As memory subsystem size and speed go up, it becomes more important that subsystems be thoroughly tested, but as size goes up, the time required for tests becomes significant. Tradeoffs must be made between component testing and subsystem testing in production situations. The time required to find all possible memory faults in a large memory subsystem is prohibitive and not cost-effective in a production environment. Thus, it is important that memory subsystems be designed so that they have adequate operating margins so that production testing does not have to weed out those units where the components are in specification but the tolerances all fall in the wrong direction. To ensure that memory subsystems are not operating on "the edge" or have pattern-sensitive problems, all memory subsystems must be thoroughly tested during the design verification process, or when major components or interconnection systems (e.g., pc boards) are changed. High-speed BiCMOS/CMOS buffers and memory outputs increase the chance for crosstalk, and high-speed memory devices, particularly SRAMs, are more sensitive to crosstalk on address, data, and write signals. Both crosstalk and ground-bounce disturbances are a function of the data pattern and are typically worse during all 1 to all 0 or all 0 to all 1 data patterns.

Testing a large memory subsystem is not easy, and the requirements and the degrees of flexibility are much different for RAM systems than for ROM or PROM systems. Little flexibility exists for changing data patterns in a ROM or PROM memory. However, ROM or PROM memory subsystems can and should be checked at the extremes of the specified power supply and temperature limits, and at the extremes there should be sufficient frequency margin left to ensure operation with worst-case process variations (see Chapter 11). RAM-based systems should be checked for timing margin and for pattern sensitivity at the extremes of the allowed operating conditions. High-speed RAMs are extremely sensitive to noise, and noise is generally a function of data patterns.

Often RAM memory tests are started with simple data patterns that check for basic functionality. For example, 1s are written to all locations and read back and then 0s are written to all locations and read back. Such a test is easy to implement, fast to run, and quickly indicates gross problems, but it will not catch data lines shorted to each other, address malfunctions, or data pattern sensitivity problems. Another simple test often run that is a little more comprehensive is an alternate "101010 - 0" and "010101 - 1" bit and word data pattern, called a checkerboard pattern, followed by its complement (see Table 13.1). A checkerboard pattern checks for adjacent data bit shorts but is still a very limited check for address or pattern sensitivity problems. All but the least-significant address line could be stuck, yet no problem would be indicated since the data are the same in all odd and even locations. Even more complex RAM tests are often misleading because RAM systems may echo back what is written regardless of where it is written. For example, if address lines are malfunctioning, data may be written to an incorrect location, but nonetheless they are written somewhere in the memory. If the data are not overwritten, when a request is sent to read the data, they come back correct because the address malfunction also causes data to be read from the (same) incorrect location. Even for the extreme case where all address lines are stuck, if only one pattern or if only one word is written and then read before another word is written, the address malfunction is not apparent. Thus, detection of address malfunctions requires special test patterns that are unique for given locations.[15] The same is true for detecting some forms of interaction, such as crosstalk, between address and data lines.

To check for address malfunctions, the test data must be a function of the address. For example, data equal to the address or some portion of it (generally the least-significant bits) can be written into the mem-

TABLE 13.1 Checkerboard and Complement Data Patterns

Address	Data
First test	
0	01010101
1	10101010
2	01010101
3	10101010
⇓	etc.
Second test	
0	10101010
1	01010101
2	10101010
3	01010101
⇓	etc.

TABLE 13.2 Address and Complement Data Patterns

Address	Data
First test	
0	000000...
1	100000...
2	010000...
3	110000...
4	001000...
⇓	etc.
Second test	
0	111111...
1	011111...
2	101111...
3	001111...
4	110111...
⇓	etc.

ory array and then checked (Table 13.2). If address lines are stuck, data will be overwritten (in most cases), and it will be obvious that a fault exists when the memory is read.

To check more than basic functionality, more complicated patterns than the ones described above are required, and the time required for the tests goes up significantly. In most cases, the time goes from a function of N, where N is the number of words in the memory, to a function of N^2 and the time required for a single test can take hours or days as memory size goes above 256K words. However, to check for data-dependent crosstalk or ground-bounce-induced problems, patterns must be run that change a word and check all other words, or some subset of other words, to see if they have been disturbed. One such test is called a "marching 1/0" pattern.[15] A marching test initializes a memory to all 0s and then sequentially writes words of all 1s into the memory and, after each write, checks all the previous locations to see that they have not been disturbed, and then sequentially writes words with all 0s and again sees that the previous locations have not been disturbed (Table 13.3). A more thorough variation of the marching test checks all locations above as well as the ones below the location where the last all 1s or all 0s word was written to see if any other location was disturbed by the write operation.

Another pattern that checks for disturbances is called the "walking 1/0" pattern.[15] In the walking pattern, the memory is initialized to all 0s or all 1s; then a word of the opposite sense, for example, all 1s if the memory had been initialized to all 0s, is written to a location, and all other locations are checked to see that they have not been disturbed.

TABLE 13.3 Marching 1/0 Data Pattern

Address	Data	Remarks
		First test
0	1111111...	
1	1111111...	
⇓	⇓	
$X - 1$	1111111...	
X	1111111...	Write address X, check all previous locations
$X + 1$	0000000...	
⇓	⇓	
N	0000000...	
		Second test
0	0000000...	
1	0000000...	
⇓	⇓	
$X - 1$	0000000...	
X	0000000...	Write address X, check all previous locations
$X + 1$	1111111...	
⇓	⇓	
N	1111111...	

That location is then returned to all 0s and the process repeated until all locations are checked with 1s and then the memory is written with all 1s and the process repeated for words of all 0s.

One of the most extensive test patterns is called a galloping pattern. A galloping pattern checks all possible read/read transitions.[15] A galloping test starts by initializing a memory to an all 0s or an all 1s state. Then a word of the opposite sense of all 1s or all 0s is written to a location and all other words in the memory are checked one at a time with the initial word checked after each of the other words is read to see that it has not changed. The process is repeated until each word in the memory has been used as the reference word. The galloping read/read pattern is useful for checking internal device interaction as well as system read problems, but it is not a good check of system write problems caused by crosstalk and ground bounce. A more useful test from a system standpoint is what is called a "galloping write-recovery" pattern. In a galloping write-recovery test, the memory is initialized to all 1s or all 0s and then a word is changed to the opposite state and a second word is changed to the opposite state. The first word is checked and the second word is changed back to its initial state and the first word is checked again. The process is repeated until all other words are checked. Then a new reference word is picked, and the process is repeated until all words in the memory have served as the reference word.

These special memory test patterns take a long time to run, but they, or variations of them customized to specific applications, must be run during the design verification process when high-speed BiCMOS/CMOS RAM memory devices are used. The faster the RAM memory chips, the more likely crosstalk, ground bounce, and other interaction are to cause problems. Noise, glitches, or other disturbances on write signals to high-speed SRAMs are a particular problem since most SRAMs have asynchronous write inputs and are sensitive to very narrow pulses at any time. Data sheets may call out a minimum write pulse width, but under typical conditions much narrower pulses may cause writes. SRAMs that are specified for 20-ns minimum write pulses often can be written with 3- or 4-ns pulses under typical conditions. Thus, high-speed RAM memory subsystems must be checked for possible data-dependent interaction problems even though long times are required for the tests. It is much less costly to find problems before a large number of systems are out in the field. As a minimum both the enhanced marching 1/0 (where all other locations are read to check for changes) and the walking 1/0 tests should be run. Ideally, the galloping write-recovery test should be run, but in large memory systems (for example above 256K) the time required is often prohibitive. Table 13.4 shows the number of memory cycles or operations required for the various test patterns where N is the number of words in the memory subsystem.

The time required for a particular test is determined by multiplying the number of cycles for a test by the time required for a memory cycle or by the time required for the test system to complete a cycle. Often a test setup has significant overhead and cannot exercise a memory at

TABLE 13.4 Number of Memory Cycles per Memory Test

Test pattern	Number of cycles where N = number of memory locations
All 1s and then all 0s	$4N$
Checkerboard and complement	$4N$
Address pattern and complement	$4N$
Marching 1/0	$N + 2N!$
Improved Marching 1/0	$N + 2N^2$
Walking 1/0	$2(2N + N^2)$
Galloping 1/0	$2(N + 4N^2)$
Galloping write-recovery 1/0	$2(N + 12N^2)$

its normal operating speed. Note that to read or write each location in a 1M memory with a 100-ns test cycle time per operation (i.e., a read or a write) requires 104.86 ms (1,048,576 × 100 ns) which is not long from a system test standpoint, but an improved marching 1/0 test for the same memory requires 2.199×10^{12} cycles which translates to 2.199×10^{5} s or 61 h. Thus, the more complicated patterns are only practical for small high-speed memory subsystems, such as high-speed cache memories.

13.4 Summary of High-Speed Memory System Design Techniques

1. High-speed memory address and data signals must be isolated from each other by reference planes and extra-wide horizontal spacing to prevent cross coupling. Running memory address and data signals at right angles may not be adequate when high-speed devices are used.
2. Memory control lines, such as chip selects, out enables, and write signals must be isolated from address and data lines and other noisy signals.
3. Memory write lines require the highest possible degree of isolation from crosstalk. They must be isolated by reference planes and by extra-wide line-to-line spacing from other signals, particularly other memory chip selects and address and data buses.
4. When prototyping boards, such as welded-wire or wire-wrap boards are used, address and write lines must be run at right angles to data lines.
5. Use a decoupling capacitor of the value recommended by the manufacturer for each high-speed BiCMOS/CMOS memory device (typically recommendations are from 0.1 to 0.22 μF).
6. When a large number of memory devices are colocated and are accessed at the same time, use bulk storage capacitors that are as large as physically possible near where power enters the board and around the perimeter of the memory devices.
7. Minimize signal path lengths for all memory address, data, and control signals to minimize interconnection delays.
8. Protect electrically alterable nonvolatile memories from false writes during transient power-supply conditions (see Chapter 12).
9. Do not tie memory data inputs or outputs directly to backplane or system buses. Memory devices typically do not have a great deal

of drive, and they are often sensitive to damage from transient conditions.

10. Make sure there is no bus contention on memory data buses. Particular care is required for common I/O memory devices.
11. Do not mix write signal buffers with buffers driving address, chip select, data, or other heavily loaded signals in a common package. Ground bounce may cause interaction.
12. Study data sheet timing cycles thoroughly. Ensure that all setup and hold times as well as all ac operating conditions are met. *Read the requirements over and over and over.*
13. Prepare a complete timing diagram that shows all *maximum* and *minimum* times for each component and interconnection segment in all memory address, data, and control paths to ensure that adequate timing margin exists.
14. Terminate or dampen all address, control, and data paths to minimize signal settling times in very high speed applications.
15. Do not allow contention between memory outputs and data bus drivers.

13.5 References

1. Altnether, Joseph P.: "High-Speed Memory Takes Fast Chips and Little Delay," *Electronic Design,* July 10, 1986, pp. 135–140.
2. Chritz, Jeff, and Al Reddy: "The 10- and 5-ns Solution to SRAM Contention," *Electronic Products,* April 1, 1987, pp. 43–46.
3. Leonard, Milt: "Density and Speed Drive Static RAM Technology," *Electronic Design,* December 8, 1988, pp. 63–70.
4. Leong, Raymond M.: "Purge RISK-Based Systems of Wait States," *Electronic Design,* February 9, 1989, pp. 69–72.
5. Myrvaagnes, Rodney: "SRAM Suppliers," *Electronic Products,* February 1989, pp. 45–52.
6. Bursky, Dave: "Advanced Self-Timed SRAM Pares Access Time to 5 ns," *Electronic Design,* February 22, 1990, pp. 145–147.
7. Rogers, Kathy: "SRAMs Are Self-Timing," *Electronic Engineering Times,* March 5, 1990, pp. 44–45.
8. Leonard, Milt: "IEDM Tackles Fresh Design Approaches," *Electronic Design,* November 23, 1989, pp. 41–46.
9. Mekhiel, Nagi: "Speed System Memory by Interleaving DRAM Accesses," *Electronic Design,* October 12, 1989, pp. 65–72.
10. Bursky, Dave: "Choices Abound for Nonvolatile Memories," *Electronic Design,* April 26, 1990, pp. 39–52.
11. Springer, John: "Designers' Guide to: Semiconductor Memory Systems," *EDN,* September 5, 1974, pp. 49–56.
12. Gunn, Lisa: "The Problems of RISK-Based Designs," *Electronic Design,* November 23, 1989, pp. 69–74.

13. Sokal, Nathan O: "Check List Helps You Avoid Trouble with MOS and Memory ICs," *EDN*, November 27, 1986, pp. 229–235.
14. *Bus Contention Considerations, Application Note #5*, INMOS, Colorado Springs, Colorado, December 1982.
15. Marshall, Martin: "Through the Memory Cells—Further Exploration of IC's in Testingland," *EDN*, February 20, 1976, pp. 77–85.

Chapter

14

Using PLDs, FIFOs, and Other LSI Devices

Programmable logic devices (PLDs), first-in first-out (FIFO) devices, dual- and multiport memories, and a host of other advanced CMOS and BiCMOS LSI devices available today have greatly simplified the system designer's task. They make it possible to keep parts count, cost, and power down while increasing performance. They provide flexibility that allows designers to develop new equipment faster and get products to market sooner. However, as with all new devices, they bring new problems. Large-scale devices exacerbate most of the electrical problems inherent in high-speed BiCMOS/CMOS applications. Ground bounce, transmission-line effects, and noise all are worse and cause more problems when LSI devices are used. The use of LSI devices, particularly PLDs, often leads to neglect of important decisions in the early phase of designs. Often it is assumed that system requirements can be met with PLDs or some other LSI devices when in actuality they cannot. It is important to not fall into the trap of assuming a function can be performed by an LSI device until it is certain that it can.

14.1 PLD Application Tips

Programmable logic devices provide system designers a great deal of flexibility and in many applications are an excellent alternative to standard SSI/MSI logic devices. PLDs reduce parts count and save board area. They offer a means of upgrading existing systems where board space is limited.[1] They may, if properly applied, optimize performance and improve reliability and testability. They shorten the design cycle and reduce the time needed for systems to be put on the market. Large programmable logic arrays (PLAs) provide a minimal risk alternative to application-specific integrated circuits (ASICs) in many lower gate count applications (up to approximately 5K of usable gates).

Today PLDs are available in a great assortment of sizes and logic structures sold under a confusing hodgepodge of names such as programmable array logic (PAL), programmable logic array (PLA), field programmable logic array (FPLA), and erasable programmable logic device (EPLD).[2] The purpose here is not to describe the fine points of the various PLD implementations or to provide a survey of the devices available but to list some general system application concerns that are universal to all PLD applications.

Speed and size are two of the foremost concerns when PLDs are considered for an application, and speed is often more of a limiting factor than density in high-performance system applications.[3] High equivalent gate count is always desirable, but if a PLD will not function at the speed needed, it is of no use. In applications where large FPLAs are used, there must be a tradeoff between speed and utilization. Interconnection delays become significant when several levels of logic must be interconnected. Careful evaluations of actual routed delays are essential before committing to an FPLA or any PLD solution.

In general, PLDs should not be used to synchronize asynchronous signals. PLDs tend to have a potentially wide window of susceptibility to metastable operation (i.e., the differences between their worst-case setup and hold times tend to be large),[4] and there is always the danger that an asynchronous input will inadvertently be routed to two or more flip-flops. Attempting to synchronize an asynchronous signal in two or more places introduces the risk that the signal will not be captured at the same time at all locations. No matter how good the metastable characteristics of PLD flip-flops may be, there is always the chance that different flip-flops will react in a different manner when their setup and hold time requirements are violated. Another reason for not synchronizing asynchronous signals in PLDs is that asynchronous and synchronous signals must be mixed in the same device and in the same board area which is not a good practice from a noise and cross-coupling standpoint. If a PLD is used to synchronize asynchronous inputs, the logic configuration must be controlled so that asynchronous signals only go to one clocked element (beware of automatic logic optimization), and PLDs must be selected with good metastable recovery characteristics. Advanced CMOS PLDs tend to have good metastable recovery characteristics[5,6] but regardless of how good the metastable recovery characteristics are, asynchronous inputs must not be allowed to change two clocked elements on the same clock edge. Gray code techniques and state machine implementations must be used to prevent multiple state changes in state machines with asynchronous inputs.

Care must be taken to properly initialize PLDs. Do not assume a power-up state. Some manufacturers design PLDs to power up with all internal registers in a given state and others do not. Where an asynchronous reset or initialization input is available, it should be used in conjunction with the system reset signal to initialize PLDs and reduce the chance for output contention or other possible abnormal start-up conditions that could damage parts (see Chapter 12).

See Section 14.3 for other important system design considerations when using PLDs.

14.2 FIFO Application Tips

Asynchronous FIFO devices greatly simplify the exchange of data between different frequency domains. Asynchronous FIFO devices allow data to be loaded and read at the same time, under control of signals that have no relative phase or frequency relationship. FIFOs are available in a number of configurations today. Most modern FIFOs are 8 or 9 bits wide and 64, 512, 1K, 2K, or 4K words deep. Most early FIFOs, of the 64-words-deep or smaller variety, were of the register fall-through type. Register fall-through FIFOs have the disadvantage, from a test standpoint, that data availability at the output is not fixed in time relative to a clock. Data fall-through time (i.e., the time data takes to move from the input port to the output port) varies with temperature and other environmental conditions as well as inherent device speed and depth. Fall-through register designs are not practical for large FIFOs, so most larger FIFOs, 512 words and above, are memory-based with pointers to write and read data from a memory array. Memory-based FIFOs with pointers have a fixed fall-through time independent of FIFO size which is usually no longer than a write and read cycle. As with PLDs and other LSI devices, FIFO speed is often of more concern than size or depth. The larger FIFOs have trouble running at the speeds needed for most high-performance CPU and processor applications, but higher-speed advanced CMOS and BiCMOS FIFOs are constantly being introduced.[7]

Most large FIFOs have write and read pulse requirements that are difficult to meet on a clock-to-clock basis in a single-phase synchronous system running at 20 MHz or higher. Most have write and read pulse width requirements that exceed one-half of a high-frequency clock period which means a gated clock cannot be used to generate write or read pulses. Thus, either multiple clock periods must be used for data transfers which is usually not acceptable, or special circuitry must be added to generate properly shaped write and read pulses. Figure 14.1 shows one scheme for generating proper write and read pulses using a delay line.

When generating write and read pulses, care is required to avoid data hold-time violations on FIFO inputs and at FIFO output data receivers. When loading data into FIFOs, the write pulse must go away before the input data goes away (check data sheets for hold-time requirements—some FIFO data inputs have significant hold-time requirements). Thus, any special circuitry required to shape write pulses must not cause extra delay. Write pulses must go away with the clock or slightly before (see Figure 14.1). Read pulses have the opposite requirement; read pulses must not go away before the active clock edge occurs at the receiving device since the output state of the typical large FIFO is controlled by the read pulse. Most FIFOs are designed so that outputs are active when the read pulse is present and inactive or are in a high-impedance (tristate) mode when the read pulse is not present (some FIFOs have output enables, but most do not). In most applications, read pulses are not present at all times, which means FIFO outputs are in a high-impedance state most of the time. Unless pull-ups are used or the output data lines are connected

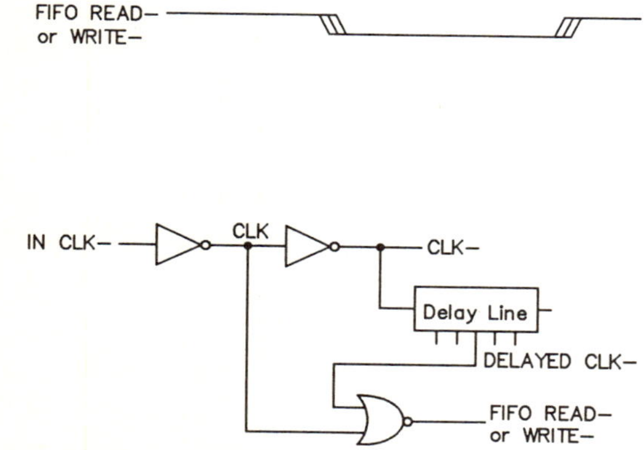

Figure 14.1 Delay line scheme for generating FIFO write and read pulses of proper width for a single clock time write or read in a system with only a single phase and frequency clock.

to other sources that are present and active, they will cause the same problems that floating lines cause in high-speed systems (i.e., noise and perhaps damage to the receiving circuits). In most cases, pull-ups are required on FIFO outputs to prevent the outputs from floating when not selected (i.e., the read pulse is not active) and during start-up or fault conditions.

See Section 14.3 for other important system design considerations when using FIFOs.

14.3 Important Considerations When Using LSI Devices

14.3.1 Ground bounce and output pin interaction

Ground bounce and power-supply droop are of particular concern when PLDs, FIFOs, and other LSI devices are used. Most such devices have a large number of outputs that can switch at once, but most LSI devices do not have sufficient ground or V_{cc} pins to support heavy simultaneously switched loads without severe output pin interaction. Center ground pins are now used in some of the new high-speed PLDs to reduce ground-bounce effects, but center ground pins only reduce ground bounce—not eliminate it. The only recourse system designers have is to limit output loads and control the mix of signals in LSI devices where that is possible (e.g., in PLDs). In general, most of the rules for controlling ground bounce in SSI and MSI devices covered in Chapter 4 must be followed when using PLDs and other LSI devices. For example, the following classes of signals should not be sourced from a common LSI device:

1. Signals going to asynchronous inputs and signals driving heavy loads
2. Clock enable signals and signals driving heavy loads
3. Write enable signals to RAMs and other heavily loaded signals
4. Clock signals of different frequencies

In general, PLDs should not be used as clock drivers because of potential ground-bounce problems and other possible interaction.

Byte-wide FIFOs and other devices with byte-wide outputs (where it is possible for all bits to change at once) must not be used to drive large buses. Large loads will cause ground-bounce problems. If outputs of byte-wide LSI devices must go to more than one or two loads, they should be buffered and redriven to prevent ground bounce from disrupting internal operation. FIFOs are of special concern. Most

FIFOs are asynchronous devices and are extremely sensitive to spikes on ground or V_{cc}. Outputs of byte-wide devices must be lightly loaded to guard against ground-bounce upsets.

14.3.2 Proper connection of unused inputs

Proper connection of unused inputs on PLDs and other LSI devices is often neglected, but when LSI devices with CMOS input stages are used, it is even more important to prevent inputs from floating than it is on SSI and MSI devices. Floating inputs cause two problems:

1. Floating input stages often go into oscillation which causes noise which may disturb the operation of the devices as well as the operation of other nearby devices.
2. Floating inputs that drift into the intermediate region between a valid logic *high* or *low* may cause excessive V_{cc} to ground current (both input MOSFETs may turn ON) which may cause overheating and parts failure (see Chapter 3).

Unused PLD inputs, particularly those that serve as either inputs or outputs, are often neglected. Unused PLD outputs that serve as feedback paths or as alternate inputs must be tied to a valid *high* or *low* logic level, or the associated output must be programmed to an active *high* or *low* logic level. Programming unused outputs that serve as alternate inputs so that they are active (i.e., not in a tristate mode) is usually the safest approach. Tying outputs to a common pull-up or pull-down voltage introduces some risk—an output might be inadvertently programmed to an opposite level which would cause contention and possible damage.

14.3.3 Proper termination of inputs

Many advanced CMOS LSI devices, including PLDs, do not have effective input clamp diodes and as a result input signals ring and take much longer to settle than may be expected. If input signals ring and take a long time to stabilize, inherent device speed may be lost. However, input dynamic characteristics are usually not specified on data sheets, which leaves the system designer with one of two alternatives where signal settling time or waveshape are of concern: (1) Parts must either be characterized for dynamic input characteristics and parts with good input clamps selected, or (2) steps can be taken at the system level to prevent ringing.

If it is decided to characterize inputs, device input dynamic characteristics can be determined with a simple test setup as shown in Chapter 7. The test consists of observing input waveforms while inputs are driven by a buffer or gate (from the logic family being used to implement the system) through a 2- or 3-ft length of coax (of approximately the characteristic impedance of the system interconnection system). If the device under test has good input clamps, input signals will overshoot a small amount (1 or 2 V) but will not undershoot back into the threshold region. If the device under test does not have good input clamps, input signals will ring and undershoot back into the critical threshold region. Testing devices for input characteristics is usually not a good long-range solution. It is expensive and time-consuming, and there is no guarantee that future parts will behave as those tested. Manufacturers often change designs and processes, and functionally identical devices from different manufacturers can have very different clamping characteristics. Often the best overall solution is to assume that LSI devices will not have clamps and eliminate the chance for ringing (on critical lines) by the use of conventional line-terminating techniques or by ending critical signal lines at devices with effective clamps. Lines feeding multiple devices (including devices without clamps) can be arranged so that critical lines always end with a device with an effective input clamp. Signal strings without devices with effective input clamps can be clamped by the addition of an SSI buffer, such as a FACT 54/74AC or ACT 240-244 or a TI BiCMOS 54/74BCT240-244, with an effective input clamp located at the end of the string. In some cases, the buffer may serve no useful purpose other than to clamp the line, but in some cases clamping buffers can be used to drive signals to test ports or test connectors. Conventional termination techniques are often less desirable than clamps for several reasons: Termination resistors may cause logistic and handling problems in surface-mount applications; source termination slows signal settling time depending upon the location of loads relative to the source, and load termination increases power dissipation.

The longer the signal lines and the faster the signal source switches, the greater the chance for system operational problems due to unclamped inputs. Very short lines, those less than the critical line length (see Chapter 7), may not cause problems. Long lines, such as at board interfaces, are of particular concern. Devices without input clamps should not be used to receive signals at board or system interfaces. Not only is there danger of operational malfunctions due to ringing, there is also danger of damage to devices. Excessive ringing, such as might occur on long lines at board or system interfaces, may exceed input level specifications and damage device inputs.

Clamping of all signals is not always required. In most applications, some signals are more critical than others. For example, clock signals to PLDs are generally much more critical than data or control signals.

Read and write controls to FIFOs are generally more important than data inputs. Clock and read-write signals must not ring, but data inputs can ring if there is time for them to settle and the amplitude of the ringing is not of sufficient magnitude to damage parts. Thus, clock and read-write signals should be clamped in most applications, but clamping of data and control inputs depends on the application.

14.3.4 Decoupling needs of LSI devices

Large-scale ICs have special decoupling needs. They often have a large number of outputs that switch at once, and they may have large internal switching currents both of which can cause large transient currents in power and ground. To avoid potential troubles due to large transient currents, care must be taken to follow manufacturers' recommendations as to the amount and placement of decoupling capacitors. System designers must not make assumptions as to the decoupling requirements. Data sheets for LSI devices usually specify the decoupling requirements, and those requirements must always be followed. If a data sheet for a device does not specify the required decoupling, the manufacturer should be consulted for recommendations. In addition, regardless of what the manufacturer recommends, practical engineering judgment should be used when specifying decoupling for LSI devices. The worst-case simultaneously switching load capacitance must be determined, and a value of decoupling capacitance must be used that will keep the local V_{cc} at a proper level (see Chapter 5). The general rule of thumb is that the local decoupling capacitance should be at least 100 times the simultaneously switched load capacitance.

14.4 Summary of Design Techniques for Using PLDs, FIFOs, and Other LSI Devices

1. Clamp or terminate all critical input signals to LSI devices that do not have input clamps.
2. Do not use LSI devices without input clamps to receive signals at board or system interfaces.
3. Do not use PLDs to synchronize asynchronous signals.
4. Follow manufacturer's recommendations for decoupling LSI devices.
5. Do not let unused PLD inputs or output feedback paths float.
6. Do not let FIFO outputs float.

7. Do not drive large buses or otherwise heavily load LSI device outputs.
8. Do not mix signals going to asynchronous inputs and signals driving heavy loads in PLDs or other LSI devices.
9. Do not mix clock enable signals and signals driving heavy loads in PLDs.
10. Do not mix write enable signals to RAMs and other heavily loaded signals in PLDs.
11. Do not use PLDs as drivers for clock signals of different frequencies.
12. Do not use PLDs as clock drivers because of potential output skew, ground-bounce problems, and other interaction.

14.5 References

1. Nass, Richard: "ASICs: The Latest Alternative," *Electronic Design,* October 12, 1989, pp. 51–60.
2. Greer, David L.: "A New IC Classification Act," *Electronic Engineering Times,* February 26, 1990, pp. 37, 72.
3. Baker, Stan: "Silicon Bits—Beyond the Gate-Count Wars," *Electronic Engineering Times,* January 1, 1990, p. 30.
4. *PAL Device Data Book,* Advanced Micro Devices Inc., Sunnyvale, Calif., 1988, pp. 3-164 to 3-169.
5. *The Programmable Gate Array Data Book,* Xilinx Inc., San Jose, Calif., 1988.
6. *GAL Data Book,* Lattic Semiconductor Corp., Hillsboro, Oreg., 1989.
7. Rogers, Kathy: "FIFO Race Revs to 15 ns," *Electronic Engineering Times,* January 29, 1990, p. 48.

Chapter

15

ASIC Application Tips

The purpose of this chapter is not to delineate all the advantages and disadvantages of application-specific integrated circuits (ASICs), most of which are well known.[1] Nor is the purpose here to survey ASIC types and availability or cover general ASIC applications. ASIC devices and technology change so rapidly that even current magazine articles often miss several manufacturers' latest announcements. Instead, this chapter discusses several topics that may not be obvious to those new to ASIC applications,[2] and covers several system-level concerns that are generally not covered in ASIC device manufacturers' data books or application information. For specific gate array, standard cell, or other ASIC device family capabilities, characteristics, and application information, readers are referred to ASIC manufacturers' most recent data books and application manuals. In all cases, manufacturers' application information must be studied closely and followed precisely.

The great majority of ASICs used today are fabricated with advanced CMOS processes, but BiCMOS ASICs are displacing CMOS ASICs in many high-performance applications. BiCMOS ASICs typically make possible at least a 50 percent improvement in operating speed over CMOS-only circuits with similar features.[3] However, at present BiCMOS processing costs more than CMOS processing, but the differential is decreasing. Even with the cost differential, BiCMOS technology is expected to quickly displace CMOS technology for most high-performance applications.

BiCMOS ASICs with bipolar output stages offer a level of performance that is not possible with CMOS-only devices. Bipolar output circuits reduce both transient switching currents and propagation times, and propagation times do not degrade as much with environ-

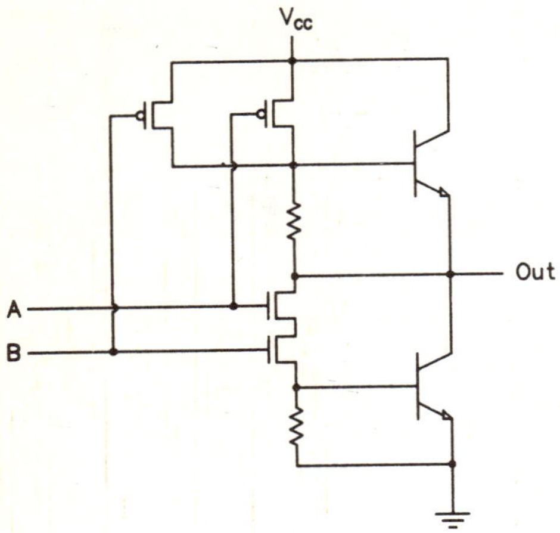

Figure 15.1 Schematic of AMCC Q24000 series BiCMOS two-input NAND showing bipolar output circuit used with internal CMOS logic cells. (*Reprinted with permission of Advanced Micro Circuits, Corp.*)

mental conditions such as high temperature, low supply voltage levels, or high capacitance. A major benefit of bipolar outputs is that transient switching currents are reduced by as much as 65 percent relative to CMOS-only output stages.[4] Lower transient switching currents mean that fewer ground and power pins are required or alternately that outputs can have higher drive. Bipolar output stages make it possible to interface directly to buses without requiring an impractical number of power and ground pins. BiCMOS ASICs are available with drive capability of up to 72 mA per output.[5]

Not all BiCMOS ASIC vendors combine CMOS and bipolar circuits in the same manner. Some arrays are fabricated with CMOS-only cores surrounded by bipolar I/O circuits. Other vendors combine CMOS and bipolar circuits throughout the array[6] (see Figure 15.1) or at least have internal sections that have an option of using bipolar output stages for driving critical signals such as clocks and buses. Some BiCMOS ASICs have an option on a cell-by-cell basis that allows internal bipolar output stages to be activated or not activated as needed for drive. Bipolar internal drivers are seen as the key to the next level of performance. In CMOS ASICs, heavily loaded internal interconnections often contribute a significant portion of signal delay and limit overall performance. In contrast, BiCMOS internal cells experience little performance degradation as the load is increased. Internal bipolar drivers significantly reduce fan-out propagation time degradation of heavily loaded internal interconnections[7] (see Figure 15.2 for comparison of CMOS and bipolar internal interconnection delay).

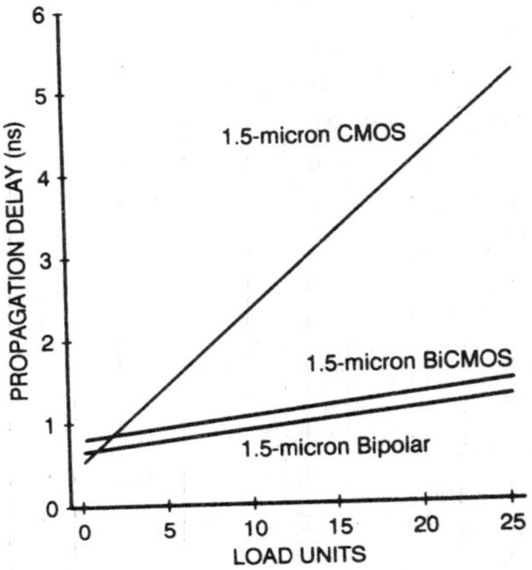

Figure 15.2 Fan-out degradation comparison of internal BiCMOS and CMOS drivers as a function of loading. (*Reprinted with permission of Advanced Micro Circuits, Corp.*)

15.1 Logic Structures to Avoid in ASICs

Special design techniques are often required for special devices, and ASICs are no exception. Certain logic structures that offer many advantages in discrete implementations may cause problems in ASIC implementations. One example is tristate buses. Tristate buses often simplify circuit board or motherboard wiring and reduce parts count. However, tristate buses usually complicate ASIC internal signal routing. Buses that go to a large number of locations tend to use up a great deal of the available on-chip signal interconnection resources and block other needed interconnections. Buses also tend to have slow response because long interconnections and many loads mean large line capacitance. Internal drivers often do not have enough drive to quickly drive large capacitance loads. Large buses also cause internal power to go up. Slow signal transitions through the region where both MOSFETs in complementary input pairs are ON cause increased V_{cc} to ground current. Tristate buses also offer the potential for internal bus contention and possible damage to parts. In general, multiplexers are a better choice than tristate buses for combining signals in ASICs. In discrete implementations multiplexers often require an excessive number of packages, but in ASICs they may be a more efficient use of silicon (than tristate buses). If tristate buses must be used, they should be controlled with an internal decoder circuit that only allows one node to be ON (the bus) at any given time. If a decoder circuit is not possible, bus drivers should be controlled with circuits with asynchronous reset inputs that are connected to the system reset or volts good signal. However, asynchronous reset control can only ensure proper initialization. Unless an exclusive decoder is used, there is no guarantee that erroneous operation will not cause bus contention.

Tristate circuits also offer the potential for problems at ASIC interfaces. Contention can also occur at ASIC interfaces, and it can be more damaging than internal contention since output drivers typically have much more current capability than internal drivers. Interface bus contention is often more difficult to control (than internal contention). Often it may not be possible to use a decoder to enable only one source. Bidirectional interface ports are special cases of tristate buses and offer the same potential for contention as other tristate buses as well as additional complications (see Section 15.2.2).

Buried circuits with internal feedback paths need to be avoided. If problems occur in buried feedback circuits with no visibility, it may be impossible to determine the cause of the problem. Ideally, feedback should be connected externally, but external feedback is often not practical in high-speed application—ON and OFF chip times are too long. If internal feedback must be used, for example in high-speed state machines, all storage element outputs should be brought out so that they are visible, or provisions should be made to disable feedback paths by external controls.

Circuits that require a long time to initialize should be avoided. They complicate test and in extreme cases make test prohibitively expensive. All ASIC storage elements should utilize asynchronous reset circuitry to ensure rapid initialization.

15.2 Guidelines for Selecting Input-Output Characteristics

15.2.1 CMOS or TTL Levels?

Most of today's ASIC vendors offer some choices for interface levels. CMOS ASICs are typically available with CMOS or TTL interface levels. BiCMOS ASICs are typically available with ECL and TTL levels and sometimes CMOS levels. When the choice is between CMOS or TTL levels, it is usually best to select TTL levels (in very high speed applications, the choice should be ECL).[8] TTL levels generate less noise than CMOS levels and are compatible with most standard devices and test equipment.[9] CMOS levels offer more noise margin (than TTL), but they create more noise and cause compatibility problems with other parts. Many LSI devices, such as PLDs, FIFOs, RAMs, and ROMs, are not available with CMOS levels. In some critical signal cases, such as clocks, CMOS levels may provide some advantage from a noise margin standpoint. However, CMOS levels complicate terminating clock lines (see Chapter 8), and clock signal waveshape must be controlled.

15.2.2 Do not let BiCMOS/CMOS ASIC inputs float

Damage to CMOS or BiCMOS ASICs due to open or floating inputs is a serious concern. Most CMOS and BiCMOS ASICs have complementary MOSFET input circuits that may conduct excessive and damaging current (between V_{cc} and ground) if inputs are allowed to remain in the indeterminate region between valid *high* and *low* logic levels (see Chapter 3). The danger is greatest during initial system test or when parts are first being checked. Test routines or test equipment may not operate as expected during the initial check. Proper signals may not be present because of software or hardware errors which may allow inputs to float (with damaging results). A careful check of component test setups prior to first installing expensive ASICs is essential to prevent floating inputs. The check must ensure that all inputs, including bidirectional ports functioning as inputs, are driven with valid logic levels at all times during test. Bidirectional ports present a special problem during test. The question is often, When is a bidirectional port functioning as an input or an output? If the direction of a bidirectional port is controlled by an external signal, control of direction is straightforward, but if the direction is controlled by an internal signal, it is difficult to prevent illegal conditions (either floating or bus contention). When a bidirectional port is controlled internally, a signal should be brought out to indicate port direction for use of testers or external buffers that might have to be used to buffer the port for test loads or other reasons.

It is not a good practice to connect ASIC inputs directly to signals that originate on other circuit boards or that originate from external sources. Expensive ASICs should be buffered and isolated from potentially damaging transients. Transients may cause direct damage or initiate latch-up that can cause damage (see Chapter 3). If ASICs must be connected directly to board or system interfaces, pull-up resistors should be used on all lines where there is a chance for the signal source to be disconnected. Series current-limiting resistors should also be used to limit injected transient current to prevent latch-up or other damaging effects (see Chapter 9 for recommended board or system interface circuits). Pull-up and current-limiting resistors add parts and complicate board layout. To avoid adding pull-up resistors to board or test stations and to reduce the chance for damage due to improper operation of interface signals either during test or during normal in-circuit operation, in most cases it is best to specify internal pull-up resistors or current sources for all ASIC inputs and bidirectional ports. Most BiCMOS/CMOS ASIC vendors offer low-current pull-ups or pull-downs. Some vendors offer only one value, but some offer two or three choices. For example, one ASIC vendor offers current sources of 5, 95, and 400 μA.[10] For most applications, where the objective is to prevent floating inputs during the initial check or under other similar circumstances, 50-kΩ pull-ups or pull-downs or equivalent current sources are appropriate. Pull-up or pull-downs must have enough current capability to provide worst-case input current requirement but should have no more current capability than is necessary to limit power dissipation. *Caution:* Current source pull-up or pull-down current values are not precise. They typically vary four-to-one with normal temperature, voltage, and process variations. Because of the possible large variation in characteristics, internal pull-ups or pull-downs should only be used to prevent floating inputs and should not be used for dynamic signal pull-up operation (unless the possible large variations in pull-up or pull-down times can be tolerated).

15.2.3 Output drive selection

Most BiCMOS/CMOS ASIC vendors offer output drivers with various current and slew rate limits. In general, output drivers with the lowest current and slowest slew rate compatible with system requirements should be selected.[11] More current capability than is necessary increases the chance for ground bounce and other system upsets due to transient switching currents. More slew rate than is necessary increases transient currents plus increases crosstalk and transmission-line effects. Package ground and power pin requirements are a function of output driver current and slew rate. Power and ground pin requirements increase significantly for high-current high-slew-rate output buffers, and package pins are usually at a premium.

15.3 Establish the Number of Ground and Power Pins Early

It is important to establish the number of package power and ground pins early in a design. Speed is a function of output drive, and the number of power and ground pins is a function of output drive characteristics and the number of outputs that switch at once. Transient switching currents cause fluctuations in internal power and ground levels that cause spikes on unswitched outputs (which can cause system upsets) and may also disturb internal logic. It is easy to underestimate the number of power and ground pins required in high-performance applications. Often optimistic guesses are made that severely limit design options later on. High-speed high-current outputs (for example 12 mA or higher) often require a ground and power pin per two outputs and in some cases may require a power and ground pin per one output. Unless a proper allowance has been made for power and ground pins, a larger package may have to be used, which may cause a number of problems, or performance may have to be decreased. Slower lower-current outputs that require less power and ground pins may have to be used which means less performance.

Most ASIC vendors have strict rules as to power and ground pin requirements and will not fabricate devices that do not meet their requirements. Thus, it is important to understand the requirements early in a design so that a proper allowance is made for their requirements. Vendor application information must be thoroughly studied. However, many system application issues also affect power and ground pin requirements. The ASIC vendor may not understand the system issues and should not be relied upon to establish the requirements. The system designer is in the best position to understand all the issues and must take responsibility for establishing the required number of power and ground pins. Even if the ASIC vendor takes the initiative to establish the number of pins, the system designer should check the results.

Important system application issues that affect the number of power and ground pins required are

1. Number of simultaneously switching outputs
2. Output speed and current driver requirements
3. Package pin inductance
4. Ground bounce and power-supply droop tolerance

Items 1 and 2 above are system issues that the ASIC vendor has little knowledge of until the later stages of ASIC designs. They must be determined by the system designer. Package pin inductance is usually available from the ASIC vendor unless a custom package is used. If a custom package is used, it is important that either the user or the ASIC vendor determine the package lead inductance. Interconnection inductance between package power and ground pins and circuit board power and ground planes must also be considered. Ideally, package power and ground pin connections should be made directly to power and ground planes so as to not add additional inductance, but that is sometimes not possible with certain surface-mount techniques. If it is not possible to make a direct connection to circuit board power and ground planes, the inductance of interconnecting traces must be added to the package pin inductance and the total used to calculate power and ground pin requirements.

15.4 Guidelines for Interface Timing

Timing of ASIC input-output signals relative to circuit board or system clock signals and system signals is a major issue that is often overlooked. ASIC vendors and their software tend to concentrate on internal signal relationships and ignore system issues as is to be expected. The system designer must take the initiative to ensure that all system components including ASICs have the ability to communicate where required. To simplify communication between ASICs and other components, all ASIC interfaces should be synchronous. The more complicated the device, the more important it is to follow synchronous design practices (see Chapter 10). Furthermore, most ASIC vendor software only supports synchronous design which means if asynchronous interfaces are used, they often cannot be simulated.

When synchronous ASIC interfaces are used, the input signal timing parameters of most interest are setup and hold times relative to the local board or system clock signal. Thus, it is important to establish ASIC input timing parameters with respect to the clock-signal phasing at the ASIC package interface, not to some internal clock phase.

Setup time requirements are usually straightforward. Input signals must propagate to internal clocked devices in sufficient time to be captured by the next clock edge. The less setup time required, the better. Setup times are usually worse at high temperature where CMOS and BiCMOS devices slow down the most.

Hold-time requirements cause most ASIC input interface problems. Inputs to clocked devices tend to have long hold-time requirements because internal clock networks tend to have long propagation paths. Long clock paths mean internal clocked devices are clocked sometime later than when the clock edge occurs at the package interface. Unless input signal paths have an equal delay, they must be held until the internal clock edge occurs. Because of the uncertainty of minimum propagation times of discrete devices or ASICs it is usually impossible to guarantee a given amount of hold time in synchronous systems. Thus, ASICs should be designed to have zero or negative hold-time requirements as most MSI and commodity LSI devices do. ASIC vendors do not like to design for zero hold times because it is an extra compilation for them (and board-level interface problems are not their concern). Zero hold-time interfaces also simplify second sourcing of ASICs. System operation is less dependent on device-to-device match of input-output propagation times.

Hold-time problems occur most often at cold temperature and maximum allowable operating power-supply voltage in systems using advanced BiCMOS/CMOS ASICs. Hold-time problems are most apt to occur where devices are fastest, and CMOS and BiCMOS devices speed up as temperature goes down and as power-supply voltage goes up (see Chapter 11).

15.5 Summary of System Application Tips for ASICs

1. Use output drivers with the lowest current and slowest slew rate compatible with system requirements.
2. Use output circuits with TTL levels where possible to reduce transient switching currents.
3. Inputs to clocked elements should be designed to have zero or negative hold-time requirements.
4. Avoid tristate internal connection.
5. Avoid buried feedback circuits that have no visibility.
6. Use internal pull-ups or pull-downs on all inputs and bidirectional ports.
7. Use internal and I/O storage circuits with asynchronous reset capability to allow rapid initialization.
8. Provide an output signal to indicate the direction of internally controlled bidirectional interface ports.
9. Buffer signals to and from external board or system destinations or sources to protect ASIC inputs and outputs from transients and ESD.

15.6 References

1. Nass, Richard: "ASICs: The Latest Alternative," *Electronic Design,* October 12, 1989, pp. 51–60.
2. Heckman, Alan: "Designing ASICs: Be Prepared for Changes," *Electronic Design,* January 11, 1990, pp. 133–140.
3. Bursky, Dave: "Digital ICs in the 1990s: Nearly Unlimited On-Chip Resources," *Electronic Design,* January 11, 1990, pp. 97–106.
4. ASIC Action: *The ASIC Newsletter,* LSI Logic Corporation, Milpitas, Calif., January 1990, p. 6.
5. "BiCMOS Arrays Boost Density, Speed, I/O Drive, and Usable Gates," *Electronic Products,* April 1, 1988, p. 60.
6. Lin, Liang-Tsai, and Richard Spehn: "Fast, Low-Powered Logic Array Unites CMOS and Bipolar," *Electronic Design,* April 16, 1987, pp. 82–88.
7. *Q24000 Series BiCMOS Logic Arrays Device Specification,* Applied Micro Circuits Corp., San Diego, Calif., 1990.

8. Chester, Michael: "Action Brews in BiCMOS Arrays," *Electronic Products,* June 1, 1987, pp. 22–24.
9. Baker, Stan: "Extending TTL," *Electronic Engineering Times,* October 16, 1989, pp. 41, 52.
10. *TSC500 Series 1-μm CMOS Standard Cells Data Manual,* Texas Instruments Inc., Dallas, Tex., 1988.
11. Runner, J. Scott, and Larry Roffelsen: "The Pitfalls of ASIC Interfaces," *Electronic Engineering Times,* November 28, 1988, pp. T30–T32.

Appendix

A.1 Conversion Factors

$$1 \text{ ft} = 30.48 \text{ cm}$$
$$1 \text{ ft} = 0.3048 \text{ m}$$
$$1 \text{ in} = 2.54 \text{ cm}$$
$$1 \text{ mil} = 2.54 \times 10^{-3} \text{ cm}$$
$$1 \text{ mil} = 10^{-3} \text{ in}$$

A.2 Definition of Symbols and Acronyms

ac	alternating or dynamic current
AC	designation used for advanced CMOS logic
AC11	designation used for advanced CMOS logic with center ground and power pins
ACL	advanced CMOS logic
ACT	designation used for TTL-compatible advanced CMOS logic
ACT11	designation used for TTL-compatible advanced CMOS logic with center ground and power pins
ALS	designation used for advanced low-power Schottky TTL logic devices
ALU	arithmetic logic unit
AMD	Advanced Micro Devices Inc.
AND	logic circuit whose output is a 1-state only when every input is in the 1-state
AS	designation used for advanced Schottky TTL logic devices
ASIC	application-specific integrated circuit
AWG	American wire gage
b	distance between reference planes in printed circuit boards
BC	designation used for Motorola and Toshiba TTL-compatible BiCMOS logic
BCT	designation used for TTL-compatible BiCMOS logic
BiCMOS	bipolar and complementary metal-oxide semiconductor
Bi-CMOS	acronym used by Motorola and Toshiba for semiconductor devices that combine bipolar and CMOS technologies

b/s	bits per second
C	designates a CMOS device
C	designates a capacitor
CAD	computer-aided design
CAS	column access strobe
CAS-	column access strobe, active *low*
C_d	decoupling capacitance
CE-	chip enable, active *low*
C_{IN}	input capacitance
C_L or C_{load}	load capacitance
CLK	designates a clock signal
CLK-	designates an inverted clock signal
C_{LOAD}	total load capacitance which includes signal track plus device input and output capacitance
C_m	mutual capacitance between lines
CMOS	complementary metal-oxide semiconductor
C_{pd}	internal device capacitance used for power calculations
CPU	central processing unit
CS-	chip select, active *low*
°C	degrees Celsius
ΔC	difference in specified and actual load capacitance
d	diameter of a wire conductor
D	designates a diode
D	designates the *drain* terminal of a field-effect transistor
dc	direct or static current
DIP	dual in-line package
DRAM	dynamic random-access memory device
ECL	emitter-coupled logic
EEPROM	electrically erasable programmable read-only memory device
EPROM	electrically programmable read-only memory device
EIA	Electronic Industries Association
EPLD	erasable programmable logic device
ESD	electrostatic damage
ESL	effective series inductance
ESR	effective series resistance
f	frequency (hertz)
F	node toggle frequency
FACT	Fairchild advanced CMOS logic

FACT QS	Fairchild advanced CMOS logic with output slew rate control and other features that reduce ground bounce
FAST	Fairchild advanced Schottky TTL logic
FBT	designation used for fast BiCMOS TTL-compatible logic devices
FCT	designation used for fast CMOS TTL-compatible logic devices
FCT-T	designation used for fast CMOS TTL-compatible logic devices with TTL output levels
FCT-A	designation used for very fast CMOS TTL-compatible logic devices
FET	field-effect transistor
FIFO	first-in first-out
FITs	failures in test; per billion device hours
F_{MAX}	maximum toggle rate of a clocked logic device
FPLA	field programmable logic array
G	designates the gate terminal of a field-effect transistor
GND	ground
h	height of a conductor above a reference plane
HC	designation used for high-speed CMOS logic devices
HCT	designation used for TTL-compatible high-speed CMOS logic devices
I_b	base current
IC	integrated circuit
I_c	collector current
I_{cc}	device supply current
I_{D}	drain current
IDT	Integrated Device Technology Inc.
I_{IH}	rated (maximum) input *high* current
I_{IL} or $I_{\text{in low}}$	rated (maximum) input *low* current
$I_{\text{in total}}$	total input current
I_{IN}	input current
I_L	load current
I_o	output current
I/O	input and output
I_{OD}	dynamic output current
I_{OH}	rated output *high* current
I_{OL}	rated output *low* current
I_{OS}	output short-circuit current
I_p	peak internal feedthrough current

ΔI	change in current
K	a constant
K_C	capacitive coupling coefficient
K_L	inductive coupling coefficient
l	unit length of a transmission line
l, w, t	length, width, and thickness
L	self-inductance
L	designates an inductor
LCC	leadless chip carrier
LED	leading-edge detector
L_m	mutual inductance between lines
L_p	package pin inductance
L_p	parallel inductance
L_s	source inductance including power source inductance
LS	low-power Schottky TTL logic
LSI	large-scale integration
MOSFET	metal-oxide-semiconductor field-effect transistor
MSI	medium-scale integration
MTBF	mean time between failures
n	semiconductor material with an excess of electrons
N	number of turns in an inductor
N	number of words in a memory subsystem
NAND	logic device whose inputs must all be in a 1-state to produce a 0-state output
NOR	logic device where any one input or more having a 1-state will yield a 0-state output
NOVRAM	nonvolatile random-access memory
OE-	out enable, active *low*
OR	logic device where any one input or more having a 1-state is sufficient to produce a 1-state output
p	semiconductor material with a deficiency of electrons
PAL	programmable array logic
pc	printed circuit
PCT	designation for Performance Semiconductor Corporation's high-speed TTL-compatible CMOS logic
PCT-A	designation for Performance Semiconductor Corporation's very high speed TTL-compatible CMOS logic
P_d	dynamic power dissipation
PLA	programmable logic array
PLCC	plastic leaded chip carrier

PLD	programmable logic device
P_q	quiescent power dissipation
PROM	programmable read-only memory
Q	designates a transistor
Q_f	final charge
Q_i	initial charge
R	designates a resistor
RAM	random-access memory
RAS	row access strobe
RAS-	row access strobe, active *low*
R_{CL}	current-limiting resistor
RF	radio frequency
R_L	load resistance
R_o or R_{out}	output resistance
ROM	read-only memory device
R_{ON}	bipolar transistor collector-to-emitter or field-effect transistor drain-to-gate ON resistance
$R_{\text{pull-up}}$	pull-up resistor
R_S	source resistance
S	designates the *source* terminal of a field-effect transistor
S	designates a switch
SCR	silicon-controlled rectifier
SRAM	static random-access memory
SSI	small-scale integration
t	time (seconds)
t	thickness of printed circuit board conductors or planes
T_{A}	ambient temperature
TED	trailing-edge detector
t_f	fall time
t_{HL} or t_{PHL}	propagation time from an input change to an output *high-to-low* transition
t_h or t_{hold}	time that a signal to a clocked device must be stable after application of the active clock edge
TI	Texas Instruments Inc.
t_{LH} or t_{PLH}	propagation time from an input change to an output *low-to-high* transition
t_{LINE}	propagation delay of a line
t_p	propagation delay of a designated length of conductor
t_{pd}	intrinsic propagation delay of the media

t'_{pd}	effective propagation delay of loaded conductor
t_{prop}	propagation time of a device
t_r	rise time
t_s or t_{setup}	time that a signal to a clocked device must be stable before the arrival of the active clock edge
t_{SH}	time that a *high*-level signal to a clocked device must be stable before the arrival of the active clock edge
t_{SL}	time that a *low*-level signal to a clocked device must be stable before the arrival of the active clock edge
TTL	transistor-transistor logic
Δt	change in time for an event
U	designates an integrated circuit
UV	ultraviolet
UV EPROM	ultraviolet erasable electrically programmable read-only memory
V_B	voltage amplitude of backward crosstalk
V_{BE}	bipolar transistor base-to-emitter voltage
V_{cc}	positive logic device supply voltage
$V_{cc\ min}$	minimum rated supply voltage
V_{CE}	bipolar transistor collector-to-emitter voltage
$V_{CE(Sat)}$	bipolar transistor collector-to-emitter voltage when fully ON
V_f	final voltage or diode forward voltage
V_F	voltage amplitude of forward crosstalk
V/I	voltage and current characteristics of a node
V_{IH} or $V_{IH\ MIN}$	rated minimum input *high* voltage
V_{IL} or $V_{IL\ MAX}$	rated maximum input *low* voltage
V_{IN}	input voltage
V_{IN}/I_{IN}	input voltage and current characteristics
V_L	voltage at the load
V_{loss}	voltage drop across a ground or power plane
VLSI	very large scale integration
V_{OH}/I_{OH}	output *high* voltage and current characteristics
V_o or V_{out}	output voltage
V_{OH}	rated output *high* voltage
V_{OL}	rated output *low* voltage
V_{pp}	programming voltage
V_{STEP}	voltage step magnitude when a wave is launched into a transmission line

$v(t)$	time-varying voltage
VTC	VTC Inc. of Bloomington, Minn.
V_x	voltage at a distance x on a transmission line
ΔV	change in voltage level
ΔV	signal swing
w	width of a square section of a plane
W- or WE-	write enable, active *low*
x	a variable, for example, distance from a point
XTK	acronym for Quad Design's crosstalk calculation program
Z_o	characteristic impedance
Z'_o	effective characteristic impedance

Greek Symbols

ϵ	relative dielectric constant
ϵ_o	dielectric constant (permittivity) of free space (8.85×10^{-12} F/m)
μ	permeability
ρ	resistivity (the resistivity of copper is $1.724 \times 10^{-6} \Omega \cdot \text{cm}$ at 20°C)
ρ_L	load reflection coefficient
ρ_s	sheet resistance
ρ_S	source reflection coefficient
ϕ	magnetic flux (webers)

A.3 Trademarks

FACT is a registered trademark of National Semiconductor Corporation.

FACT QS is a registered trademark of National Semiconductor Corporation.

FAST is a registered trademark of National Semiconductor Corporation.

PAL is a registered trademark of Advanced Micro Devices, Inc./Monolithic Memories, Inc.

ABOUT THE AUTHOR

James E. Buchanan is an advisory engineer at Westinghouse Electric Corporation's Electronic Systems Group in Baltimore, Maryland, where he serves as a technical advisor in digital and analog circuit design, high-speed logic applications, and memory systems. Buchanan holds thirteen patents, most of which cover analog-to-digital and digital-to-analog converter circuits. He is a frequent contributor to professional literature in electrical design and is the author of *CMOS/TTL Digital Systems Design*. Buchanan earned his BSEE at the University of Tennessee and his MSEE at the University of Maryland.

Index